Introduction to
Neuromarketing &
Consumer Neuroscience

Introduction to
Neuromarketing &
Consumer Neuroscience

Thomas Zoëga Ramsøy, PhD

Published by Neurons Inc 2015

www.NeuronsInc.com

Published in Denmark
ISBN 978-87-997602-0-6

1st edition, published 2015, version 1.2

Neurons Inc ApS
Trollesvej 6
4581 Rørvig
Denmark

www.NeuronsInc.com

Contents

Prologue

How do we make decisions on what to buy and how much to pay for it? How can it be that we are affected by brands when we like and choose products? Why do some people love brands, and others develop pathological gambling or "shopaholism"? Traditional approaches to such questions have relied on the behavioural sciences. Today, we see a dramatic shift in our understanding of consumption behaviours. Recent advances in modern neuroscience, and how it combines with economics and psychology, have allowed us to study of how different brain functions serve consumer behaviour. A commercial industry is emerging that offers novel ways to assess consumer attention, emotion and memory.

This book offers a comprehensive introduction to the workings of the brain and its mind, and how this knowledge can inform our understanding of consumption behaviours. The book offers both basic and front-end academic insights, and includes chapters on sensation and perception; attention and consciousness; emotion and feeling; memory; motivation; preference; and decision making. It also offers up to date and comprehensive insight about how the tools of neuroscience can be applied to assess consumer cognition and emotion.

The book is based on my years of experience in doing research and on the brain and our decisions, and in teaching courses on neuromarketing, neuroscience of branding, neuroeconomics, neuroleadership and behavioural economics at the Copenhagen Business School. It is also based on my resignation on finding a suitable textbook for my neuromarketing courses.

This book is therefore my own take on consumer neuroscience and neuromarketing. My own background is versatile: I started studying business, went over to psychology and ended up as a clinical neuropsychologist. After working a few years working in clinical neurology and psychiatry, I did my PhD in neurobiology and neuroimaging at the Danish Research Centre for Magnet Resonance, studying the effects of ageing on cognition and emotion, under the supervision of Prof Olaf Paulson and Prof Terry Jernigan. During the same period I undertook several studies on consciousness, object perception, decision making, and drug abuse, and got involved in methods ranging from psychometrics and personality testing

to structural MR, to PET scanning and Transcranial Magnetic Stimulation.

At the end of my PhD I was invited by Professor Emeritus Flemming Hansen at the Copenhagen Business School to explore the possibility of combining economics, psychology and neuroscience, both in specific research projects as well as teaching. Since 2008 I have been employed at the Department of Marketing at Copenhagen Business School, establishing my own lab (the Decision Neuroscience Research Group) and finally establishing my own research centre: the Center for Decision Neuroscience, still at the Department of Marketing at CBS. Here, a number of PhD students, senior researchers and undergraduate students work hard on exploring the brain bases of decision making, using a multitude of methods – we study everything from the effects of the ovarian cycle on women's response to erotic ads, to the brain bases of social decision making.

This book is a result of all of this: my experience from economics, psychology and neuroscience. But it is also a result of what started almost incidentally as a few talks here and there and has grown all the way to the company Neurons Inc (www.NeuronsInc.com) which is an applied neuroscience company that studies the ways in which consumers are responding cognitively and emotionally to communication, products and other kinds of communication. Today, Neurons Inc runs monthly copy testing omnibuses, has developed an automatic image and video analysis tool, and tests consumer responses in a variety of challenging conditions, ranging from in-store behaviour to human-robot interaction, and everything in between. Whenever we need to understand how we respond to an event, there really are no limits to the use of applied neuroscience.

Lately, my work has turned even more towards innovation and how our understanding of the human mind can help us better understand consumers' responses and successful or failing adoption of novel products and services, as well as how we can use our understanding of the brain to improve creativity and innovation. By becoming involved in Singularity University and Lowe's Innovation Lab, my work is increasingly turning towards understanding both the process of innovation as well as our responses to the end result of innovation. If anything, applied neuroscience is innovation, and although it may seem like a no brainer that it should be used to understand consumers, this is no guarantee that employees at leading companies see the same no brainer. Innovation can only truly be successful when it is adopted broadly. Consumer neuro-science and neuromarketing are currently facing this challenge, and it is my hope that this book can help define the science, limit unwarranted overclaims, and consolidate our understanding on consumption and the brain.

About this book

When writing an introductory book to a topic, many decisions have to be made. Most notably, there is a struggle between balancing the need for being absolutely correct, updated and nuanced on every single topic, and the didactical for making the reader understand the topic. As we are dealing with an intro book, I have chosen to move as far as I have dared towards fulfilling the didactical objective, although I have to admit that my nerdy side sometimes has won over me and allowed details, digressions and nuances that for many people will be "overkill".

Nevertheless, I believe that one important feature of this book is to show that neuromarketing and consumer neuroscience is nothing like pulling a software solution off the shelf and just start running studies. Applying neuroscience requires hard work, scientific and methodological scrutiny, and a substantial portion of knowledge of the brain, hardware, software and statistics. There is nothing like "just running a study" in this field.

My second concern has been on balancing the need to address academic readers and students who need to focus on insights into the causal mechanisms of consumption behaviours, and the need of others to address how neuroscience insights and methods can be used in more commercial applications. While both accounts are valid, I feel that it has been important to start from a more academic standpoint, and move towards commercial applications. As popular science books and articles abound, there is an outspoken need for a more balanced and less sensational approach. Also, since I want to see consumer neuroscience and neuromarketing becoming an established discipline on universities and business schools, there is a need for a standard introductory book that can qualify to the stringent academic needs.

My focus in this book is in many ways inspired by my background in neuropsychology, and in cognitive and affective neuroscience. Thus, you will see that the book is divided into topical chapters, just as many cognitive neuroscience textbooks. This, I believe, is a good start at understanding the brain bases of consumer choice. People come to this field from many different backgrounds: economics and marketing; psychology and the social sciences, engineering, natural sciences, and much more. Thus, we all have our own takes at what particular words and concept mean.

My own experience is that the best approach is to start by debating the general concepts we are studying: attention, memory, choice, and ask ourselves: what do I actually mean by this concept? Are there any nuances in how I can understand this concept? This book is an attempt at providing a foundation for how we should think of and discuss those concepts, and to show that our often folkpsychological concepts do not hold up for scrutiny. While my own philosophical approach is what can be seen as an "inter-theoretic reductionism" (yes, you should google that, but also look at http://philpapers.org/archive/CHUIRA.pdf (PDF file), I have tried to provide a more soft approach here, in showing that our general concepts and terms are better described by mechanisms rather than overall and mongrel concepts.

As part of this book, I have also chosen to add several opinion pieces. These are taken from my many years as a blogger, where I have shared my own views on the brain and the mind. You can take these pieces as digressions, but there are notable learning points here, too.

I have chosen to make all figures and illustrations myself. Everything presented here is, unless otherwise noted, my own creations: having direct access to brain scans and raw data, and the tools and experience in working with the data, has allowed me to make illustrations and figures that I feel are explaining my points well, and represent the science well. My nerdy approach with respect to neuroanatomy and years of drawing Regions Of Interest on brain scans has allowed me to have a high certainty in pointing to brain regions, and make the illustrations as pedagogical as possible.

Finally, allow me to make a plead: I am not ignorant of the interest that this topic spurs, nor the ease at which one can copy digital materials. If you have an illegal copy of this ebook, you have made the conscious choice of not paying for it. Of course, if you like the book, I hope you will change your mind and pay for it. However, an alternative that would please me equally, would be if you donated the same amount of money to charity. If you do not have the money, why not spend a few hours helping such a purpose? The value of this book can take on many forms, and I hope that in many ways it can help make this place a better home for us all.

Acknowledgements

This book has been written everywhere: at the train to and from work, on planes and airports during travels, at home with kids running around, during boring conferences and talks, in alleys and yards, hotel rooms and garden alike. Writing this kind of book has indeed been a selfish and almost solipsistic endeavour. But it has also been the result of two main threads through my recent life. One academic and one personal.

Academi cally, I could not have been without the great inspiration I have had from my many great colleagues in psychology, economics and neuroscience, and beyond. The many great people I have met in academic and commercial uses of neuroscience have always been impressive and inspiring. One special thanks goes posthumously to Professor Flemming Hansen, for his ability to see the oncoming merging between neuroscience, psychology and economics, and inviting me over from basic neuroscience research and into the fields of economic behaviour. Plenty inspiring discussions have been with some of the leading scholars of this multidisciplinary arena, including Steve Genco, Antoine Bechara, Antonio Damasio, Hilke Plassmann, Baba Shiv, Carolyn Yoon, James Rowe, Erik Du Plessis, Oliver Hulme, Bryan Knutson, Maurice Ptito, Ale Schmidts, Torsten Ringberg, Bernard Baars, Shaun Gallagher, and Morten Overgaard. One of the true inspirations I have had – and still have – is through my Adjunct Faculty membership at Singularity University, the many mind-blowing discussions at the SU HQ at NASA Ames Research Park, and the so many completely crazy and innovative people being or just passing through this place.

I could never have been where I am now without the original and lasting PhD guidance from Prof Olaf Paulson and Prof Terry Jernigan, the hardcore science approach taken by Prof Hartwig Siebner, the extremely detailed neuroscience insight and philosophical approach taken by Martin Skov, the pragmatic approaches taken by Jon Wegener and Jesper Clement, the highly inspired discussions and initiatives from my PhD students PhD students Dalia Bagdziunaite, Lou-ise Koch, Nausheen Niaz, Morten Friis-Olivarius and Sofie Gelskov, and the enormous help from research assistants Khalid Nassri, Samir Karzazi, Dimo Beloshapkov, Catrine Jacobsen, Clara Zeller and Johanna Bruehl, as well as the numerous highly inspired and motivated graduate students that have spent late nights and early mornings collecting materials and data in my lab. This book is a special thanks to all of those colleagues, whom I am always grateful for having been so fortunate to meet and be able to spend time with.

A special thanks goes to three people who have provided me with the guts to pursue my goals on combining neuroscience with today's challenges: Kyle Nel, Director of Lowe's Innovation Lab, who runs this planet's most innovative, daring and successful innovation lab and incubator consortium, and whom I have had early, long and continuous discussions with about how neuroscience can inform today's challenges in market research and beyond; Ari Popper at Scifutures and A2O, has been an enormous inspiration for thinking beyond the horizon and seeing opportunities and futures where most people stagger at thinking about just the next day; Leonard (Lenny) Murphy at GreenBook and Gen2 Advisors, for his awe-inspiring take on innovation in market research, his breath of insights and connections and not the least his incredible help and words in some difficult times. Together, this trio brings what I believe is something any field needs in these times of innovation: a willingness to think different, a daring to live it out (with the potential to fail), and at the same time smile and have a blast while doing it!

All of the work leading up to this book could never have been done with the love, devotion, trust, and support from my family. My wife and love of my life, Majken, has been the support and thoughtful help I could never have dreamt of and knowingly never deserved; our children Mike, Frederik and Sophia-Lilje are such an amazing life inspiration every day. Writing this kind of book is selfish, but without the support from my family this could never have been possible. My love for them is eternal.

CHAPTER **1**

Introduction

How do we make decisions every day? What makes us remember prior experiences? Why are we influenced by contexts such as brands, or merely believing that others are observing us? Why do we have different preferences? Do we really have a free will?

These and many other questions are among those that are dealt with in this book. In our attempt to understand the consumer and how their decisions are made, we must start from the beginning. We must start with the basic questions of what it means to have emotions, to remember, to make choices. We must start with ourselves.

In recent years, there has been an exponential rise in interest for neurosciences in different kinds of business. Today, we see a plethora of emerging disciplines, such as neuroeconomics, decision neuroscience, neuromarketing, consumer neuroscience, neuraesthetics, neuroethics, even neurocinematics and neurofinance that have made it to people's awareness.

What these terms actually mean may sometimes be less obvious. Suffice to say at this point that the addition of "neuro" signals a significant change from business-as-usual. At the very least, it entails the understanding that the use of neuroscience – both knowledge and tools – may provide new insights into what has usually been the domain of economics, psychology and other social sciences.

As the chart below shows, terms such as "neuromarketing" has seen an immense growth during the past few years. In academia, we find a strong growth in the number of publications in peer-reviewed journals using the term. Leading journals in both marketing, consumer behaviour, neuroscience, economic, and psychology regularly publish articles that fall within the domain of neuromarketing and consumer neuroscience. In addition, new journals dedicated to

this multidisciplinary field have emerged. Table 1 lists some of the most notable journals.

A second trend is found in business. On the one hand, several emerging companies are registering under the heading "neuromarketing" or related terms. Such companies typically offer tests of neural and physiological consumer responses to ads, commercials, even in-store situations and virtual environments. In addition, many major corporations employ teams that dedicate their work to neuromarketing. Today, we even see that many of the traditional "neuromarketing tools" are becoming standard tools in market research and other disciplines.

Today, we also see emerging organisations of such multidisciplinary efforts, including the Association for Neuroeconomics, Association for NeuroPsychoEconomics and the Neuromarketing Science and Business Association (NMSBA. While the first is pri-

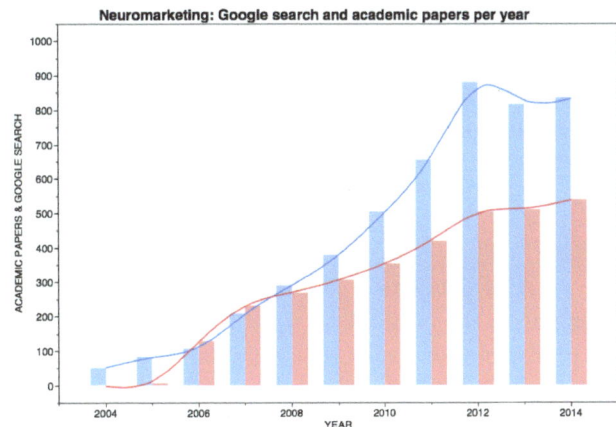

Figure Neuromarketing over the years. RED: Google hits (number is divided by 100) and show a stead incline after 2005. BLUE: number of academic papers, showing a peak in 2012 and a tentative stabilisation about 800 articles per year. Please note that 2014 is an estimate based on collect data for the first 6 months of the year.

JOURNAL NAME	PRIMARY CLASSIFICATION
Biological Psychology	psychology
Frontiers in Decision Neuroscience	consumer neuroscience & neuroeconomics
Frontiers in Human Neuroscience	neuroscience
Frontiers in Psychology	psychology
Journal of Cognitive Neuroscience	neuroscience
Journal of Consumer Behavior	marketing
Journal of Consumer Psychology	marketing
Journal of Consumer Research	marketing
Journal of Economic Psychology	marketing
Journal of Economics, Psychology & Neuroscience	consumer neuroscience & neuroeconomics
Journal of Marketing	marketing
Journal of Marketing Research	marketing
Journal of Neuroscience	neuroscience
Nature	general science
NeuroImage	neuroscience
Neuron	neuroscience
Neuropsychologia	neuroscience
PloS ONE	general science
Proceedings of the National Academy of Sciences	general science
Psychological Science	psychology
Science	general science

Figure   A small selection of notable academic journals where one can find relevant publications.

marily. American based, and the second mostly European based, the NMSBA has quickly become the global representative for everyone practicing work in the field of neuromarketing and related disciplines. Today, the NMSBA has representatives from every corner of the world, and with the count increasing steadily (see below). In addition, other major organisations in psychology, neuroscience and economics are actively engaging in similar works. Notable contributions on consumer neuroscience and neuromarketing is found

each year at meetings in the Association for Psychological Science, Association for Consumer Research, Society for Neuroscience, Organisation for Human Brain Mapping, just to mention a few.

We should not forget a final trend, which is the extreme interest that neuroscience has mobilised in the community as a whole. Today, not a single day passes without a new press release, a media discussion or a public statement that concerns the relationship between the brain and the mind. Neuromarketing

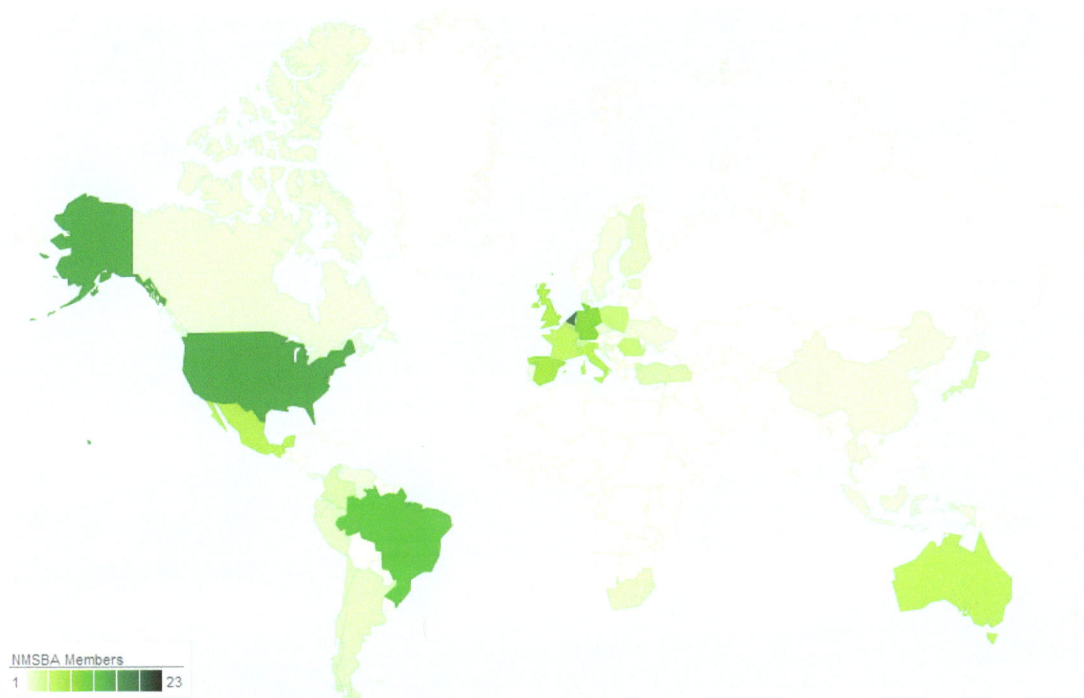

Figure   Visual representation of the number of NMSBA members. Courtesy of NMSBA.

and consumer neuroscience is just a reflection of this development. Indeed, the inclusion of neuroscience and neurobiology is disrupting every way in which we speak about the human mind, with such a diverse areas as child rearing, finance, couple therapy, politics, didactics, and leadership.

CONCEPTUAL CLARIFICATIONS

This book works with two concepts: "neuromarketing" and "consumer neuroscience". Both terms are used relatively interchangeably throughout the book, but let us clarify from the get-go that we should in fact think of the two terms as referring to two different takes on how neuroscience is used to understand consumers and communication effects.

What is "neuromarketing"?

The term "neuromarketing" is believed to be first coined by Ale Schmidts, now a Full Professor at the Department of Marketing at the Rotterdam School of Management in Holland. The term itself is defined as a part within marketing that studies the effect of marketing stimuli on consumers' sensimotor, cognitive and affective (emotional) responses. Indeed, the term itself is of course the result of "neuro", referring to neuroscience (or more likely "cognitive neuroscience") and

"marketing", which can broadly be defined as "the activity, set of institutions, and processes for creating, communicating, delivering, and exchanging offerings that have value for customers, clients, partners, and society at large" (from AMA).

Today, neuromarketing is often seen as the commercial use of neuroscience insights and tools that companies can use to better understand consumer responses to different kinds of brand-, product- and service-related communication efforts.

What is "consumer neuroscience"?

Many scholars have felt that the term "neuromarketing" has been given a bad reputation after several cases of commercial overselling, under-delivering and erroneous use of neuroscience methods and insights in the name of commercial applications. Therefore, some scholars have suggested the alternate term "consumer neuroscience" to better coin the academic approach of employing neuroscience methods and insights to study and understand consumer psychology and behaviour (Hubert 2008; Yoon 2012; Plassmann 2012).

From this tradition, "consumer neuroscience" is a combination of the academic study of consumer psychology and consumer behaviour, which are more focused on the way in which consumers respond and act. As such "consumer neuroscience" is also closely

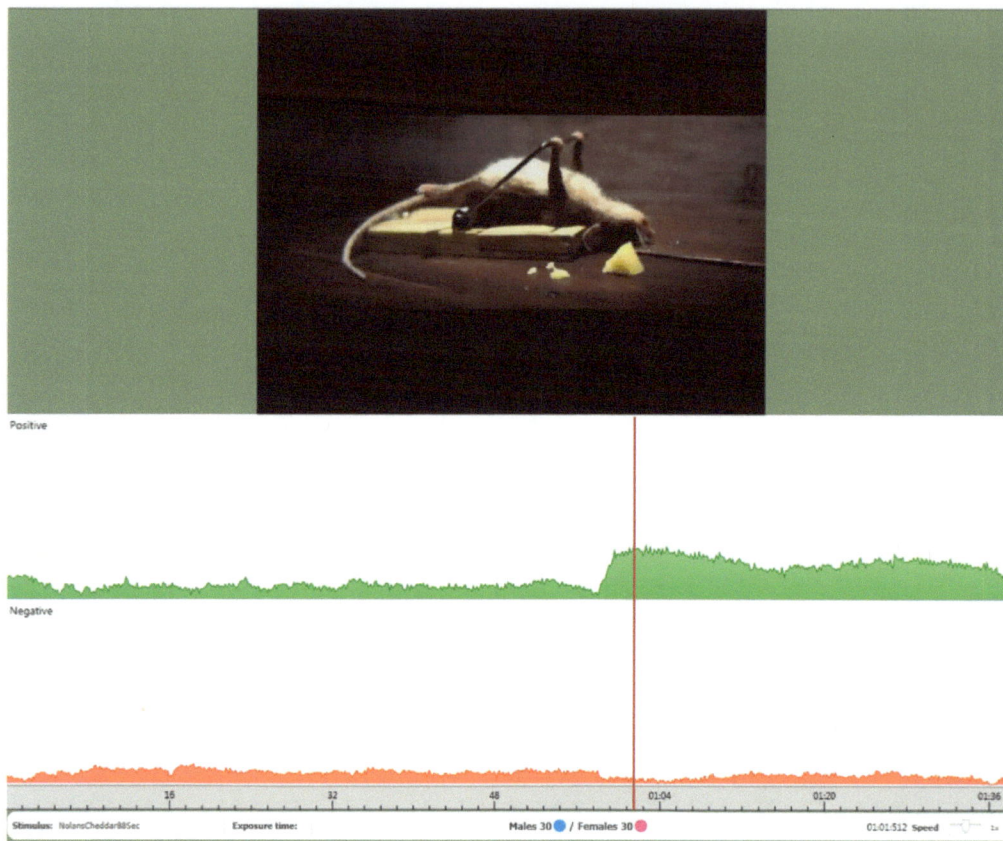

Figure   Using neuroscience and related methods to assess direct and often unconscious responses to ads and other consumer situations can allow a better understanding of communication effects. In the sample above, using automated recognition of facial expressions it may be possible to determine exactly when emotional effects occur in an ad, such as when people display positive emotional responses. Courtesy of iMotions (www.imotionsglobal.com)

linked to "decision neuroscience" and even "neuroeconomics", in which researchers attempt to understand how decisions are made, and what the causal brain mechanisms of our choices can be.

More generally, what is "applied neuroscience"?

The use of neuroscience in understanding consumers relates to a broader attempt to link neuroscience to understand thought and behaviour. Going under the heading "applied neuroscience", we can say that the use of neuroscience changes the way we try to understand human thought and behaviour substantially.

In addition to using classical methods such as interviews and focus groups, neuroscience methods and insights offer a completely different – and often provoking – take on what drives and affects our choice. Therefore, the term "applied neuroscience" can in many ways be described as the more broader attempt to link the role of the brain in the many different walks of life. A few examples could be:

- Cinema goers' emotional and cognitive responses to the movie (more -recently called "neurocinematics")
- The study and improvement of human–robot interaction
- Assessment and influence on creativity by way of neuroscience methods and insights
- Brain training – using neuroscience tools and insights to boost your cognitive capacities
- Managerial decisions and leadership (sometimes called "neuroleadership")

What is neuroscience?

In many ways, it is completely wrong to use the term "neuroscience" in the way we are using it throughout this book. The term "neuroscience" covers a much broader approach to the brain that we use here – it covers the study of single receptors of cells, single cells, sea slugs, leeches, reptiles, mammals, primates in addition to humans. It uses a whole range of tools, from chemical analyses of cells in a petri dish to structural and functional imaging made with Magnetic Resonance scanners. So the term is actually much too broad.

There are other related disciplines that are more narrow in their scope: for example "cognitive neuroscience", which is a term that focuses on the brain bases of cognitive functions and processes. "Affective neuroscience" is a discipline that focuses more on the brain mechanisms underlying emotional responses. "Decision neuroscience" is, as the name says, a discipline that focuses on the neural mechanisms that are responsible for our choices. "Social neuroscience" focuses on our social life and the brain structures supporting or disrupting this part of our lives. "Neuroeconomics" is the study of the neural bases of economic behaviour, broadly defined.

Thus, the term "neuroscience" is in many ways a misnomer, as it is too broad for the very narrow use of insights and methods that are used in neuromarketing and consumer neuroscience. Nevertheless, we cannot avoid using the term "neuroscience" as a common reference to point to the approach of looking at the brain for clues to our understanding of consumers. In any case, if you meet the term "neuroscience" in this book, it will only be as a vague and general term.

What about "neurology"?

Some authors and speakers often speak of "neurology" when we are looking at how the study of the brain can inform our study of consumers. However, I will argue that the term "neurology" is a misnomer, too. Specifically, neurology is the study of brain disorder. Neurologists study the brain bases of degenerative diseases such as Alzheimer's and Parkinson's disease. They study the gross effects of stroke on human thinking and behaviour. At all times, they are focusing on *disorders* of the mind, and never at anything even remotely related to consumer behaviour.

Being a clinical and theoretical neuropsychologist by training, I find it puzzling how the term "neurology" has snuck its way into these arguments about consumers. Neuropsychologists, by contrast, look less to the organic changes that underlie brain disorder and injury, and focus more on the *mental* changes accompanying brain changes. Neuropsychology is the study of the relationship between brain injury and disorder, and changes in cognition, affect and behaviour. It is still a focus on the disorder of the brain, but at least it is a focus on how particular regions relate more closely to mental changes.

So if anything, let's call it neuropsychology, not neurology. As we will see in the chapter about consumer aberrations, this approach can indeed be fruitful. However, let us leave "neurology" and "neuropsychology" to what they are best at: understanding brain disorders and injuries, and let us not use these terms in our understanding of the brain bases of consumer choice.

Two approaches: methods and insights

There are two gross ways in which the study of the brain can be used to study consumer behaviour. First, cognitive and affective neuroscience offers a whole new set of tools that can be used to study our direct, unconscious responses to stimuli, that can supplement or even sometimes replace other methodologies. This use of neuroimaging is what we can call the *methodological application* of neuroscience.

Second, the study of the brain itself offers several insights that can be key to understanding consumers. Reading a textbook on cognitive neuroscience can reveal new insights into how we should understand consumers. This take is what we can call the *theoretical application* of neuroscience.

Both accounts offer tremendous new ways in which we study and understand consumer behaviour and choice. In this book we will provide both accounts, hopefully in a balanced way, that can allow a broader and more comprehensive understanding of what drives our choices as consumers.

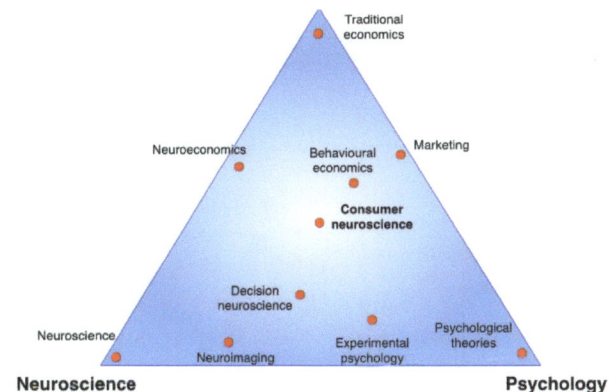

Figure   An overview over the cross-disciplinary mix that consumer neuroscience and neuromarketing entails.

THE EMERGENCE OF NONINVASIVE NEUROSCIENCE

Modern neuroscience is in an innovative phase, and is moving from a sole reliance on huge labs with equally large brain scanners and an accompanying highly trained and skilled team of engineers, physicists and other technical staff. The innovation we see today is the result of the combination of an increasingly maturing neuroscience and technical inventions. Today, we find that the methods with which you can study and interact with the brain are exploding and appearing in new and unprecedented ways.

Some of these methods are no more than a fad, and will probably go away over time. Other methods are a new step into what we can call the "mobilisation" of neuroscience.

Faster, cheaper, better

One of the big developments is price. While a lab with EEG and eye-tracking would earlier have costed you the better part of a quarter of a million US dollars, today the same lab can be obtained for a fraction of the cost.

Similarly, methods such as functional Magnetic Resonance Imaging (fMRI) are highly costly. A regular scan can often cost around USD 2,000 or more per person, and you would normally test at least 20 people for a decent study. Beyond this, analyses of neuroimaging data are both time-consuming and highly complex, even to the most trained statistician.

We see new eye-trackers being released now that cost as low as $99, and new head band with EEG sensors costing only $200. While their actual use and application to neuromarketing still needs to be explored, this is a development that will soon be available for testing. If they turn out to be usable, it will ultimately change the face of neuromarketing and market research alike.

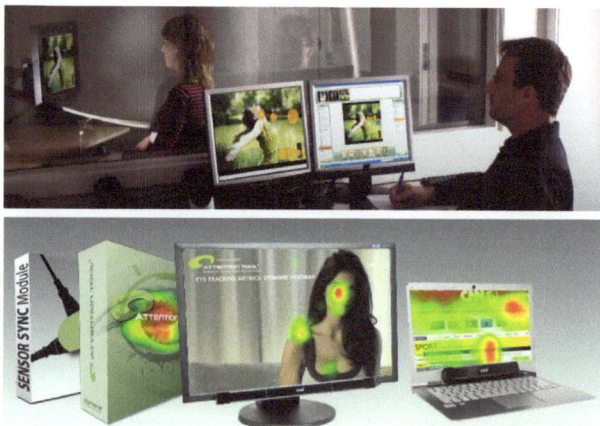

Figure   Methods for studying the unconscious drivers of consumer behaviour are becoming more accessible, as both the technology is becoming more reliable and affordable, and the science is becoming more robust. The images demonstrates a setup from iMotions (www.iMotionsGlobal.com) which allows complete integration of eye-tracking and a palette of biosensors and facial coding. Reprinted with permission from iMotions.

From stationary to mobile

With technological and scientific advances it is now possible to move our studies out of lab environments and into more realistic consumer settings. With many technologies, such as fMRI, this is not possible. But some methods, such as eye-tracking, GSR and EEG, it is possible to move out of the lab and into the real world. This is not to say that the move is straightforward, as any physical movement of the body is typically associated with profound measurement errors and noise in the data. The analysis of imaging data in mobile environments still require a deep insight into noise correction, data quality and other data preprocessing methods.

The example below shows the use of Tobii Glasses and the EPOC Emotiv 14 channel EEG from a study performed by Neurons Inc on consumer responses to specific retail environments. This tool, while suboptimal for many purposes, allowed us to assess movement via accelerometry, which could be used to correct data artefacts in the EEG analysis.

Figure   Mobile solutions can provide novel and more realistic insights into consumer behaviour. Courtesy of Neurons Inc (www.neuronsinc.com)

WHY NEUROSCIENCE IN CONSUMER BEHAVIOUR?

One of the hallmarks of neuromarketing relies on two major developments. On the one hand, the development and sophistication of tools in neuroscience and neurophysiology now allows us to assess human responses to different kinds of stimuli and conditions, and the making of a "neuromarketing toolbox."

The second development is the realisation that human decision making deviates substantially from a normatively rational, deliberate and conscious process. Here, a major contribution was the cross fertilisation between psychology and economics, epitomised by the 2002 Nobel prize in economics awarded to the psychologist Daniel Kahneman, based on his research with Amos Tversky in uncovering how human deci-

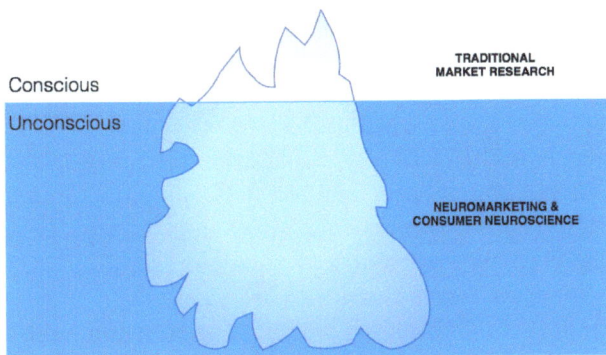

Conscious

Unconscious

TRADITIONAL
MARKET RESEARCH

NEUROMARKETING &
CONSUMER NEUROSCIENCE

Figure   The iceberg analogy highlights how consumer neuroscience and neuromarketing offer new ways to understand the unconscious bases of consumption behaviours. If indeed a lot of our mental processes are operating unconsciously then traditional research methods – such as interviews, surveys and focus groups – fall short of addressing this aspect of consumption behaviour. The tools and insights from neuroscience offers ways that we can better understand the unconscious and conscious drivers of our choices.

sion making is neither optimal or formally rational. That is, even though we feel that we are making controlled, conscious and fully informed choices, research throughout the decades has demonstrated how our choices deviate from what can be considered optimal choices, our decisions are often influenced without our knowledge, and we rarely if ever decide after considering all options. This suggests that we need to understand human choices in other ways than merely asking people.

In general, there are three notable effects that argue for a need for assessing unconscious effects in consumer choice:

1.   Our choices are often based on unconscious processes and influences. Framing effects and other contextual cues affect our choices without our knowledge of being affected.
2.   Emotions affect our choices substantially – we do not decide after carefully and "rationally" calculating the effects of each option.
3.   Decisions are not made after complete information is obtained – rather, our choices happen almost instantly and mostly after just receiving a fract ion of all available information about our choice options.

This all means that we are not rational (in the traditional sense), not fully informed, we are influenced by others, and our choices are not purely conscious. By this token, can we then trust what people respond when we ask them why they chose as they did? This is, indeed, a problem. A major portion of traditional market research and consumer psychology is based on surveys, interviews, focus groups and other que-

ries to people's thoughts and feelings about their behaviours. Some of the notable research focuses on consumer narratives and social constructivism, in which we study how consumers create meanings about their thoughts and behaviours as consumers.

However, if we cannot say that consumers are on top of their own choices and decision processes, then any assessment relying on personal narratives can only be a highly limited account of consumer behaviour and psychology. Yes, it does make sense to understand consumers' personal narratives, but if consumer choice is at least somewhat driven by unconscious processes, then we need another set of tools to understand what drives consumer choice.

The models of choice and how we can assess causal mechanisms of consumer choice, is what this book is about. This encompasses models and methods that focus on both the conscious AND the unconscious mind, of rational AND irrational choices. For a full understanding of consumers, we need to have both feet to stand on.

CONSUMER BEHAVIOUR AND MARKETING IN NEUROSCIENCE?

An equally important question is the use of consumer behaviour and marketing in neuroscience. This is a largely forgotten and ignored topic, as neuroscientists tend to view the approach of consumer behaviour and marketing as an unwanted commercial application of their basic research. Often, you may hear that consumer neuroscience and neuromarketing is like "being in the pocket" of the commercial industry. Often this comes from proponents that receive funding – and forget they do – from the medical industry...

The claim, even at it's core, is nonsensical. It suggests that "real research" should be done inside the lab in highly controlled environments and under optimal measurement conditions, and that any attempt at applying neuroscience in studying consumer behaviour outside the lab is "false" and "erroneous". To this, on can ask: what is the purpose of any research, maybe in particular lab research? Basically, this kind of lab research tries to understand human decision making, and the causal (brain) mechanisms underlying such behaviours. Indeed, this research is trying to understand human behaviour, and try to translate their lab based findings back to the real world. What worth does lab research have that cannot make us understand human behaviour outside the lab?

This leads to a challenge: how much can we translate the findings in a highly controlled lab environment to the real world? The more controlled and artificial the lab environment is, the less we can make any direct extrapolation from the findings to the real world. After all, how often do you see a product first for 4 seconds, then the price for 4 seconds, and then choose during 4 seconds? How often are you exposed to very controlled stimuli for about 40 milliseconds, before you make a risky choice?

The problem with cognitive and affective neuroscience research today is that it does lack the external validity – the way in which we can say that lab based findings can explain our human behaviours. However, interestingly, technological and scientific advances today allow us to take those steps out of the lab and into the real world, with many of the same tools available. We can indeed assess brain activity and physiological responses in more real life situations. Nobody are saying that it is easy, but the challenge is actually interesting in itself. Today, we can allow ourselves to make lab based studies, and then test the findings in a more "noisy" environment such as allowing people to freely roam and choose inside a store environment. Indeed, one may argue that this is a natural next step in understanding the brain bases of human choice and behaviour. It does not rule out the basic research, but allows a valid bridge for us to study, test, and learn about the roots of human decisions.

So does neuromarketing work?

In a recent blog post at ESOMAR, there was a critical remark about neuromarketing. I found the remarks too simplistic and caricaturing applied neuroscience, and made a response which I share here:

- Neuromarketing is not a unique and novel application of neuroscience outside it's domain of origin. Psychology has used neuroscience for decades now, aka neuropsychology, and with great success in understanding and predicting behaviour. Neuroscience and physiology was the very part of the origin of psychology since the times of Fechner and Wundt more than a century ago. Why should this not be the case for understanding consumers and communication effects?

- We understand much more than the basics, but even for the basics, there is much added value. For example 1) knowing where people actually look (we are poor at knowing ourselves); 2) how we respond emotionally (also often unconsciously); and 3) how such initial responses predict likelihood of purchase/click/behaviour-of-choice. These are very straightforward questions that are "easy" to answer with neuroimaging and related measures, and yet can have profound insights to marketers.

- Whether neuromarketing works is actually not something that should be determined as a beauty contest. It's an empirical question. Today, we see an increasing number of studies showing that neuromarketing predicts actual behaviour.

- On criticism from Wilson & Trumpickaite, it is true that many measures are bivalent, i.e. cannot tell us whether an elevated response is due to positive or negative responses. However, novel measures now allow better determination of this; and even for traditional measures, we usually operate in the neutral-to-positive scale, rarely we see customers run away screaming... This means that arousal responses are typically a signal of the positive relevance and appeal that a person ascribes to a stimulus

- Thinking neuro informs your psychology: the way you use terms such as attention, memory, preference and choice are highly informed by the combined efforts of economics, psychology AND neuroscience. For example, there is solid evidence from neuroscience that we have (at least) two motivational systems with distinct speeds and processes. Not exactly the same as dual process theories, but then again converging evidence as such.

- Fishy studies abound, but more than anything, this demonstrates an honest appeal to rigorous methodology in neuroimaging measures, not something that is problematic for neuromarketing only. Then again, if we are left with surveys, interviews and focus groups, then let's take the discussion of validity here, too. We know that is a contentious topic

- Finally, while neuromarketing comes very much across as an assessment toolbox, it is so much more than this. When used properly, it is a strategic tool to shape the way information is conveyed to the recipients, how a brand is construed, and the way companies communicate. Think of neuromarketing not as something different from marketing, but a new leg to stand on that is based on rock solid science

I may be coloured on this aspect since I have spent the better parts of my life devoted to these questions. But I firmly believe that when we are able to sort out the snake oil and false promises in neuromarketing, and other places where neuro is used, we can focus on the true insights that can be gained. It's a learning process for all sides, and I find that added value can be made on every step of the way for all participants.

A basic model of consumer choice

To better understand consumer behaviour, we need a framework within which we can place all information and discussions. In this book, we will use a model that was first introduced in Plassmann, Ramsøy & Milosavljevic (Plassmann 2012). In this model (see the figure below) we can distinguish between different subprocesses in consumption behaviour, and especially in how contextual effects such as brands affect decision making. These steps are characterised by four separate stages:

• Representation and attention
• Predicted value
• Experienced value
• Remembered Value & Learning

Representation & attention

At this level, there are three basic subprocesses – representation, and two types of attention: bottom-up and top-down.

Representation denotes the way in which needs and desires are presented to us. Imagine that you are thirsty and how that feels. The sensation of thirst, and the mental preoccupation of seeking out items that can quench the thirst are good examples of this first step. The representation of thirst, both as a feeling and bodily urges for something to drink, is something that signals a vital step in an organisms well-being.

Many consumers will report having a strong desire for a particular product – such as the latest iPhone, a Snickers chocolate, or a new dress from Dolce

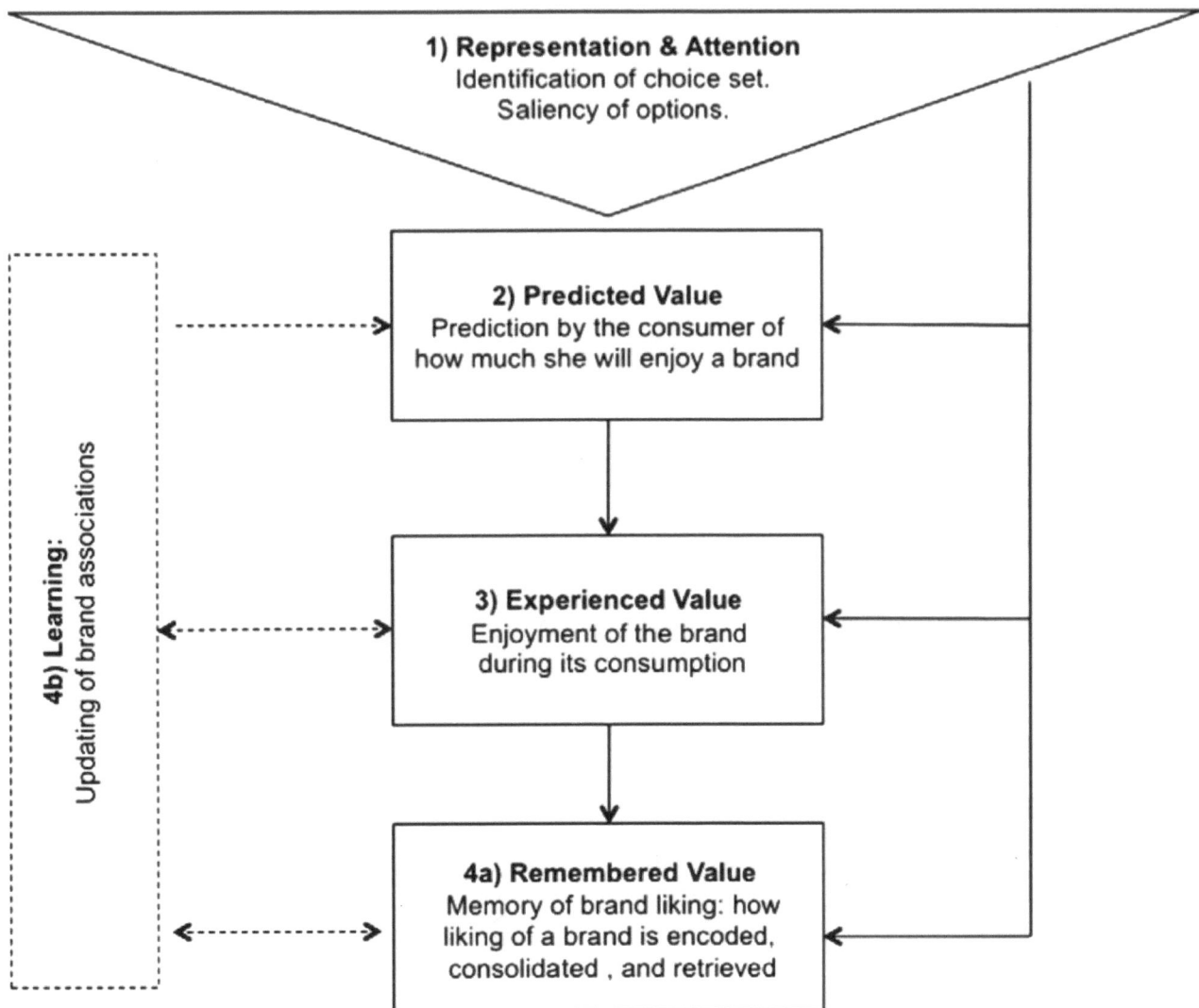

Figure   The Consumer Neuroscience model of branding, showing the cognitive and emotional steps related to branding effects. From (Plassmann 2012)

& Gabbana. People can report feeling urges equally strong to being sexually attracted or even "hungry" in the literal sense. These are all representations of needs, physical or psychological.

By "representation" we therefore refer to the way in which your mind treats desires and needs – from those that ensure your survival (such as oxygen) to those that are absolutely superfluous for survival (e.g., a new pair of expensive jeans). As we will see, your brain treats aspects of those desires in very much the same way, using the same or similar neural structures to calculate their values – although there are also some notable differences.

We will treat the representational aspects of consumer behaviour throughout this book, but especially in the chapter about emotions and feelings and the chapter about wanting and liking. As we will see, representations will show large differences in their neural foundation, depending on which representation we study.

Attention comes in two forms. On the one hand, **bottom-up attention** is the situation in which your mind gets tuned to events that occur either outside or inside yourself. A sudden explosion, a shape that is different from its surroundings, or a stomach ache all force their way into your conscious mind and simultaneously lead to changes in your behaviour, such as fear response, turning one's head or eyes, and holding a hand over one's stomach and crouching. Bottom-up attention is the process in which attention is automatically drawn to events or stimuli, driven by the properties of the information itself. Bottom-up attention is when you are ruled by your senses.

Attention is often related to specific brain systems of the frontal and parietal cortex, but how these regions are recruited can be very different. In the case of bottom-up attention, we will see that processes much earlier in the system – down to subcortical structures such as the thalamus and even the brainstem – can determine the likelihood that something will be attended or not.

The second form of attention is **top-down attention**, in which you actively choose to focus on certain aspects of the world. You could start looking for red things in your surroundings, or focus on the feeling of sitting in a new chair, or force yourself to read a difficult (but possibly important) passage of this book! Top-down attention is typically situations in which you actively search for particular kinds of information, be it something edible, a particular word in a book, or a face in a crowd. In this way, top-down attention is when you rule over your senses.

When you actively choose to focus your attention, we know that regions of the prefrontal cortex are engaged. In this way, top-down attention is very similar to many kinds of decision-making. Indeed, we will see that "paying attention" is a kind of a choice itself, and that the brain computes such decisions very differently independently of whether the attentional mechanism is bottom-up or top-down.

Attention is treated as one of the core aspects – and yet a much forgotten aspect – of consumer psychology and behaviour, in the chapter on attention and consciousness. Here, we tease out both the concepts of attention, how they rely on different brain mechanisms, and we will show that attention is a limited resource for the consumer mind.

Predicted value

Predicted value is the value that an individual assigns to different options before they are chosen, be it consciously or unconsciously. Predicted value can be seen as both a mathematical calculation, a neural process, and as a subjective feeling. Economists talk about the expected utility of a choice option; neurobiologists talk about the engagement of activity in the basal ganglia of the brain; and psychologists may talk about the feeling of anticipation or even gut feeling towards the outcome of a decision.

In fact, the talk about predicted value (as contrasted to experienced value) is essential to our understanding of consumer choice, and how neuroscience can contribute to this. In many ways we can say that the predicted value of an option is the actual driving force of choice, and that the assessment of brain mechanisms at this stage can predict actual purchase, long before the consumer has any conscious experience of making such a decision.

Experienced value

Once a choice is made and acted upon, we face the consequences of our choice. We experience the pleasure of eating a delicious cake, and (sometimes) the negative emotions associated with guilt of eating the cake rather than staying healthy. We feel the joy of purchasing a long desired product, or the disappointment of the product not living up to our expectations. In a large part, our experience of pleasure or displeasure – what we can call *hedonic experience* – relies on a set of brain mechanisms that are quite different from the mechanisms driving predicted value. As we will see, such brain activations are often "online" after a choice has been made. It is an after-the-fact matter in consumer behaviour.

This fact is worth noticing: if our conscious experience of making choices and their outcomes are something occurring after the fact, our consciousness is bound to be an imprecise proxy to understand consumer choice. If our choices are to a large extent driven by unconscious processes, why ask the conscious about why we choose how we choose? This is a core theme for consumer neuroscience and neuromarketing, and it even extends well into the ethical domains. The more researchers are exploring the buying brain, the more we see that early brain responses predict consumer choice, long before the decision maker is privy to such processes.

Remembered value & Learning

All choices have consequences. Even a choice that does not lead to anything is a consequence in itself – the choice may not have had the desired outcome (e.g., adding coins to a machine to buy some snacks and then nothing happens), or the choice may have taken something away (e.g., taking a pain killer and the headache fortunately goes away). In any case, choices and their outcomes allows us to learn from experience. We are not likely to add coins into the snack machine if it didn't give us any snack last time we tried; we are very likely to take painkillers the next time we have a headache, if painkillers worked last time we had a headache.

Decision making is unavoidably intertwined with learning: we learn from our mistakes and successes alike. Thus, every choice comes with it's own full dynamics, and as shown in the model displayed in the figure above, such learning can influence subsequent processing at all stages in a consumer decision making process.

The memory aspect of choice comes in two steps. First, the remembered value of a choice is our experienced, episodic and semantic memory of the event – what you did, where it happened, what and who was involved, what the outcome was. Second, outcomes influence subsequent choice at all stages through learning – the containment of information for some time until it is utilised. As we will see, such influences of memory can operate both consciously and unconsciously.

Together, the model proposed here serves as both an up-to-date model for understanding consumer behaviour, but also s a general pedagogical tool for guiding you, the reader, through the different sections of the book. In time, we will hopefully challenge and extend the model.

Neuromarketing and reverse inference

Using neuroscience to study consumer behaviour (and other applied approaches) has indeed captured the imagination and creativity of business and the media. There is no single day in which we cannot read about a newly discovered "centre for" a particular brain function, or some use of neuroscience to determine whether a person is lying, is conscious or not, and what the person is thinking about. This is epitomised by recent popular science books, which may indeed spur interest, but nevertheless make extreme logical mistakes. Before even embarking on our journey to understand the brain and the consumer, let us first kill the simplistic view of the mind and brain.

In his famous book "Buyology: truth and lies about why we buy" (Lindström 2010), Martin Lindstrøm wrote the following:

"A brain-imaging experiment I conducted in 2006 explains why antismoking scare tactics have been so futile. I examined people's brain activity as they reacted to cigarette warning labels by using functional magnetic resonance imaging, a scanning technique that can show how much oxygen and glucose a particular area of the brain uses while it works, allowing us to observe which specific regions are active at any given time.

We tested 32 people (from Britain, China, Germany, Japan and the United States), some of whom were social smokers and some of whom were two-pack-a-day addicts. Most of these subjects reported that cigarette warning labels reduced their craving for a cigarette, but their brains told us a different story.

Each subject lay in the scanner for about an hour while we projected on a small screen a series of cigarette package labels from various countries -- including statements like "smoking kills" and "smoking causes fatal lung cancers." We found that the warnings prompted no blood flow to the amygdala, the part of the brain that registers alarm, or to the part of the cortex that would be involved in any effort to register disapproval.

To the contrary, the warning labels backfired: they stimulated the nucleus accumbens, sometimes called the "craving spot," which lights up on f.M.R.I. whenever a person craves something, whether it's alcohol, drugs, tobacco or gambling."

So what is really the claim here? Let's summarise the logic:

1. Viewing warning signs led to stronger activation of the nucleus accumbens in smokers
2. Prior research has linked the nucleus accumbens to craving.
3. Therefore, the activation in the nucleus accumbens means that smokers who see the warning signs are led to crave cigarettes...right?

Do you spot the fallacy? It's called **reverse inference**, and the "logical" structure is as follows:

1. We observe X
2. X has previously been linked to Y
3. Therefore, our observation of X must mean Y

This probably highlights the problem a bit more. The mistake happens when we assume that prior research can be blindly applied to our current finding. Yes, it is true that the nucleus accumbens (also known more generally as the ventral striatum) is reliably engaged during reward expectancy and craving (Knutson 2001; Adcock 2006; Knutson 2007; Ikemoto 2010). So from this perspective, the logic seems sound.

But what if we had evidence showing that the same region was involved in other functions? For example, in a study by Levita and colleagues, it was found that activity in the nucleus accumbens was high for both positive and negative outcomes (Levita 2009). This destroys the simple argument: if we cannot reliably claim that the activity in one brain structure is related to one function, then we cannot just "look at" a particular brain activation and say what the person is thinking or feeling!

In a more recent book, "Brandwashed: Tricks companies use to manipulate our minds and persuade us to buy" (Lindström 2011), Lindstrøm makes a similar statement:

> ""Using fMRI for my new book Brandwashed we discovered that [when iPhone users watch images of an iPhone] there's a flurry of activation in the brain's insula-- which is connected to feelings of love and compassion – in short the subjects loved their iPhones; responding the same way they would respond to their boyfriend, girlfriend, niece, nephew, or family pet."

Again, the logic is flawed:
1. The insula is activated when people look at their iPhones
2. Prior studies have found the insula to be engaged when people look at people they love
3. This must mean that the insula activation must mean that people love their iPhones...right?
4.
Do you spot the fallacy?
Indeed, there are many studies linking the insula to reward (Preuschoff 2008; Hare 2010; Sescousse 2010; Rutledge 2010; Smith 2009). But there are equally many – and probably more – studies linking the insula to numerous other functions, including negative emotions such as disgust (KrolakSalmon 2003; Wright 2004), arousal (Decety 2004; Berns 2010; Dreher 2007), avoidance behaviours (Canessa 2013; Nitschke 2006; Kuhnen 2005; Dreher 2007), and consciousness (Craig 2009; Christensen 2006). Again, this kills the argument. We **cannot** just look at a particular brain response and infer a person's mental state. This was clearly stated in a blog post from Russ Poldrack, a highly reputable neuroscientist:

> "Insular cortex may well be associated with feelings of love and compassion, but this hardly proves that we are in love with our iPhones. In Tal Yarkoni's recent paper in Nature Methods, we found that the anterior insula was one of the most highly activated part of the brain, showing activation in nearly 1/3 of all imaging studies! Further, the well-known studies of love by Helen Fisher and colleagues don't even show activation in the insula related to love, but instead in classic reward system areas. So far as I can tell, this particular reverse inference was simply fabricated from whole cloth."

Still, reverse inference is not false. If we, indeed, had a perfect understanding of the brain and the mind, reverse inference would be perfectly legitimate. If we were able to discern some very specific brain response that was only related to one mental operation, then finding that particular brain response would indeed be indicative of the mental operation.

So the logic behind reverse inference needs to be clarified:
1. We observe brain response X
2. Brain response X is active if and only if mental function Y is present
3. X must mean Y

So if we have a perfect understanding of the brain, and this understanding allows these inferences, then we can make those statements. As long as we do not have a perfect understanding of the brain and mind relationship, we cannot allow ourselves to make 100% reverse inferences.

CHAPTER 2

The Brain

The brain is by far the most complex organ we know. Today, we can distinguish between the different layers of the brain, and measure brain activity by the millisecond. To the newcomer it may seem unsurmountable to get a grasp of the many different regions and processes that the brain consists of.

So we need a basic brain model. Below you can find a model illustrating the surface of a brains seen from the (left) side: To the untrained eye the brain is a chaos of folds, hills and valleys of organic tissue. However, as we will see, just as we can use a map to orient ourselves in the external world, we have both directions and landmarks to orient ourselves in this "inner" space.

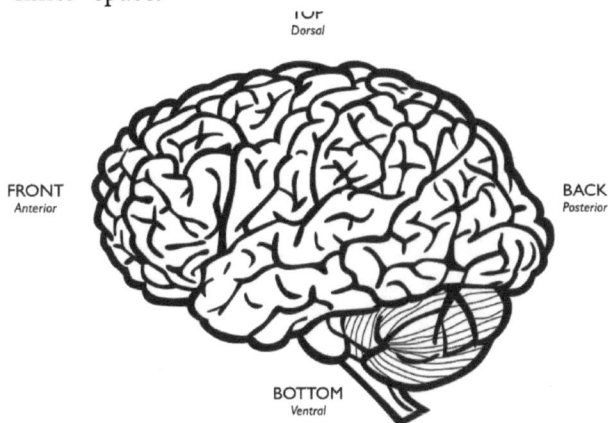

Figure The Brain Compass model to understand brain orientations.

THE BRAIN COMPASS

To orient ourselves, we need a set of rules. Here, we introduce the terms dorsal, ventral, posterior, anterior, medial and lateral. They all bear specific meanings, and as we will see, we can use those meanings to orient ourselves on where in the brain we are.

Imagine that you are reading an article reporting increased activation in the "ventromedial prefrontal cortex" when participants report enjoying the taste of wine. Here, we will show you exactly how to find that region.

The different orientations we use to orient ourselves in the brain come along three orthogonal planes:

- **Anterior – posterior**: this is the terminology for what is forward (anterior) and back (posterior) in the brain. To some extent, we can also use terms like rostral – caudal.
- **Dorsal – ventral**: this is used to point to regions that are either upwards (dorsal) or downwards (ventral) in the brain
- **Medial – Lateral**: this is the dimensions assigned to things more towards the middle (medial) or sides (lateral) of the brain. This dimension is not shown in the figure above.

With this in mind, we now know that for a region entitled "ventromedial prefrontal cortex", we can already guess that within this "prefrontal cortex", we should look towards the bottom (ventral) and middle (medial) parts of this structure. As we will see, the regions we use are gross and helps us understand where we are in the brain.

LATERALITY AND STRUCTURAL DOUBLING

Most brain structures are bilateral, meaning that we have two samples of each, basically due to the way in which our nervous system is organised. Just as you have two sides of your body, and a dual set of most body parts, so the brain also has a dual – or bilateral – side. Each cortex should thus actually be referred to as cortices, and individual structures should be described using their rightful plural name – amygdalae, thalami, nuclei accumbens etc. However, to avoid such complicating terminology, we will refer to each brain structure using only the singular term – amygdala, thalamus, nucleus accumbens – unless there is good reason to treat hemispheric differences.

THE OCCIPITAL LOBE

If there is one thing that the human brain is focused on, it is visual input. Of the five senses (sight, sound, touch, taste and smell), our brain shows an unparalleled use of "brain real estate" for the visual sense. The occipital lobe is placed at the back of the human brain. In fact this is a tell tale of the fact that our brains have evolved rather than being carefully crafted and designed to optimise information flow. After all, why put the visual brain as far away from the eyes as possible? Despite it's many virtues and amazing properties, the brain is not well designed, as we shall see plenty of times throughout this book.

Visual input from the eyes travel through the visual radiations through deep brain structures such as the thalamus and projects further back to the primary visual cortex. Here, the visual signals are processed more thoroughly in smaller functional units. The visual input are undergoing a "divide and conquer" processing: the visual cortex has several distinct modules that have specialised functions, such as processing edges and contours, others for colour, and yet others for motion.

Processing in the visual cortex is further projected to other regions such as the parietal and temporal cortices for the processing of position/action and object identify, respectively. The visual system is a good model for understanding many general properties of the brain, including how it processes information and creates a representation of the world in itself.

What does the region "do"?

In many ways the occipital cortex is a "simple" structure, although it is still way too complex for us to have a firm grasp of how it works. Yet, the occipital cortex is relatively simple to understand in terms of its functions, as it is solely dedicated to vision. As we will see, this is not the case for the other brain lobes.

Functions associated with the occipital cortex include:

- Primary visual sensing
- Bottom-up attention
- Specialised functional units: contour, colour, movement etc.

Damage to this region can lead to:

- Full or partial cortical blindness – an inability to see consciously but with paradoxical residual visual capacity (blindsight)
- Inability to see colours – not the same as colour blindness
- Inability to see movements – the world looks like still images for seconds that change every few seconds

Functions related to consumer behaviour

- Bottom-up attention – how density, contrast, brightness, colour, orientation and other visual properties automatically attract attention.
- Basic functions of brand and product recognition – including response to specific features such as red ("I'm looking for a Coke").

THE PARIETAL LOBE

The parietal lobe of the brain is located above (dorsal) to the occipital lobe, stretching towards the top of the brain. It covers the sides (lateral) of the brain as well as the middle (medial) sides of the brain on each hemisphere.

The parietal cortex receives input from certain parts of the occipital cortex, in particular information related to position and movements. It also receives much information from the other senses. In particu-

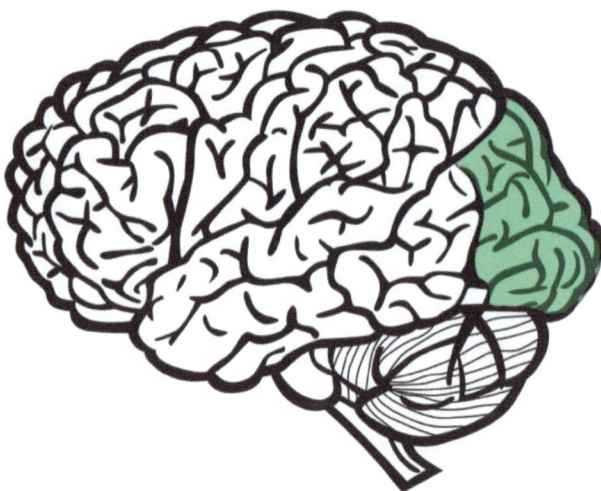

Figure   The primary visual cortex is positioned at the back (posterior end) of the cortex.

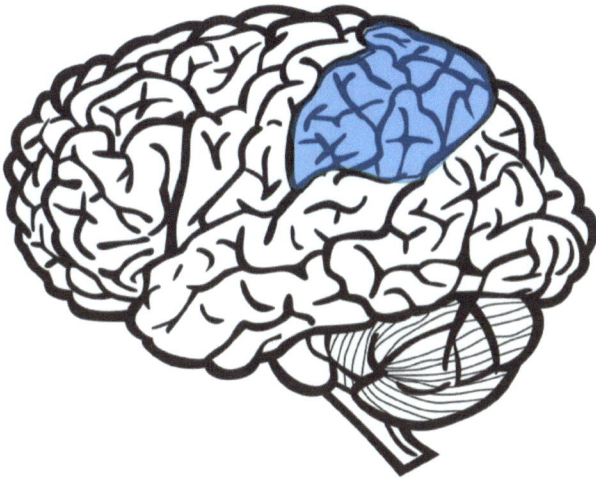

Figure The parietal cortex is positioned at the back-top (dorsal) end of the human brain.

lar, the most anterior parts of the parietal cortex has a neat 1:1 representation of our body, allowing us to map different parts of the body just from observing the location of brain activation along this strip.

The parietal cortex has been implicated in a variety of functions. First, there is the inescapable and hardwired 1:1 relationship between the body and brain found at the somatosensory strip. Beyond this, it is well known that different regions of the parietal cortex have been implicated in a host of different mental functions, including attention and consciousness, self-awareness and social reasoning such as empathy and mentalising, as well as in more rudimentary processing of objects in space and how to deal with them.

What does the region "do"?

- Spatial processing ('where')
- Object action ('how')
- Navigation
- Self-awareness
- Body sense and body representation
- Attention & consciousness

Damage to this region can lead to:

- Loss of specific body senses
- Navigation problems
- Apraxia – inability to know how objects are used
- Unilateral neglect – inattention towards one side of the body and world
- Anosognosia – inability to recognise own problems and limitations (often accompanies unilateral neglect)

- Attentional dysfunction, including lack of attention control

Functions related to consumer behaviour

- Top-down attention – our ability to wilfully focus our mental effort on certain items, such as an active decision to search for particular products or brands.
- Product handling – our ability to learn to operate products, such as a smartphone, operating system, and dishwasher.
- Product navigation – assistance on navigating real or virtual stores, web pages and magazines.

THE TEMPORAL LOBE

At the bottom part of the cortex we find the temporal cortex. Along the posterior–anterior axis it spans from the visual cortex and ends at the temporal pole. The medial side of the temporal lobe is highly intertwined with del structures of the brain such as the basal ganglia and the diencephalon.

The temporal lobe receives an abundance of information from a large portion of the brain, and thus serves as one of the brain's "convergence zones", where information from different senses aggregate and are processed conjointly.

The temporal cortex is involved in a host of different functions. Most notably, it has been show to play a central role in visual processing of object identity, where vision goes from crude processing (in the occipital cortex) to more sophisticated object processing and cataloging. Indeed, the temporal lobe – especially it's bottom/ventral parts – is highly connected to our

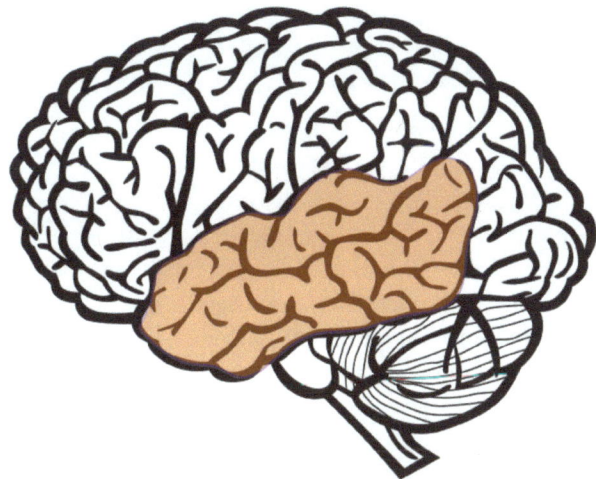

Figure The temporal lobe make out the bottom (ventral) parts of the brain.

ability to recognise a brand and a product, as well as other things such as items, places and faces. In the same way, through increasing levels of processing, the temporal lobe is crucially implicated in certain kinds of memory, such as our ability to remember events and have factual knowledge. Parts of the (dorsal) temporal lobe also receive auditory inputs, and are thus critical to our ability to hear and understand speech. Moreover, the temporal lobe has been implicated in certain aspects of social reasoning and behaviour.

What does the region "do"?

- Recognition of objects, places and faces
- Hearing and perception of sounds
- Certain kinds of memory, such as declarative, conscious memory
- Social perception and reasoning

Damage to this region can lead to:

- Agnosia – inability to recognise objects, places and/or faces
- Hearing disabilities
- Language disability, especially understanding of spoken and written language
- Amnesia – loss of episodic and semantic memory (but preserved procedural and other memory types)
- Deficient social cognition

Functions related to consumer behaviour

- Processing, learning and conscious memory of brands and products
- Understanding of communication (e.g., via ads)
- Social framing effects

THE FRONTAL LOBE

The frontal lobe is by far the largest portion of the human cortex. It occupies the frontal 1/3 or so of the cortex, and extends from the top-most (dorsal) parts of the brain and skull, to the bottom-most (ventral) parts that neighbours the deep brain structures and almost resting upon the eyes; from the frontal-most (anterior) parts of the brain resting just behind your forehead, to its back (posterior) end that is set apart from the parietal lobe through a deep valley entitled the central sulcus.

Due to its sheer size, the frontal lobe can be expected to be involved in a number of different functions. Indeed, by looking at a connectivity map of the

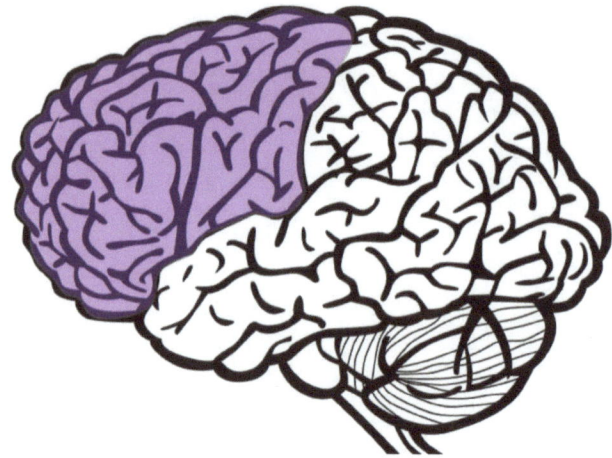

Figure   The frontal lobe occupies a large portion of the brain.

brain, which displays the highways of information across the brain, we see that the frontal lobe very much operates as a convergence zone for the abundance of sensations, emotions and thought

In terms of functions, the frontal lobe can very much be seen as a region implicated in all walks of decision making. Just as the parietal cortex has a nice 1:1 mapping of the sensory input from each body part, so the frontal cortex has a "motor strip", just adjacent to this "sensory strip", that has a finely tuned 1:1 mapping of the body. If one stimulates this part of the brain, it is possible to induce jerking of an arm, a chin etc, just by moving the stimulation physically across the motor strip. From here, more anterior regions are engaged in increasingly complex behaviours, such as motor sequences (like playing a piano), ultimately to long-term planning. Crucially, the frontal lobe has an important role in behavioural control – both initiation and inhibition – through the "executive system". Other parts of the frontal lobe are implicated in hedonic experience – the conscious perception of pleasure – and thus also in certain aspects of motivated behaviours. The frontal lobe has also been implicated in certain kinds of attention, especially top-down attention and in consciousness, as well as in certain kinds of memory, such as working memory. Finally, the frontal lobe has been implicated in social behaviours, such as empathy, social reasoning and social choice.

What does the region "do"?

- Crude and sophisticated motor skills
- Planning and executive control
- Top-down attention and consciousness
- Working memory
- Hedonic experience
- Social cognition and behaviour

Damage to this region can lead to:

- Paralysis
- Impulse control problems
- Lack of initiative
- Loss of working memory

Functions related to consumer behaviour

- Consumer choice execution
- Long-term planning of actions
- Impulse control (or lack thereof)
- Manual handling of objects
- Hedonic experience – pleasure/displeasure perception – of products, events, services, brands etc.

DEEP STRUCTURES OF THE BRAIN

For a long time, the relationship between the brain and the mind was highly focused on the cortical surface of the brain. Anything beyond and especially below the brain's brim was neglected, thought of either as less important with respect to actual human mind and behaviour, or as too complex and inappropriate to study. Neuropsychologists thought that their primary brain region was the cortical mantle, leaving neurologists to study what they thought of as the boring and rudimentary functions of the subcortical underworld. In many ways, the view about the relationship between the brain and the mind was 'cortiocentric' – the cortex was the most important parts of the human brain.

With a few exceptions, science had to wait for the advent of modern neuroimaging to face new methods to study the entire brain, especially during it's deeper parts. In early studies, we can still read the unwillingness to report the strange subcortical engagement to higher order functions. How could it be that the basal ganglia were involved in decision-making? Why do we see cerebellum activity to cognitive tasks? How can it be that the thalamus is involved in awareness?

It took a decade or so to realise that the cortex was not the only important part of the human brain – deep regions had also evolved and were involved in human's complex processing capacity. Despite this realisation, we are still not at the point of understanding these functions properly, and it is a field of active research.

Here, we will treat some of the regions that are most crucial to consumer neuroscience and neuromarketing. As such, this means that we are leaving out several regions such as the cerebellum, habenula, and pons, to focus our attention on regions such as the midbrain, basal ganglia, thalamus, amygdala, hippocampus, cingulate cortex, and insula.

Some of the important subcortical regions are shown in the following figure:

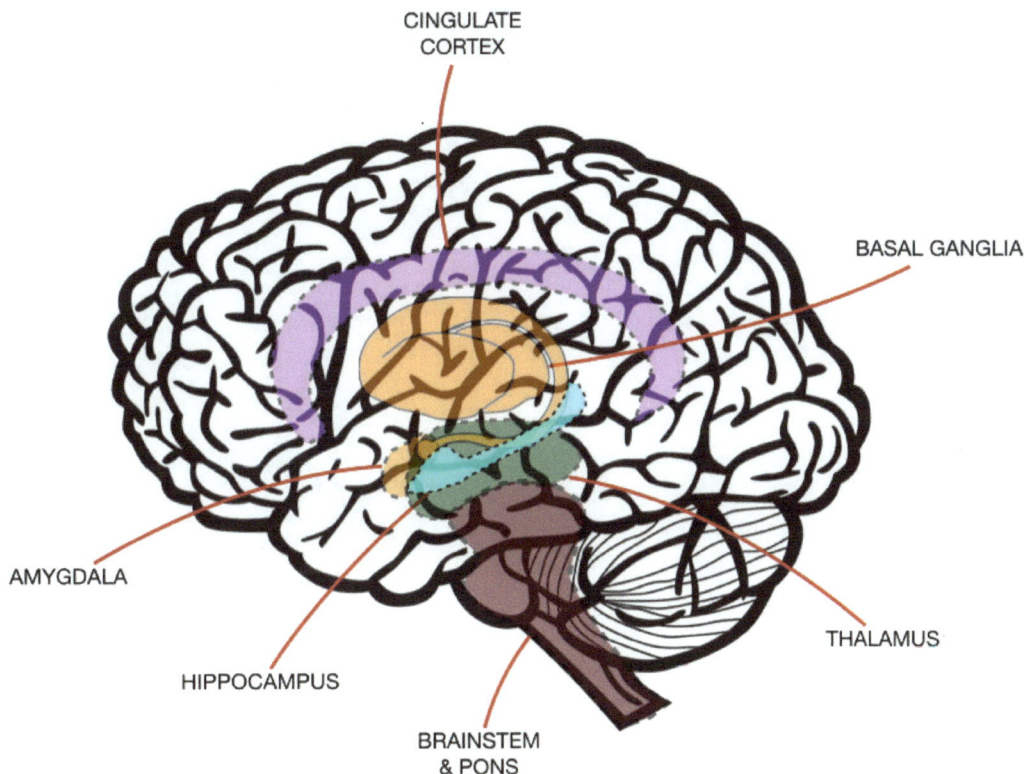

Figure Subcortical structures relevant to consumer behaviour (and beyond).

The autonomous nervous system

Before looking at these systems, we need to take look at an even more basic brain region: the autonomous nervous system. This system crucial to most of the emotional responses we will be treating in this book. The autonomous nervous system is divided into three parts: the sympathetic, the parasympathetic and the enteric nervous system.

The sympathetic and parasympathetic nervous system work in opposition in activating and relaxing the body, sometimes making it ready to respond and act, and other times allowing the body to relax and recoup. Below, you can see the main distinctions between the two systems and their effects on the body:

FUNCTION	SYMPATHETIC	PARASYMPATHETIC
Pupil	Dilation	Constriction
Salivation	Inhibits	Stimulates
Pulse	Increases	Slows
Respiration	Increases	Decreases
Digestion	Inhibits	Stimulates

As can be seen from this table, the two systems work in opposition. As we will see, emotional response activate sympathetic responses, which is associated with an inhibition of digestion and increase in "body preparation." The absence of danger or other arousing events is related to increases in digestion and other "body relaxing" processes, including lower respiration and pulse.

Brainstem and upwards

From the spinal cord, the most basic parts of the brain is the brainstem, pons and midbrain. These regions maintain several basic functions of the brain and human mind, and survival is almost impossible upon lesion to these regions. For example, certain regions of the brainstem are responsible for respiration and pulse, and all projections between the brain and the body go through this region, ensuring bodily sensations and movement control at all levels.

Certain brainstem nuclei are also found to regulate different levels of awareness, and studies have demonstrated that stimulation or inhibition of the brainstem can control whether an animal is conscious or not. This suggests that one basic function of the brainstem region is to control awareness.

Regions of the midbrain are also well known to be central to the production of certain neurotransmitters – signal substances that brain cells use in communi-

cation. In particular, it is all known that dopamine is synthesised and processed in three different regions of the midbrain (ventral tegmental area, substantia nigra, and retrorubral area), but the actual mechanisms and distinct roles are still being mapped.

In general, the brainstem regions are undergoing much reconsideration after decades of treatment as "simple" and basic function regions of the brain and central nervous system. Today, we see that these regions in no ways are "fixed in time", but rather have co-evolved with the increasing complexity of mammals. We also see that many of these regions are engaged in what is usually considered complex psychological mechanisms, such as social behaviour and other kinds of decision making.

Thalamus

The thalamus operates as one of the most central gateways of information in the brain. It rests on top of the brainstem region (especially the anterior most part, i.e., the midbrain). It is a collection of multiple nuclei that are still arranged in a systematic way to provide a global outreach.

One function of the thalamus is to convey information from the outer senses to the more sophisticated processing of the cortex, while in doing so also providing basic processing of such input. For example, when relaying visual information, thalamic substruc-

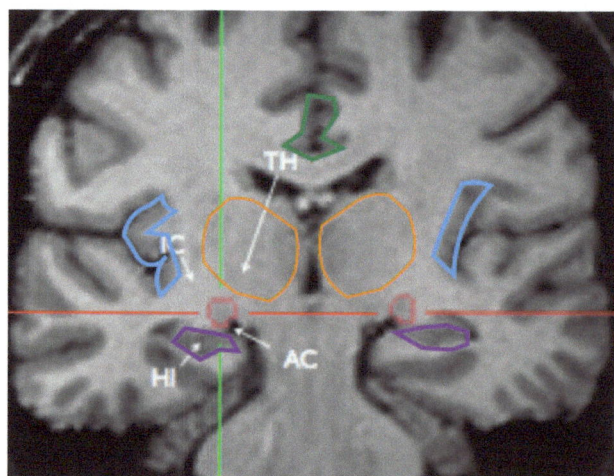

Figure Example of the Lateral Geniculate Nuclei (LGN) in a coronal slice of the brain (red circles). The slice also allows identification of other brain regions such as the hippocampus (HI, purple) and the vaguely defined thalamus (TH, orange), as well as the insula (blue) and cingulate cortex (green). As one can see, the LGN are at a relative distance from the thalamus yet is considered a part of the extended thalamus. Structural landmarks such as the internal capsule (IC) and ambient cistern (AC) are used mainly to determine the LGN location and extent, and not of relevance here.

tures such as the lateral geniculate nucleus (LGN) are known to have subcomponents that process either high-resolution information (allowing for detailed processing) or more gross macroscopic properties of visual input (allowing detection of gross changes and movements).

Another example is the pulvinar (a nucleus within the thalamus), which is well known to provide crude "relevance" determination of the outside world, and its responses are well known to be associated with strong emotional responses such as shock (behavioural "freeing"), fight and flight, in which the pulvinar exercises its effect through the amygdala and other regions.

The thalamus also serves another important function: it distributes information across the brain. In many ways, the thalamus is a hub that allows "globalisation" of information in the brain, hinting at a role for consciousness in behavioural flexibility and higher degree of control over subprocesses. Several models of consciousness, including yours truly, claims a central role of the thalamus, including in disorders/distortions of consciousness (e.g. epileptic seizures, coma, general anaesthesia, deep sleep).

Basal ganglia

As with the thalamus, the term "basal ganglia" covers a host of different structures in the deep regions of the brain, including the striatum, pallidum, substantial nigra, and subthalamic nuclei. The functions of the basal ganglia have long been known to include motor behaviour, and until only recently it was believed that this was the only functions. But in the more recent years it has been realised that this area plays important roles in other functions, such as reward and anticipation, as well as certain motivated behaviours.

One of the most prominent structures we will see in this book is the striatum and its subcomponents – caudate nucleus, putamen and nucleus accumbens (NAcc), as shown in the figure.

The NAcc is a structure that is engaged every time we are talking about motivation and choice in consumers. As we will see later in this book, the NAcc is a structure that is not only highly related to emotional responses, or to prediction of outcomes, but also to a very specific kind of motivation – what we will call a *wanting* motivation.

Amygdala

If there is one single subcortical structure that has fame, it has to be amygdala. Located almost lying on

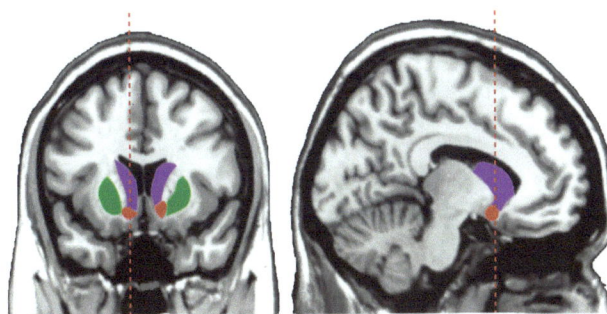

Figure The basal ganglia shown in both a coronal section (left) and a sagittal section (right). The basal ganglia substructures are shown, including the caudate nucleus (purple), putamen (green) and the nucleus accumbens (red). Dotted red lines indicate sections, to provide a common reference.

top anteriorly and dorsally relative to the hippocampus, it receives input from all senses, both directly from the thalamus (pulvinar) but also through cortical processing.

All popular accounts of the amygdala stress it's role in emotions, especially negative responses such as fear and anxiety. In fact, for a long time we have thought that the amygdala was a pure fear structure.

Today's accounts of the amygdala are much more nuanced. First, we see it not as a structure, but a conglomerate of many substructures. Also, several studies have now implicated amygdala in a host of non-emotional functions, such as novelty processing and object processing. Even within the emotional domain, it has been found that the amygdala both processes positive and negative outcomes, and the anticipation of positive and negative outcomes. This suggests that the amygdala might operate more as a "relevance barometer" rather than a purely fear-based response mechanism.

Hippocampus and the medial temporal lobe

At the innermost (medial) side of the temporal lobe, we find a tightly bundled bunch of structures commonly referred to as the medial temporal lobe (MTL). Even despite its collective label, the MTL cannot be defined as a cortical (or subcortical) region per se. It is rather to be considered something of a mix between the cortical and subcortical regions.

Being the most prominent member of the MTL, the hippocampus has long been known to be involved in the kinds of memories that we can explicitly state that we know – who is the current president of the US; remembering the last movie you watched. But through a better understanding of the MTL and the informa-

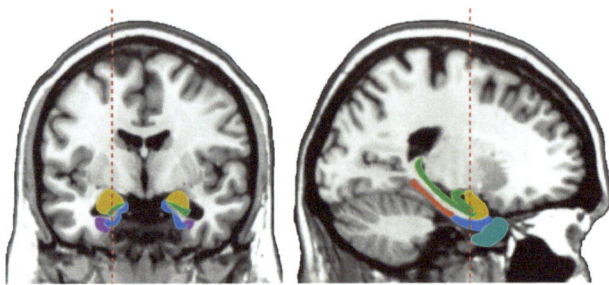

Figure Illustration from the coronal section (left) and saggital section (right) of the MTL region showing the amygdala (yellow), hippocampus (green), entorhinal cortex (blue), perirhinal cortex (purple), parahippocampal cortex (red) and temporopolar cortex (cyan). Dotted red lines indicate sections, to provide a common reference.

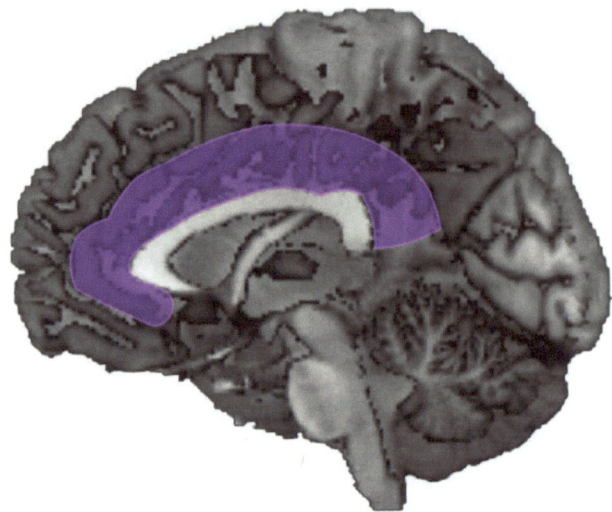

Figure The cingulate cortex, which is often subdivided into an anterior and posterior end.

tion flow to and from this region, we now understand that it is one of the important convergence zones of the brain, where many kinds of sensory information are brought together to compose coherent *episodes* of ourselves and the world around us, and join many different kinds of associations together. As we will see, our ability to recognise brands, products, places and people depends on this intricate set of regions.

The MTL is usually noted to consist of the following structures: hippocampus, entorhinal cortex, perirhinal cortex, parahippocampal cortex, and many researchers also include the temporopolar cortex and even the amygdala (see figure).

Together, the structures encompassing the MTL can often be considered a functionally integrated unit, especially when it comes to processing integrated "events" as whole episodes, and thus encoding and remembering them as such, especially the hippocampus. The MTL has also been implicated in a number of other functions, such as novelty processing, semantic processing, object and place processing, as well as emotional responses (in particular the amygdala).

Notably, the entorhinal cortex is an important structure in the processing of odour, and with its direct connection to the hippocampus, this can be said to be the structural foundation for odour's ability to automatically trigger memories.

Cingulate cortex

This part of the cortex is typically considered a part of the more ancient parts of the brain, but it also has a tight relationship with "higher" cortical regions. The cingulate cortex is typically divided into anterior and posterior (and sometimes a middle) sections.

Several studies have demonstrated a role for the anterior cingulate cortex (ACC) in decision making,

including monitoring and resolution of emotional and choice conflict (Botvinick 2004; Braver 2001), but also in functions such as action-outcome prediction (Alexander 2011), as well as error detection, conflict monitoring, stimulus-response mapping, familiarity, and orienting (Wang 2005).

The ACC has also been implicated in social and moral reasoning, as in a study by Greene and colleagues (Greene 2004a) in which it was found that both the anterior and posterior cingulate cortex, in conjunction with the dorsolateral prefrontal cortex, were strongly engaged when participants were solving difficult versus easy moral problems.

In a study of consumer preference for brands, Schaefer and Rotte (Schaefer 2007) demonstrated that the perception of familiar brands engaged the posterior cingulate cortex in conjunction with the hippocampus, and superior frontal gyrus. When looking at value brands, the ACC was activated together with the superior frontal gyrus.

Insula

The insula is often seen as a hidden part of the cortex, as it lies "insulated" within the folds of each side of the brain. The insula has been implicated in numerous functions, including emotions, consciousness, and value-based decision making.

In a study by Tusche, Bode and Haynes (Tusche 2010) it was reported that even when products were not attended, activation in the insula, together with the medial prefrontal cortex, was predictive of subsequent product choice. In a study by Knutson and colleagues

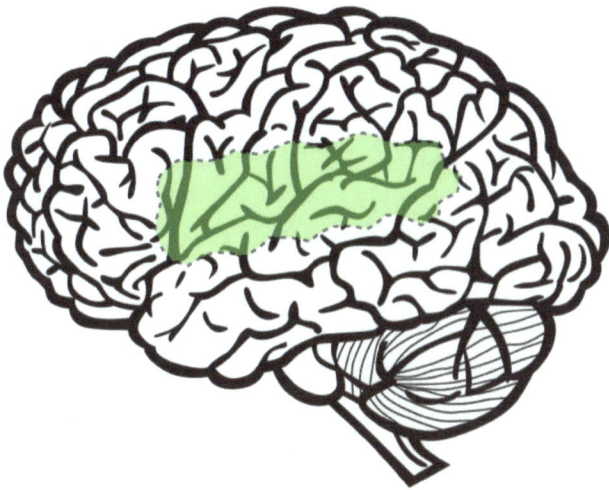

Figure The insula is hidden from view when the brain is seen from the side.

(Knutson 2007) it was found that excessive price for a product was associated with stronger activation of the insula and lower likelihood of purchase. Thus, the role of the insula in choice is not yet fully understood, but it is possible that the (anterior) insula shows a bivalent response pattern, i.e., response to both positive and negative events.

STRUCTURES AND FUNCTIONS

While we go through different regions of the brain, the observant reader might have noticed that the relationship between structures and functions is complex. When reading the list of what a particular brain region does, you may have been struck by the many and diverse functions implicated within the same relatively small region. At the same time, you may have observed that some functions are repeated between the structures. As we will see, this hints about some basic principles about the organisation of the brain and its mind.

Compared to the not-too-ancient phrenologists who believed in a tight relationship between the brain and its functions, our current understanding of the brain is, if anything, much more complex and less straightforward. We wish that we could say that one structure was responsible for one function, just as journalists frequently talk about a "reward centre" or a "fear centre." But it is a futile wish nevertheless.

Today, we know too much about the brain to accept such simplifications. Rather, we need to embrace the complexity of the brain, not as a mount improbable, but as a dynamic, complex and tightly interconnected structure. In doing so, we may also be highly inspired by other disciplines in biology, including ge-

netics. Here, in a very parallel fashion, one have had to realise that there is no neat one-to-one relationship between a gene and its function. Rather, a specific gene may be involved in many functions, and a single function may have many genetic building blocks.

In the brain, we too see this complexity. There is no neat one-to-one mapping between a brain structure and a function. We do not have a reward structure any more than we have a moral centre in the brain. But instead of resorting to statements like "oh it's very complex and let's leave it at that", we can embrace this complexity in very specific ways. Here, I will point to two ways in which we can relate brain structure to function. As Price & Friston (Price 2005) put it:

> "Functional neuroimaging data preclude a one-to-one mapping in two ways. First, attempts to manipulate a "single" cognitive process (e.g., semantics) often elicit a distributed pattern of activation over many areas (i.e., a one-to-many mapping from function to structure). Second, the same brain region, or set of regions, may be activated by tasks with different cognitive processes (i.e., a many-to-one mapping). In short, there is a many-to-many mapping between cognitive functions and anatomical regions, with a range of cognitive processes emerging from different patterns of activation among a limited number of brain regions" (p. 262)

Thus, we may talk about two ways in which we need to understand the brain: a one-to-many mapping and a many-to-one mapping:

One-to-many mapping – When we study brain activation while a person performs a simple task, such as perceiving a brand, we observe a flurry of activation in many brain regions, not just a single brain region. This suggests that for us to perceive that brand we need to engage many brain structures. Thus, we do not have a "brand centre," but rather a network of brain regions that together lift the task of perceiving the brand. This one-to-many mapping is also called *pluripotentiality* in both neurobiology and genetics.

Many-to-one mapping – The second type of mapping exists when we observe a single brain region and realise that it can be engaged by what seems to be many different functions. For example, the amygdala was long assumed to be a fear structure. Today, studies have implicated the amygdala in reward expectation, novelty responses and object processing. This many-to-one mapping is often referred to as *degeneracy* and *redundancy*.

The distinction between these two concepts is only slight:

- **Degeneracy** – a function is performed by two or more dissimilar regions (e.g., in blind people the "visual" occipital cortex tends to respond to auditory and tactile stimuli)

- **Redundancy** – a function is performed by two or more identical regions (e.g., the control over the middle finger can be taken over by the brain region for the index finger)

FUNCTIONAL DISSOCIATIONS

Still, many studies of brain functions can still employ some basic learnings from other disciplines such as neuropsychology. Through decades, neuropsychologists and neurologists have studied the role of specific regions in mental functions, and used studies of brain lesions, accompanied by animal lesion studies, to better understand these relationships. In particular, one aim is to identify so-called dissociations, which can be summarised as two main categories:

Single dissociation – This is a case where one would find two groups (A and B) with different brain lesions, in which only one group (A) has a functional deficit, such as the inability to recognise objects. The other patient group, who do not have this lesion, do not suffer from this problem. The good thing about this contrast is that one can rule out that the functional deficit (I.e., recognising objects) is not due to just having a brain lesion, but a very specific one. Such studies may help pointing to a role for one particular brain region in a particular function.

Double dissociation – By the same token, studies of two groups may also reveal cases where what seems to be one function can actually be teased apart as two separate structure-function relationships. In the case of objects, such as brands, we know that patients with a lesion to the ventral (inferior) temporal lobe may suffer from the inability to name or recognise objects, but still with an intact ability to point to where the object is and even demonstrate its use. Conversely, another group with a lesion to the dorsal parietal cortex will show an intact ability to name and recognise an object, but be fully incapable of demonstrating what it can be used for. Thus, double dissociations are good for demonstrating how certain brain regions are important for specific functions.

MINDFUL NETWORKS

A brief historical overview of neuroscience will tell us a major part of all work has focused on the role of specific brain regions in mental functions and behaviour.

In many ways, the whole enterprise of brain scanning methodologies were run under the heading "human brain mapping", leading to the many activated blobs we still see in the popular press. Here, region X is found to be involved in function A, much in the way that we saw about the method of single and double dissociation method from neuropsychology.

During the past few years, there has again come a strong focus on the actual network of the brain. As you will be able to read in any neurobiology textbook, the brain consists of billions of brain cells (neurons), and top that off with at least twice as many "supporter" cells such as glia cells and astrocytes – actually outnumbering the "true" brain cells. Notably, each brain cell is thought to have around 10.000 connections, which in total means that the brain holds more connections in total than there are stars in the known universe!

If there is one thing that this tells us, it is that the brain is an information processing device. The brain, in other words, is extremely apt at making cells mutually influence each other. This is also why it does not make sense to talk about "brain centres" for specific functions. Rather, we should focus on which networks are engaged – locally and globally – in subserving specific mental operations and behaviour.

Today, we know much more about the functions of these networks than ever before. Modern techniques such as Diffusion Tensor Imaging (another way to use a Magnetic Resonance Imaging machine) allow us to better understand how regions are connected, and how they are the foundations of human thought and behaviour.

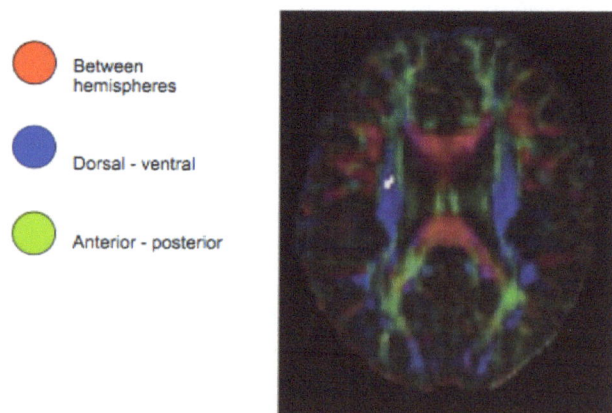

Figure Diffusion tensor imaging (DTI) allows researchers to assess the fibre tracts across distances within the brain. In this image, the brain is seen from above and the fibre tracts are running from the brain's front to back (green), sidewise (red) or up and down through the brain (blue). Changes and differences in this network has been found to play a substantial role in cognitive and emotional processes (Vestergaard 2011; Madsen 2010).

A brain in three parts?

There are many popular notions about the brain, and a few prevalent ones are important to address. One particular view stresses that the brain is divided into three major parts – the reptilian complex, the paleomammalian complex and the neomammalian complex. Essentially, this idea – labelled the Triune Brain model – is that these three parts are evolutionary and functionally divided. In this vein, the reptilian complex is thought of as a part of the brain that humans shares with reptiles, both structurally and functionally. It is as if this part of the brain is frozen in evolutionary time.

This division can be traced back to MacLean's Triune Brain theory (Maclean 1973; Maclean 1970; Maclean 1990). Notably, it is a theory that was posited decades ago, and that LeDoux famously debunked in his book "The emotional brain" (Ledoux 1998). Nevertheless, the idea is infamously used in popularised accounts of how the brain works.

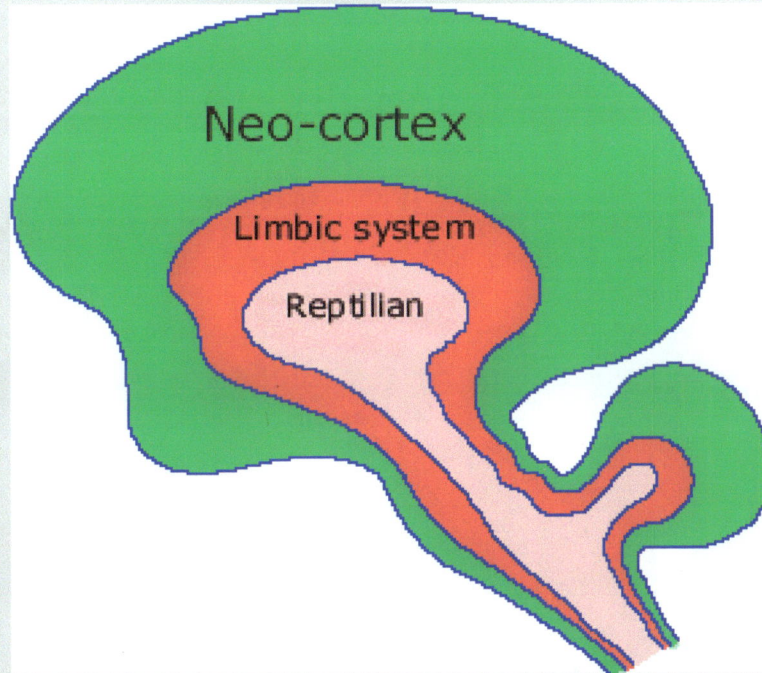

Figure The Triune Brain as illustration. The core assumption of this model is that "lower" and more primitive parts of the brain are frozen in time.

The view, despite it's fame, has plenty of problems. Most notably, regions are not "frozen" over evolutionary time, but have been found to have evolved tremendously over the ages. For example, the collection of nuclei commonly referred to as the amygdala (part of the paleomammalian brain in this terminology) would be posited to be the same (identical) in all mammals. After all, it should be structurally and thus functionally "frozen" in time.

However, this is not the case: the amygdala shows additional functions, altered connectivity and even additional nuclei in humans compared to, say, other mammals. Making inter-species comparisons of the amygdala – or any other brain structure – in rats, monkeys and humans will show both gradual as well as qualitative morphological differences. Basically, we find that regions originally put in the "deeper" regions of the brain, are not functionally or morphologically frozen in time, but have co-evolved in species with a larger cortices. The basal ganglia and amygdala show an involvement in complex functions and behaviours, and we even find substantial morphological differences between species, demonstrating that these regions are all but frozen in evolutionary time.

In other words, evolution means change, and no brain structure is immune to change. Thus, the Triune Brain theory has been thoroughly debunked (Ledoux 1998). Today, the Triune Brain is seen as an overly simplistic view of the brain, something that is briefly but well summarised at Wikipedia (http://en.wikipedia.org/wiki/Triune_brain).

See also these (going to the core with primary scientific references):

http://cms.unige.ch/fapse/EmotionLab/Publications_files/Sander_etal_RevNeuro-2003_1.pdf

http://www.ncbi.nlm.nih.gov/pmc/articles/PMC2507884/

http://neuron.nimh.nih.gov/murray/NLM.2004.pdf

NEUROCHEMICAL SYSTEMS

Neurons are *electrochemical* cells – this basically means that neurons work through electrical and chemical processes. Although the single neuron is not highly relevant to consumer behaviour, it might be good to understand the processes at the single cell level, so that one can better understand why we talk about the brain as an organ that *both* has electrical and chemical properties.

To better understand these processes, let's look at the following simplified diagram:

Neurons, despite being extremely diverse most often have a main body (soma), dendrites and an axon. In very simple terms, the dendrites are the dominant receiving part of the cell, the axon is the main transmitting part of the cell. However, it should be noted that the soma also receives signals, and that the flow of "information" within the cell is not unidirectional.

When dendrites (and the cell body) receives signals it changes their electrical and chemical properties within the cell relative to the outside of the cell. Some signals to the neuron tend to excite the cell while others can inhibit the cell. Overall, there is a constant flux in the excitatory and inhibitory signals.

When a neuron reaches a critical level of excitement, it makes a binary switch from inactive to active. At the start of the axon (the "axon hillock") started an electrical wave that travels all the way to the axon terminals. Upon reaching the terminals, vesicles within the terminal are released out of the neuron and into a small gap between this cell and another reining cell (called the "gap junction"). The vesicles, that can contain a particular substance that can affect the receiving cell, a substance we call *neurotransmitters*, simply because they transmit signals between neurons.

There are multiple neurotransmitters in the brain, but some of the best known are dopamine, serotonin and acetylcholine. They are often thought of as playing different roles in our brains, and are unevenly distributed in our brains, as the following figure shows:

Interestingly, the human brain is thought to be much more densely distributed in terms of the neurotransmitter systems. Compared to other mammals and primates,

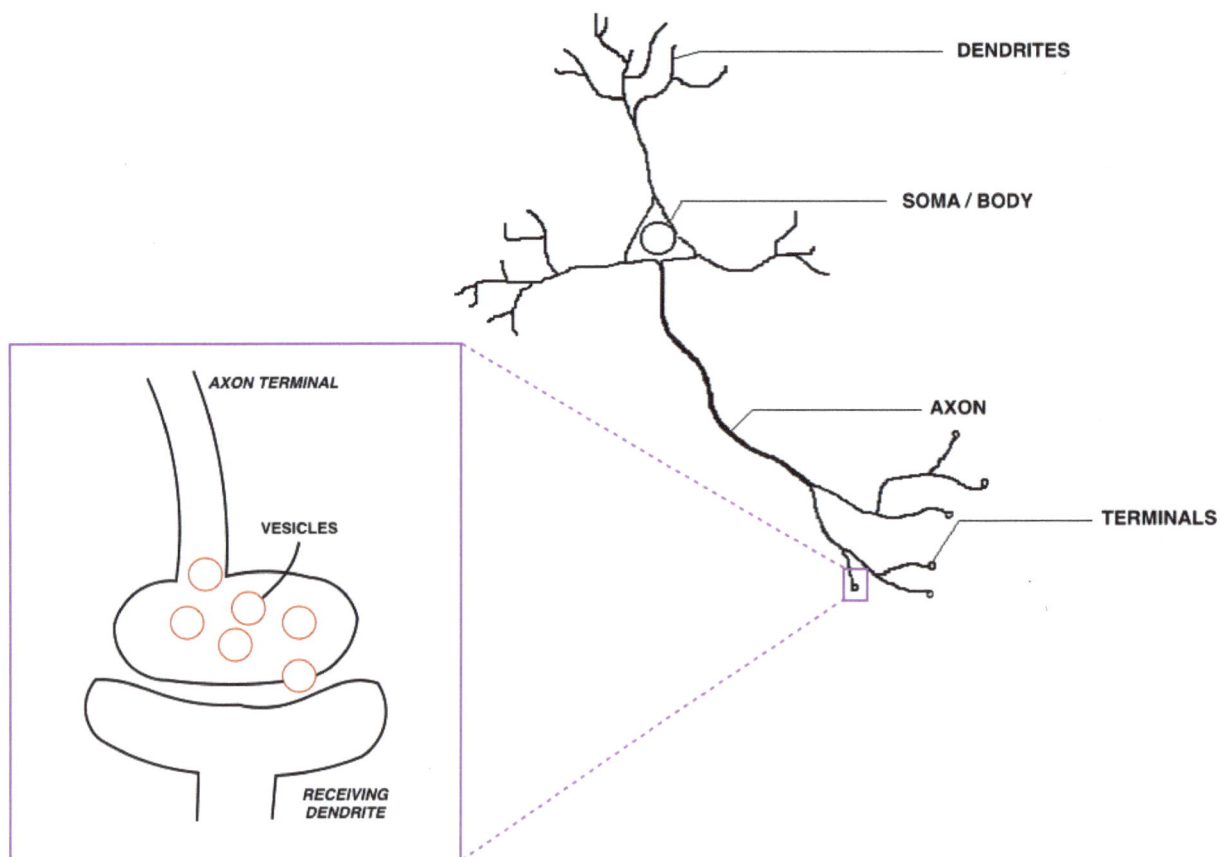

Figure The single neuron and its neurochemical process. Nerve cells typically have the receiving dendrites, a main body (soma) and an axon that ends in terminals. This graphical example is a typical pyramidal cell (the soma has a pyramid shape). When an electrical action potential travels from the soma to the axon terminals, vesicles containing neurotransmitters are released into the gap junction (inset), from which some transmitters are taken up by the receiving cell, and some are taken up again by the original cell.

DOPAMINE   **SEROTONIN**   **ACETYLCHOLINE**

MEDIAL VIEW   MEDIAL VIEW   MEDIAL VIEW

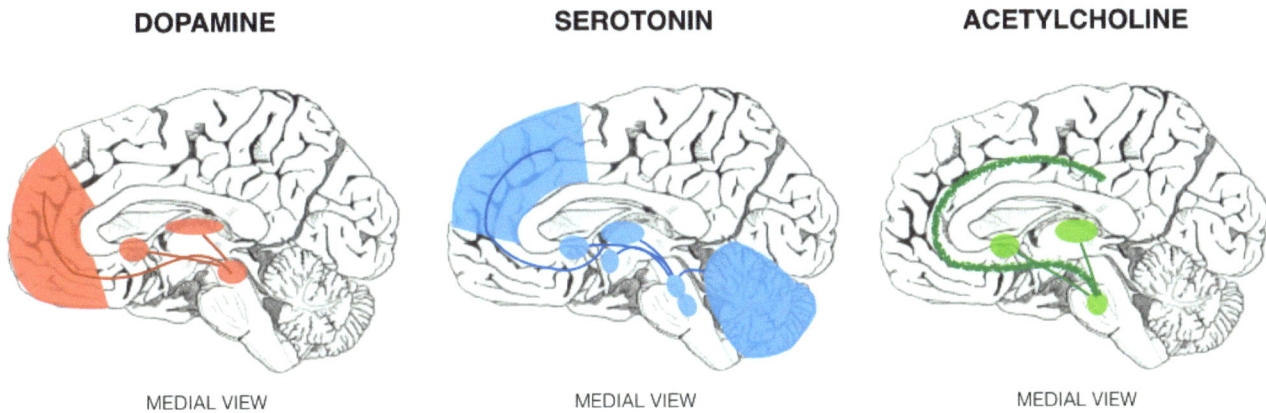

Figure Distribution of three neurotransmitter systems in the human brain. All brains are shown from the middle (medial) view. As can be seen from this simplified illustration, each neurotransmitter system originates in different deep structures and show different projection patterns throughout the brain.

the distribution of the different neurotransmitters has been found to be more overlapping in humans. The significance of this is not known, but one may speculate that it may lead to a higher degree of "cross-talk" between the many different functions of each neurotransmitter system, which in turn may provide a neurochemical basis for complex behaviours.

For the three neurotransmitter systems, we can summarise some insights in the following table:

Another neurotransmitter of note could be the opioid system, which is synthesised in deep structures of the brain and distributed to several structures of the brain, including the ventromedial prefrontal cortex. It has been suggested that opioids are related to our experienced pleasure (Chelnokova 2014; Barbano 2007) – as opposed to the possibility that dopamine is more related to our more unconscious reward seeking (Berridge 1998), and related to ventral striatum function.

NAME	PRODUCED WHERE?	EXAMPLE FUNCTION	EXAMPLE PATHOLOGY
Dopamine	Central nervous system, deep structures of the brainstem and midbrain	Associated with the "pleasure system" or "anticipation system" of the brain and promotes feelings of enjoyment and reinforcement to motivate performance	Depression Parkinson's Disease Depressed libido Learning Disorders Attention Deficit Disorder Chemical Addictions Psychosis and schizophrenia Changes in creativity
Serotonin	Central and peripheral nervous system	anger regulation, body temperature, mood, sleep, pain modulation and appetite; motility and pain modulation	Obsessive-compulsive disorders Anxiety Disorders Bipolar Disorders Loss of pleasure in interests Unable to fall into a deep, restful sleep
Acetylcholine	Central nervous system, deep structures of the midbrain and pons	Promotes excitatory actions for cognition, memory and arousal; regulates your ability to process sensory input and access stored information; controls your brain speed by determining the rate at which electrical signals are processed throughout your body	Learning disabilities Memory lapses Calculation difficulties Diminished comprehension Loss of visual and verbal memory Attention Deficit Disorder Slowed mental responsiveness Decreased creativity

Oxytocin

A whole different class of substances in the brain are ligands that can operate both as hormones and neurotransmitters. Several studies have recently demonstrated a role for the ligand oxytocin in social behaviour, where increases in oxytocin levels can lead to changes as diverse as increased interpersonal trust, pair bonding, caregiving, and hypnotisability (Barraza 2009; Bryant 2012).

ESSENTIAL SYSTEMS

To summarise the many different brain regions, below is a figure that illustrates and highlights some of the more essential brain structures in relation to neuromarketing and consumer neuroscience. Please note the following:

- The striatum is divided into the caudate nucleus, putamen and nucleus accumbens. When we are speaking about the "ventral striatum", this typically refers to nucleus accumbens
- The anterior cingulate cortex and orbitofrontal cortex are different structures, but are both part of the more general label "ventromedial prefrontal cortex"
- Certain structures, such as amygdala, hippocampus and parietal cortex, are hidden from view in this model
- The insula is a large structure but is only shown as a small strip here

In the following chapters we will populate this brain model with what these structures are actually doing.

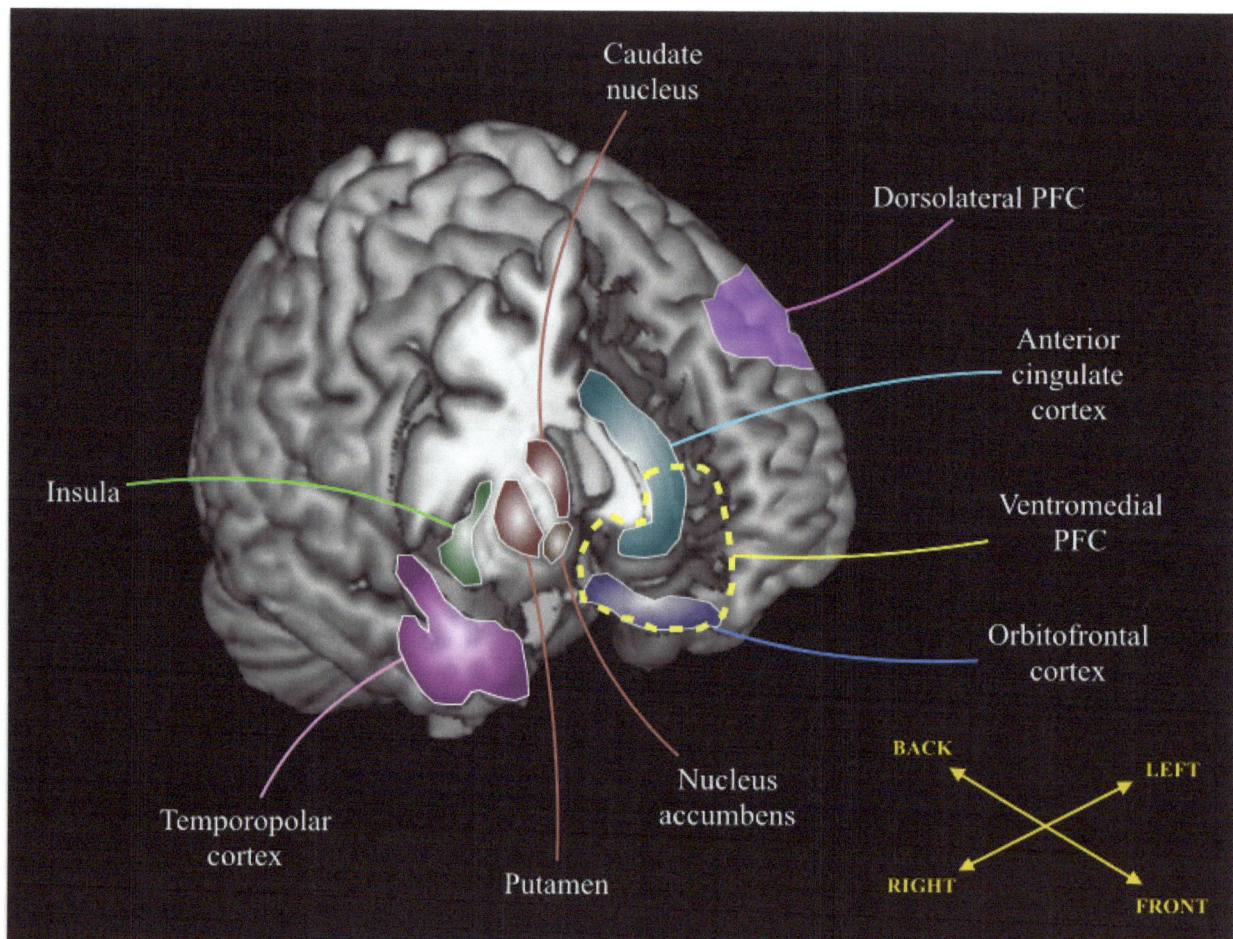

Figure   A summary of some of the important structures for neuromarketing and consumer neuroscience. For understanding the orientation of the brain, see explanation at bottom right. Certain structures such as the amygdala, hippocampus, midbrain and parietal cortex are hidden from view here.

CHAPTER 3

The Neuromarketing Toolbox

While much of what we learn in neuromarketing and consumer neuroscience concerns novel insights and views about consumer behaviour, another important aspect is to consider the different methods that can be used to explore the unconscious sides of consumer thought and choice.

Is there a single magic key method to explore the human mind? The answer is a resounding 'no'. Instead, we should think about the methods available as mutually informative and often overlapping, but not mutually exclusive. Methods may be good or bad for certain purposes, and with their own strengths and weaknesses. But there is no single method that should be considered the only one.

In this section we will explore the many different methods that can be applied to our understanding of consumer psychology and behaviour. One important aspect of this chapter is to learn how to consider pros and cons of each method. While everybody wants to discuss fMRI research, the actual work needed, the price, and the limitations in use and interpretation could and should easily encourage the consideration of other methods.

It is also worthwhile to note that the toolbox also includes many methods that do not measure directly on the brain. Indeed, we should consider what method serves our purposes best. Is it absolutely crucial to run an expensive fMRI to study visual attention, or is it more useful to use eye-tracking and possibly extend with computational neuroscience? The costs will differ enormously, but the insights might be the same. Or when does it make sense to combine measures, such as using eye-tracking simultaneously with EEG?

Here, we will build a **Neuromarketing Toolbox** for the academic and commercial study of consumers. The list will by no means be comprehensive and treat all methods available, but rather focus on those methods that are either more prevalent or useful. Please note that the methods that are most usable (in certain situations) may be those methods we believe to be "boring", and vice versa, methods that are high on our wish list may in fact not be so usable in many or most contexts.

With the toolbox in hand, we will be allowed to focus more on the actual research questions we have in mind. Therefore, this chapter will focus on the actual methods and limitations, but the toolbox will be a recurring theme throughout most chapters in this book.

CLASSES OF MEASURES

The tools available do not all bear a "neuro" name, and as neuromarketers we should consider methods that provide the necessary insight, not insist on using "neuro" every time. Thus, we may see at least three different classes of tools:

- **Self reports** – This is not a method unique to neuromarketing, but rather something any neuromarketer cannot ignore (although they easily tend to). Asking people seems like what neuromarketing methods are going against. However, on the contrary, neuromarketing should never be seen as something that should replace traditional methods, we should have good protocols for asking people. In studies of, for example, in-store behaviours a participant may report feeling frustrated in the store, but without much more elaboration to go on. Here, neuroscience methods can be used to go beyond self reports to better understand whether in-store frustration is due to information overload and stress, emotional responses or other specific events. But to demonstrate the added insights from neuroscience methods, we need to have proper methods for assessing customers' conscious perceptions and thoughts.

- **Behavioural measures** – Many tools that are used to observe consumer behaviour to infer particular mental states and responses. For example, *facial expressions* are used to understand a person's state of mind and emotional responses. Similarly, *response times* are used to understand motivation as well as choice conflict. As such, behavioural measures are the first forgotten measure in neuromarketing: without a proper assessment of what consumers are actually doing, we need to have a proper assessment of their actual choice behaviours.

- **Physiological measures** – This denotes measures in which one use different kinds of measures of bodily responses, ranging from eye movements, pupil dilation, palm sweating, respiration and pulse. Crucially, many of these measures are highly correlated: pulse, respiration, GSR and pupil dilation coincide, as we will see in this book, due to the underlying way in which our brain and central nervous system is wired.

- **Neuroimaging** – Measuring the brain is a daunting task, and anyone who begins will be faced with a plethora of options, limitations and challenges. Methods include functional Magnetic Resonance Imaging (fMRI), electroencephalography (EEG) and even optical imaging, but these tools belong to a much larger class of measures, such as structural MRI (sMRI), Positron Emission Tomography (PET), magnetoencephalography (MEG), Single Photon Emission Tomography (SPECT) and many more methods (see figure below).

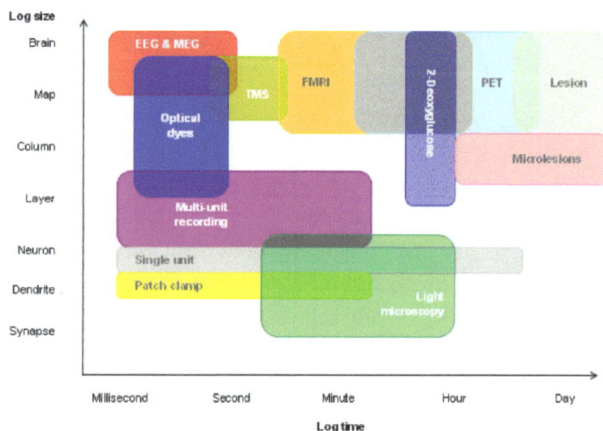

Figure Spatial and temporal resolutions of different neuroscience methods. While most of those methods cannot be expected to be a part of the neuromarketing toolbox, it nonetheless is meaningful to see how some of those measures (such as EEG and fMRI) match up to other methods. Recently, many measures, such as fMRI and EEG, have seen considerable changes in their spatial and temporal resolution. For example, fMRI can now both assess activation at individual cortical layers, and EEG can reconstruct activation in deep structures of the brain.

AFFECTING BEHAVIOUR

Besides the assessment techniques that are used to measure brain activity, physiology and behavioural responses, other neuroscience methods are used to affect behaviour in different ways. Some of these measures include the following:

- **Direct brain stimulation** – There are many different ways to stimulate the brain, although most of these methods are not used in the neuromarketing toolbox. However, as academic researchers it can sometimes make sense to use such methods to better understand the causal mechanisms at stake. Some neuroscientists use *direct electrical stimulation* to exit or inhibit specific neurons, but a more applicable method is *Transcranial Magnetic Stimulation* (TMS), in which a magnetic coil over the head can be used to induce changes in the underlying brain region's activity, thus grossly affecting the operation of this region and its effect on behaviour.

- **Chemical affliction** – Other ways to affect brain activation can be done by affecting its chemistry. This includes a whole range of methods, including the use of *psychoactive drugs* (e.g., cannabis and cocaine), *psychopharmacological/psychiatric drugs* (e.g., antidepressants and antipsychotics), and yet again other studies can use hormones (e.g., testosterone) to test for effects on choice behaviours.

- **Subliminal effects** – Other methods that belong to the neuromarketing and consumer neuroscience toolbox include those methods that operate solely as external effectors with respect to the person. These include methods in which specific external stimuli are tested for their effects on behaviour. One such method is the use of very brief exposures to stimuli (e.g., a brand name) to nudge cognitive, emotional and behavioural responses in consumers. Stimuli do not have to be brief as such, but vague and by definition not available to the person's conscious mind. By this token, things that are present within the eye of sight but not registered consciously, can still be processed unconsciously by a person, and have subsequent effects on their thoughts and choices.

- **Behavioural framing** – Even when an external factor is presented so the person is fully conscious, they may be unaware of how they are affected. If a product is shown together with a famous, well-known brand, people are more likely to report that they like the product than if the product is either shown with an unknown or less liked brand, or no brand. Similarly, products associated with a high price are reported as being better and more enjoyable than lower priced products. Furthermore, merely naming a task or a situation can affect how we think and act. In social games, our behaviours are more pro-social if we believe we are playing a "community game" than if we believe we are playing a "Wall street game."

To many, neuromarketing practitioners may seem highly focused on using "neuro tools", but I will contend throughout this book that there is much to gain from applying other measures and methods. Furthermore, the successful neuromarketer will excel on most of those measures, just as he/she will also utilise the possible combinations of data that will be available through combining measurements.

IMPLICIT MEASURES

Imagine that you are walking down a street. You are exposed to multiple sensory impressions, you notice only a few of this steady stream, and actively decide to look more at even fewer. Still, you may ask, are you affected by those things you cannot consciously say you saw? This is an area still under intense scrutiny. Can something undetected by our conscious minds actually impact on our thoughts and behaviour?

In 1957, James Vicary reported that he had successfully affected 45,699 movie goers to buy more popcorn. He claimed that he had inserted brief flashes of the text "Eat popcorn" and "Drink Coca-Cola" into the regular movie, and observed a tremendous (57.5%) increase in the sales of popcorn after the movie. The Coca-Cola sales effect was less impressive (18.1%) but still statistically significant. The story soon got press

and fame, and Vicary was figured in numerous talks, seminars, and even political debates. The unconscious persuaders soon became a theme, becoming every advertiser's wet dream and every consumer's great fear.

Only a few years later, Vicary admitted that the story was made up. He never actually did the popcorn "experiment", but was inspired by contemporary talks about the whole topic, and even led to the urban legend of hidden cinema ads. This was also later referred to in movies such as "Fight club", in which the main character inserted snips of pornographic images into children's cartoons which led the children to cry during the movies.

With the case of Vicary in mind, what is the current status of this so-called "subliminal perception1"? To put it simple: it's in good shape. Today, we see numerous examples in media, and a huge scientific literature supporting the notion that subliminal messages can have an effect on people's behaviour. In all ways, we have seen that perception can occur without a person's knowledge or consciousness, and ultimately affect their behaviour and/or thinking.

A good example can be found in magic. In a British TV show, magician and entertainer Derren Brown invited a couple of designers to design a new ad campaign for him. He offered a taxi ride through London to their meeting, where carefully placed specific items and events along their trip. In doing so, Brown demonstrated clearly that the end product of the designers' work was highly affected by the cab ride. The clip is available on YouTube, see the link below.

Figure Derren Brown's example of priming designers to make exactly the kind of design he wanted them to. The clip can be seen at: http://youtu.be/YQXe1CokWqQ

In this way, Brown demonstrated what scientists have known for a long time: that things we either don't consciously see or forget that we have seen, may have a strong impact on our thoughts and behaviour. Even brief glimpses of words may skew the way in which we perceive a product or an event.

Turning the table around, there are also ways to use this implicit method as means to test consumers' responses, even though they do not see the stimulus, or even know what aspects of their behaviour that we are measuring. In other words, we may use many methods that consumers are not privy to, as a way to understand their unconscious emotional responses.

Here, we can identify a range of methods, all ready to put into our toolbox:

Priming

Basically, one dictionary definition of priming is "*to prepare (someone) for a situation, typically by supplying them with relevant information.*" In other words, priming is way in which we provide information or other contextual aspects that can affect a person's perception, thought and action. This is a method that is readily used by researchers who want to understand branding and framing effects. In a typical setup, two or more groups are presented with a particular product, and in a preceding

There are numerous examples of how priming can be used

- **Altering drink preferences:** Samuel McClure and colleagues asked participants to taste cola while being scanned using functional Magnetic Resonance Imaging (fMRI, see the section about fMRI in this chapter). When no information was given about the brand, the researchers found stronger activation in the brain's ventromedial prefrontal cortex to be related to subsequent preference ratings. But when the participants were told that they were going to taste Coca-Cola, the researchers found both a higher rating of the drink, and a stronger response in two memory-related regions of the brain: the hippocampus and the dorsolateral prefrontal cortex. When the participants were told they were going to drink Pepsi Cola, no such effect occurred. We will look more at this in the chapter about wanting and liking.

- **Imaginary pricy wine boosts preference:** Hilke Plassmann and colleagues (Plassmann 2008a) had people fMRI scanned while they performed a wine tasting and preference rating task. Prior to each wine, the participants were informed about the price of the wine. The researchers found that price was positively related to preference: the higher the price, the higher the preference. Interestingly, the price itself was made up; there was in reality no relationship between price and the wine the participants tasted – it was all in their heads. As we will see in our treatment of wanting and liking, Plassmann and colleagues identified the neural processes underlying this hedonic response.

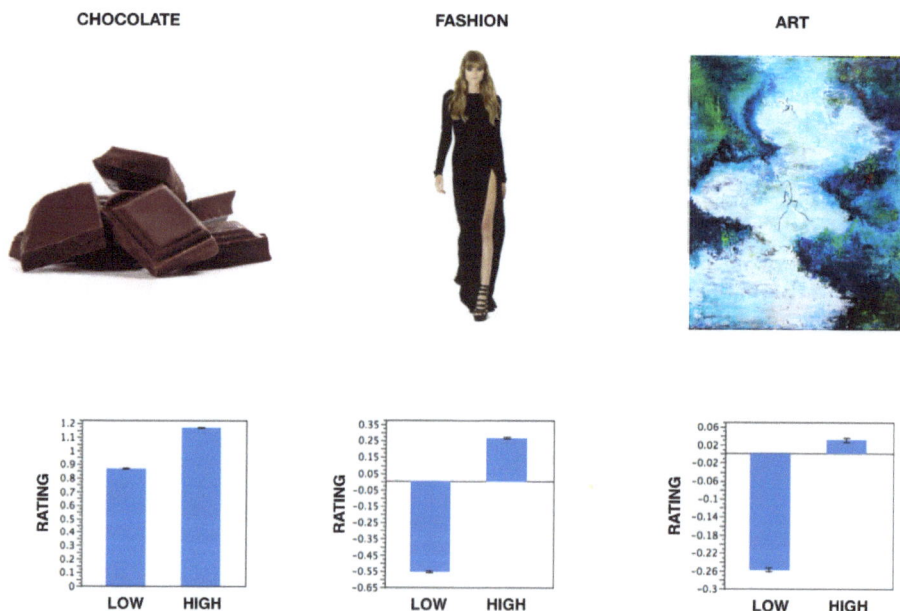

Figure   Effects of framing on liking. In this study, participants were shown either a category-relevant high-value brand or a low-value (discount) brand prior to rating chocolate taste, fashion clothing or abstract art. Here, we found a remarkably strong effect. For chocolate, ratings went from positive to more positive (probably due to the inherent positive value of chocolate); fashion clothing went from negative to positive; and abstract art went from strongly negative to slightly positive.

- **Framing across conditions**: In a recent study performed in my own lab, we had participants make many different kinds of rating: chocolate, fashion clothing and abstract art. Prior to each image or taste, participants were shown either a high-value brand or a low-value (discount) brand. As you can see from the figure below, the brand information had a strong and significant effect on people's ratings.

Subliminal stimuli

While priming studies typically use clearly visible types of information – e.g., showing the brand – other methods include the display of information below the threshold of conscious detection. Despite Vicary's erroneous claims in the past, numerous studies have today demonstrated that subliminal perception is a reality. We can show information below the conscious threshold and record changes in thought, neural activity, physiology and behaviour.

While subliminal perception can be an effective means of inducing psychological and behavioural changes in people (see the section about attention and consciousness), the method has been deemed unethical and most modern societies have banned the method by law. Nevertheless, subliminal perception can be used as a method to test consumers' responses as part of a neuromarketing test. A few such ways could include the use of *The Implicit Brand Equity Test* (TIBET). In this test participants are given tasks to rate their experience and preference for specific goods. Prior to this task, they are shown a specific brand, and one use a host of measure to test the effect of the brand on their responses, including:

- Self-reported preference, e.g., using a visual-analogue scale
- Self-reported mental preoccupation – how much time you spend thinking about something
- Willingness to buy / willingness to pay
- Response time: the time it takes for participants to respond to the task
- Response strength: the relative strength of a button press, or a hand grip
- Physiological responses, including respiration, pulse, palm sweat, and pupil dilation
- Brain responses, including engagement of the nucleus accumbens, or alteration in the prefrontal asymmetry

As we will see, most of these measures are not mutually exclusive, but rather mutually informative and can add to the overall predictive value of the test battery.

Response time

The more we like something, the faster we tend to respond. This is the simple yet powerful insight from multiple models of preference and behaviour (such as the Sequential Choice Model, which we will look at later). But as long as this relationship holds, it means we should be looking at response time as a potential index of preference.

In general, response time (RT) can be used for multiple purposes. First, the positive relationship between preference and RT has now been well documented. In conditions with two or more choices available, studies have consistently shown a relationship between RT and decision conflict: the closer in value two options are, the harder it will be for a person to decide. Even though such effects are in the order of ten to a few hundreds of milliseconds, they are easy to detect, provided one use the correct measurement tools. In recent studies in my own lab, we have found that RT can be influenced both by the level of conflict between response options during gambles, and that individual personal differences affects the ease at which we resolve conflicts, such as in social dilemmas.

One particular well-known version of the use of RT is the *Implicit Association Test* (IAT). There are different versions of the actual setup of this test, but one basic setup is as follows:

- **Word Valence Task:** Participants are trained to respond with one finger (e.g., right index) when a word is positive, and another finger (e.g., left index) when a word is negative. This allows a training of a habituated response that couples the response of one finger to the valence of the choice (positive/negative)
- **Implicit Association Task:** Subjects are then, most often interspersed between the word valence task, shown another type of stimuli. These stimuli are in fact the ones we are interested in. This can be words, brand names or faces. In one classical experiment (Mcconnell 2001), researchers used faces of people with either dark or light skin colour. The task for people in this task was to, for example, respond with the right index finger if the face had dark skin and left index finger if it was a light skinned person. The crucial component in the IAT is that the learned association between response and valence may be in conflict with the IAT test. That is, in studies using skin colour, when a the response to a man being black and a word being positive are using the same response (right index finger), caucasian participants have a significantly longer RT. This has been interpreted as a hint that we all

tend to have implicit racial biases. In a similar vein, research in my own lab has showed that a similar IAT approach could be used to predict overt brand preference2, thus suggesting that behavioural measures can be used as an index of unconscious brand preference.

Response strength

Although rarely tested or employed in consumer research, academic studies have uncovered a relationship between desire and response strength. Typically, the more you like something, the more pronounced your response is (Pessiglione 2007). Just as with response time, and its relation to the Sequential Choice Model, we see that desired items generate stronger physical responses – such as harder button presses and hand grip – than less desired items, even when the person is unaware of such differences.

Measuring response strength can be done in many ways. Certain kinds of hardware (e.g., music keyboards and computer keyboards) can record the strength of button presses. Other methods include ways to assess hand grip strength, such as a hand dynamometer. Collectively, such measures allow us to precisely measure the actual strength that a person puts into a response, often without noticing such effects.

Besides the possibility of relating response strength to models of motivation, another appealing aspect should be noted: response strength is closely tied to the basal ganglia of the brain. As we will see later in this book, the basal ganglia is involved in motivated behaviours, especially motivation that operates unconsciously.

Response strength has yet to become popular as an assessment tool, and it is likely that few have thought of this as a potent assessment tool, when the focus of neuromarketing so much focuses on neuroimaging approaches.

Thus, with recent studies demonstrating the potent role of using response strength as a measure of desire and motivation, such as the work by Pessiblione and colleagues (Pessiglione 2007), and with an appealing model to explain the phenomenon, it is likely that we can soon see new solutions that integrate response strength as one of their implicit measures.

Posture

Anecdotally, we know that when we are engaged in something, we tend to lean forward; and when we see something disgusting or unappealing, we tend to move away. In recent studies of posture, work in my own lab has demonstrated that posture, especially when combined with pupil dilation, can predict a person's subsequent product choice.

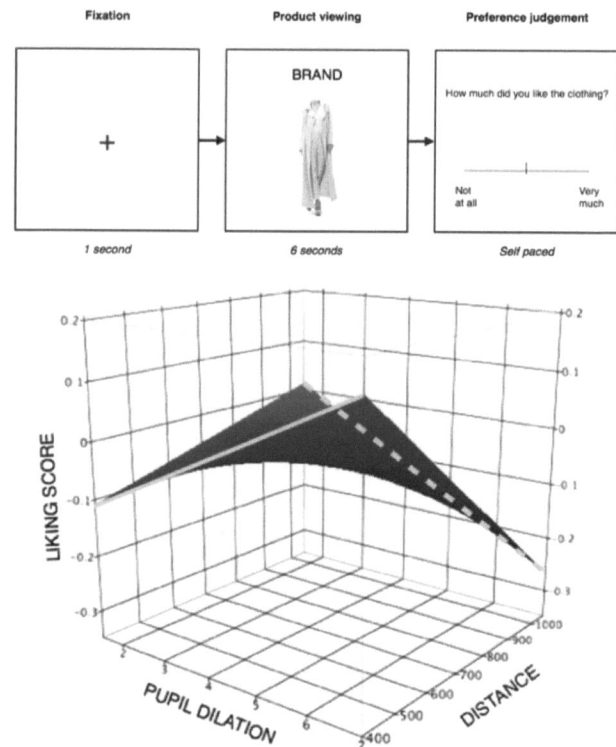

Figure Relationship between posture and pupil dilation in predicting liking and willingness to pay for a fashion item. Top panel: The study design, in which participants first saw a product, and then rated their liking of the product using a visual analogues cale. Bottom panel: Three dimensional model of pupil dilation and posture and their joint explanatory relationship to liking score. At a low distance, pupil dilation response was positively related to liking, while at a high distance the relationship was inverse, and larger pupils would be associated with lower liking.

Mental preoccupation

Many studies of unconscious motivation attempt to avoid talking to people, while in fact, psychology holds a long history of assessing needs and desires from just talking and observing people. This is famously illustrated in Shakespeare's Rome and Juliet, act 2, scene 2, in which Romeo proclaims:

But, soft! what light through yonder window breaks?
It is the east, and Juliet is the sun.
Arise, fair sun, and kill the envious moon,
Who is already sick and pale with grief,
That thou her maid art far more fair than she:
Be not her maid, since she is envious;

Her vestal livery is but sick and green
And none but fools do wear it; cast it off.
It is my lady, O, it is my love!
O, that she knew she were!
She speaks yet she says nothing: what of that?
Her eye discourses; I will answer it.
I am too bold, 'tis not to me she speaks:
Two of the fairest stars in all the heaven,
Having some business, do entreat her eyes
To twinkle in their spheres till they return.
What if her eyes were there, they in her head?
The brightness of her cheek would shame those stars,
As daylight doth a lamp; her eyes in heaven
Would through the airy region stream so bright
That birds would sing and think it were not night.
See, how she leans her cheek upon her hand!
O, that I were a glove upon that hand,
That I might touch that cheek!

Being in love is, if anything, one of the clearest signs of mental preoccupation. For branded products, we see the same effects: when you simply cannot wait for a specific upcoming movie, or when you cannot stop thinking about a particular pair of shoes, or buying an iPhone. Mental preoccupation is highly linked to motivation and the pursuit of that item. In consumer research, one should pay much attention to mental preoccupation as an index of desire, including "love brands" (Sarkar 2011) and brand loyalty (Plassmann 2012; Jacoby 1973).

Indeed, mental preoccupation might speculatively be seen as two components: a conscious and an unconscious. On the one hand, it can reflect stated and overt constant thinking about a brand and/or product; while on the other hand it might reflect actual recurring and persevering behaviours.

COMPUTATIONAL NEUROSCIENCE

Although we do not yet have a complete understanding of the brain, we have a sufficiently good knowledge about certain of its processes that we can make pretty reliable models of these. Such models can operate as mathematical algorithms that model specific aspects of, say, vision, or memory.

The approach of making such models of aspects of the brain often go under the heading "computational neuroscience" (Churchland 1993; Rolls 2002; Trappenberg 2009), which roughly can be defined as the study of the information processing mechanisms of the brain.

Computational neuroscience covers a range of topics, ranging from the modelling of single neurons and even sub-cellular processes, all the way to modelling consciousness. For our purposes, some aspects may be more relevant to the study of consumer behaviour and marketing than others.

For example, it is possible to model certain aspects of visual processes. Attention can be subdivided into two presses: top-down and bottom-up. While top-down is the aspects of actively focusing and selecting to focus on certain aspects, bottom-up attention is automatic, unconscious and driven by external cues. What actually makes us automatically look at certain parts of an image, or turn towards a sudden sound, rests on processes that are well understood.

NeuroVision

Vision is one of the most studied and best understood senses, and we have an exquisite and detailed understanding of how information goes from the eyes, project through the visual pathways, via the thalamic nuclei, and ending in the primary visual cortex, where the information is further delegated into specialised processors. It is exactly these processors that are in part responsible for what automatically draws on our visual attention, also called bottom-up attention. Changes or differences in contrast, brightness, movement, density, colours and other parameters can all lead our eyes to be automatically directed to certain parts of our visual scenery.

With this knowledge in hand, we can make computational algorithms that model these effects. One such model, NeuroVision (see http://www.neuronsinc. com/neuromarketingservices/neuromarketing/neurovision/#.UyK2muddUwo), is exactly such a weighted model of the part of vision that automatically grabs attention. This allows the making of a process that automatically analyses images and videos for exactly the same parameters. NeuroVision is therefore an online/cloud based tool wherein you can upload videos and images, and have them analysed automatically, rapidly and at a low cost.

While computational neuroscience has existed for almost as long as modern neuroscience itself (15-20 years), it has not been introduced as a proper tool in understanding consumer behaviour and marketing effects. This may soon change, with new models entering fields of more complex human behaviours, including emotions and decision making.

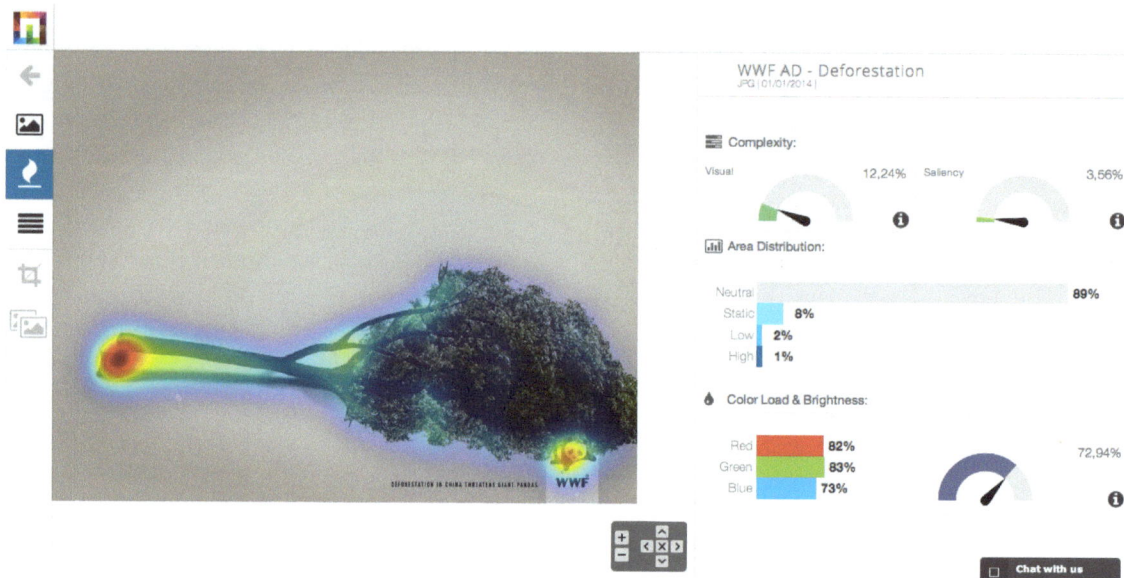

Figure The NeuroVision online interface that shows the heat map representation of the most salient regions of a WWF ad aiming to increase awareness of the negxative sides of deforestation. As can be seen, the broken tree and the panda are the most salient parts of the ad, owing to a high degree of angle and contrast, with an overall simple ad with few distractors. The image also has a low complexity and large regions of neutral parts, all ensuring that the main messages of the ad has a high likelihood of being seen. The ad requires us to see both the broken tree and the panda, and the high saliency scores point to a high likelihood that both features are likely to be seen automatically and effortlessly. Courtesy of Neurons Inc (www.neuronsinc.com)

EYE-TRACKING

To understand where people are looking, we cannot just ask. We are extremely poor at knowing, let alone reporting what we are looking at. Try this out: take a walk down a nearby street and try to notice why you are seeing. On the one hand, it's an impossible task! Noting what you are looking at becomes an insurmountable task of keeping track of your eye's every move, and we will tend to filter out the smaller fixations and focus on what you paid attention to. Furthermore, giving you this task changes your mindset from a natural strolling down the road, to an intense interoceptive focusing on where your eyes are fixating. You move from a natural behaviour to a highly unnatural one.

If you then are asked by another person to report what you saw, this changes you from a mode to perceiving to actually trying to remember what you saw. It becomes a memory task, which is different from reporting on what you looked at.

This may seem trivial, but when in fact a vast majority of market research uses self reports to address visual attention and related processes, it is clear that other tools and methods are needed. We need ways to better at passively recording eye fixations, which allows the person to act as naturally as possible.

Eye-tracking itself is an older invention, and the earliest accounts of eye-tracking can be found in 1879, where Louise Emile Javal noted that when people read,

their eyes do not run smoothly over the text, but rather jumps and stops. Inventors such as Edmund Huey constructed a device that could indeed measure eye movement, but it was extremely intrusive to use. Nevertheless, researchers soon realised that eye-movements were another source of information that could be used to understand the human mind.

But it would not be until the 1980's that eye-tracking became used in marketing and advertising (Wedel 2000; Pieters 2008). Indeed, eye-tracking allowed researchers to determine what parts of a magazine page were seen, which elements of the page were actually read, and how much time was spent on each part. This exceeded prior methods such as voice stress analysis and galvanic skin response tests significantly, and provided researchers with a tremendous wealth of information about consumer attention.

In the history of eye-tracking, other highlights include:

- **1990's** – Gallup Applied Science's eye-tracking system was used on people watching NFL games, to determine what parts of the game the typical watcher saw and missed. The setup included filming the participants' eyes, after which a computer would track where the eyes followed on the screen
- **Late 1990's** – Several companies began using eye-tracking to understand how users of the World Wide Web consumer information that was displayed on screen.

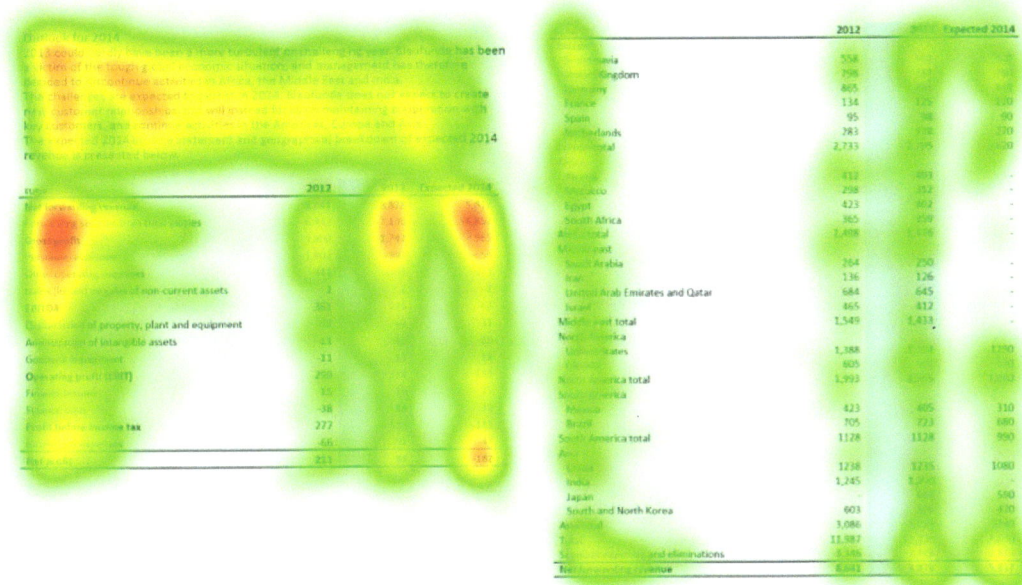

Figure Heat map representing how trained accountants are reading a financial report. Warmer colours indicate regions that receive more attention. Interestingly, key numbers such as bottom-line numbers that indicate overall business performance, were not the most attended. Courtesy of my team member Samir Karzazi, who was on the top-three of the 2015 NeuroTalent competition.

- **2006** – A study by Bunnyfoot demonstrated the effectiveness of advertising in video games, spurring a tremendous interest in how in-game ads can boost brand equity and sales.
- **From 2001 until present** – Eye-tracking has become commonplace in all major (and often minor) market research agencies, and many of the major corporations have their own labs that include eye-tracking. Academic consumer research also has seen a tremendous boost in interest for using eye-tracking to gauge visual attention as a way to better understand consumer psychology and behaviour. No serious commercial or academic researcher today can avoid using eye-tracking as a minimum requirement in their research.

How does eye-tracking work?

To track the eyes, most methods employ the use of a sender and a receiver of light in the infrared spectrum of light. A light is beamed into the eyes, and the reflection of each eye (and surrounding face region) is recorded by a set of infrared cameras. Specifically designed software recognises the eyes in the recording, and infers where the eyes are looking on the screen. All eye-tracking procedures require an initial calibration procedure, where a participant is asked to fixate on particular areas of the screen (typically 9 reference points, shown one at a time). A more comprehensive explanation can be found on the eye-tracking Wikipedia page.

When speaking of eye-tracking, we usually distinguish between stationary and mobile eye-tracking. As the names hint, each method is specialised for conditions with either static or moving states. As always, there are pros and cons for each method.

Stationary eye-tracking

So how does eye-tracking work? As described, eye-tracking involves the use of infrared cameras that record the position of each pupil. These cameras are typically mounted below a computer screen, and after a simple calibration (e.g., forcing participants to look at nine fixation points) we have a reliable estimate of where on the screen they are looking.

The Heat Map

One of the most used visualisation of eye-tracking studies is through so-called heat maps. This is a representation of how many people are looking at a particular region of the screen. Each participant's eye fixation is represented by a green dot, and the more people that are looking at the same region, the warmer the colour of this region will be. This visualisation allows us to instantly see whether visual attention is strongly guided by certain elements, and whether something is not noticed by the viewers. As such, this is a good way to provide a quick and dirty diagnosis of visual attention towards an ad or other materials.

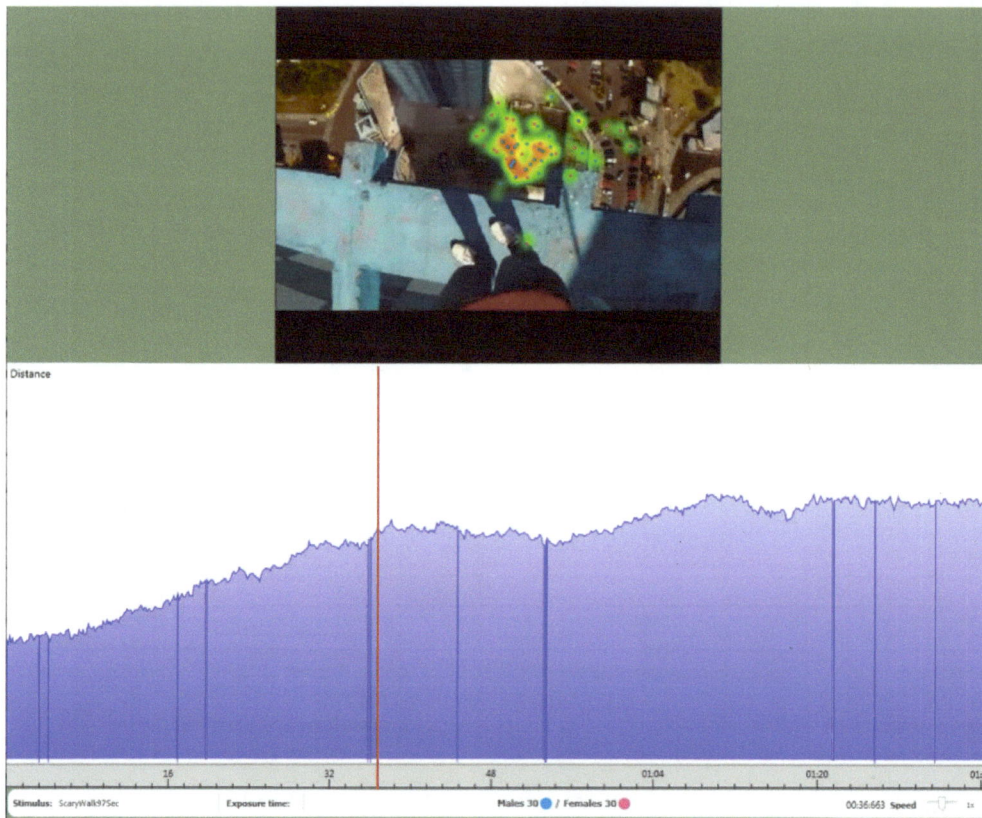

Figure   Using eye-tracking sometimes also allows the assessment of a person's distance to the screen. This example shows where people look (top: heat map where warmer colours indicate where more people are looking) and how people are moving away from the screen (bottom graph: distance gets higher) when watching a recording of a person balancing at a high altitude building. Courtesy of iMotions (www.imotionsglobal.com)

Gaze Replay

This is a visualisation method in which we replay the way in which participants – typically a single individual – have looked at the image or video. There are different ways to show this, but one powerful solution is the weighted gaze replay that is provided by iMotions' Attention Tool, where longer fixations are shown as expanding orange circles.

Figure   Stationary eye-tracking allows you to generate heat-maps that represent how groups of people look at an ad or other visual materials. Here, we see how visual attention for most people is attracted to the face of the woman in the commercial, and that few are really looking at anything else. Courtesy of Neurons Inc (www.neuronsinc.com)

Figure This gaze replay shows how a person is attracted to look at calorie-rich food, although her task has been to only focus on (and evaluate) the abstract art. In this study, we found that high-calorie food has an automatic appeal on people's visual attention, and that it cannot be overridden by instructing them to focus only on abstract art.

Gaze replay is a way to visualise eye-tracking results that allows us to understand the sequence in which people are looking. The drawback is that this works best for looking at individuals' route through a stimulus, which will only have limited usability in massive testing situations. That said, it is a powerful tool for going behind statistical analyses to understand the scan patterns that people are taking.

Area of Interest

One of the more important ways to analyse eye-tracking data is to use a so-called Area of Interest (AOI) analysis. By this method, researchers are highlighting specific regions of a display or moving stimulus. By using this method, it is possible to extract eye-fixation data, as well as other metrics that are recorded simultaneously.

When using AOI as a way to do analysis, most typical eye-tracking systems allow you to assess many different metrics related to visual attention, including:

- **Eyeball Count (EC)** – This is a count of the number of people ("eyeballs") that at any one time are fixated within the AOI (or on the screen). In some industries, such as TV, this is a highly valuable metric.
- **Time To First Fixation (TTFF)** – This is an index of the time it takes from stimulus onset to when participants are looking at a particular AOI. Shorter TTFF can be an indication of either automatic attentional processes or top-down driven searches.
- **First Fixation Duration (FFD)** – The time participants are spending during the first fixation within the AOI. This may serve as an index of the "stopping" power that a stimulus has, and thus an index of interest.
- **Total Fixation Duration** (TFD) – The total time that respondents are looking within the AOI over the entire time that the stimulus is present. This is often an index of motivated, top-down attention.
- **Fixation Count (FC)** – The number of fixations that participants make within an AOI. The more appealing an area is, the more people will tend to gravitate towards this area. As this will be highly correlated with TFD, FC is often regarded as a less robust and sensitive estimate compared to TFD.

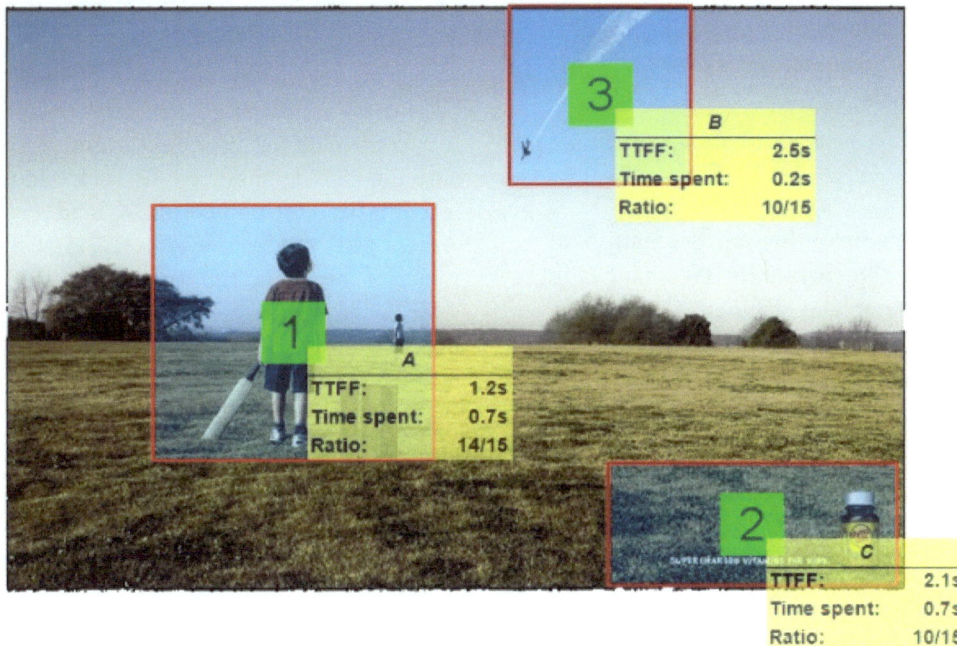

Figure Area of Interest analysis. Some eye-tracking software solutions – like iMotions' Attention Tool – allow an analysis of specific areas of the image that we are particularly interested in. How fast are customers looking at the product? How much time do they spend looking at the crucial information? In which order are they looking at the material? These and other questions can be answered by the AOI method, and provide crucial and actionable insights for advertisers. Courtesy of iMotions (www.imotionsglobal.com)

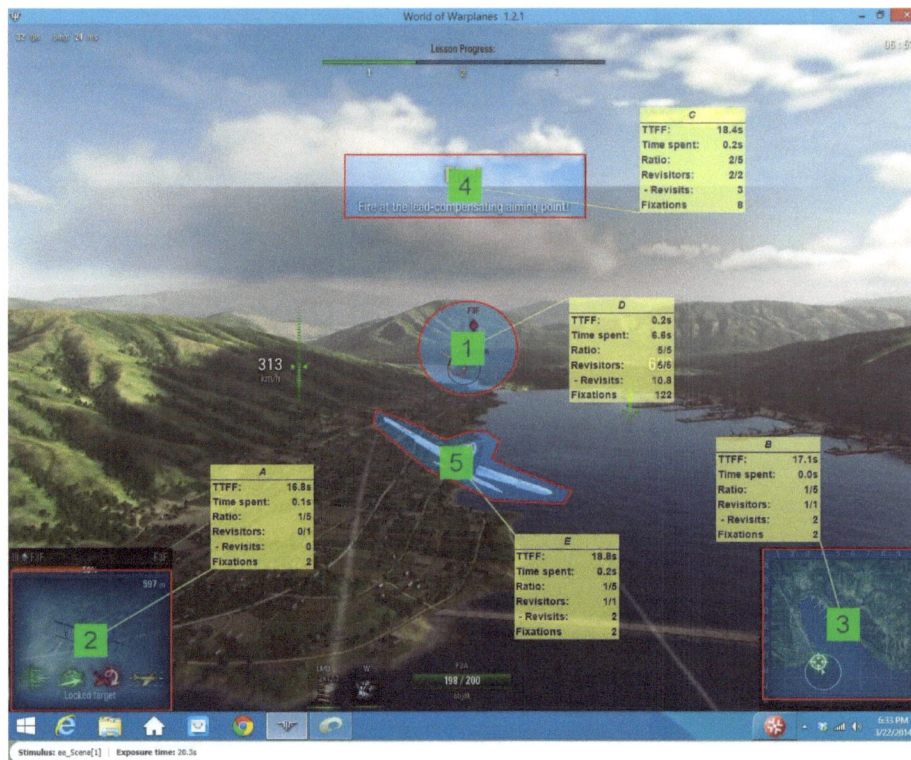

Figure Area of Interest drawing can also be dynamic and moving, which allows us to understand how we process eye-tracking data, and connected scores during movie watching, in-store movement or, as here, during gaming sessions. Such insights can prove critical in our understanding of how consumers are processing the information present in front of them. Courtesy of iMotions (www.imotionsglobal.com).

While there are many other measures, and the list can be comprehensive and exhaustive, these are some of the most used and well-known. As well, these are metrics that are most valuable for providing insights about customers' responses to ads, packaging, movies and many other materials.

Pupil dilation

Many eye-tracking systems allow an estimation of pupil size, and are able to correct for the changes in pupil size that is attributable to changes in the distance between the eyes and the eye-tracking sensors. As we will see in the chapter about emotions and feelings, pupil dilation is one of the many ways in which the body shows emotional responses, also known as "arousal".

Pupils constrict and dilate as a result of mainly three processes:

1. **Brightness and darkness** – Light constrict the pupils, darkness dilates the pupils
2. **Cognitive load** – More difficult tasks are associated with pupil dilation.
3. **Arousal** – High arousal, whether due to positive or negative events, is related to pupil dilation.

How can we tease apart these different effects? It is all about experimental control! If you want to focus on pupil dilation as an arousal measure, there are two main strategies for doing so:

- **Keep other things constant** – make sure the brightness of images are the same, or in some other way controlled for (constant brightness) and; make sure the task demands are the same across the different tasks (constant cognitive load)
- **Assess and model the effects** – assess the luminance of the image and use it as a regressor in your statistical model (controlled brightness) and; assess task demands and model in the statistical analysis (controlled cognitive load)

Through these means, it is possible to "take out" the disturbing effects of cognitive load and image brightness, and thus construct a reliable metric of arousal. In doing so, it has been found that pupil dilation is highly correlated with other indices of arousal, including heart rate variability and galvanic skin response ("hand sweating").

We will dig deeper into the relationship between pupil dilation and emotions in the chapter about emotions and feelings.

Posture

Anecdotally, we tend to lean forward when we are interested in something, and lean backwards when we are disgusted or fearful. Some eye-trackers are indeed able to track these postural changes in posture, both sideways, but critically also the forward-backward leaning (measured as the distance to the screen) in minute details.

This allows researchers to assess postural changes, and as shown previously in this chapter, we have demonstrated that changes in posture can be combined with pupil dilation changes to assess and even predict customer preference and choice. We will look more into this in later chapters.

Mobile eye-tracking

The majority of eye-tracking research has been, and is still being done, in stationary settings. Today, a combination of technological innovation and scientific interests, allow researchers to move out of the lab and into more mobile environments.

Glasses

Eye-tracking glasses have been around for some years now, but are seeing an explosion of interest these days. The basic method is similar to the one used in stationary settings, but with the need for more sophistication in solving problems associated with a mobile environment. All glasses have at least one camera pointed towards the eyes, and one camera pointing outwards, allowing us to put the eye fixations as an overlay on the video stream.

Figure Sample image from a single person's walk through the store, using Tobii Glasses, and analysed through iMotions' Attention Tool. The orange line indicates the path of the eyes, and the circle indicates the current focus. The number within the focus area denotes the length of the fixation in milliseconds. Courtesy of Neurons Inc (www.neuronsinc.com)

Visualisation the results are similar to stationary settings. It is possible to show the tracking of a single participant; AOI analyses and even heat map representations. Together, this allows researchers to assess different aspects of consumer attention, very much in the same way that one does for stationary eye-tracking.

The big difference between stationary and mobile eye-tracking is at the level automatisation and standardisation of the testing situation and data recording. **Stationary settings** allow us to standardise the presentation and duration of stimuli, and thus have a high certainty for when all participants were exposed to a stimulus. This allows us to make more automatic data analyses.

In **mobile settings**, however, this is not the case: participants are typically allowed to roam around a store, and no single person will walk the same path, turn their heads in the same way at the same time, and thus be exposed to the same items at a reliable pace. Thus, mobile settings are often associated with a substantial amount of data processing prior to the actual analysis, where one needs to define moments, draw AOIs and other methods to identify and define similar events in recordings from different people.

Figure A heat map generated from mobile eye-tracking data on 40 people. The warmer the colour, the more people have looked at this region.

GALVANIC SKIN RESPONSE

Whenever you have an emotional response, whether it is due to rewards or to danger or pain, you will start to sweat slightly more. This is particularly pronounced in the palms of your hand, as well as the arm pits and soles of your feet. Thus, finding a way to measure this effect could provide a powerful metric of arousal.

Indeed, the solution lies in basic school teaching on electronics: water conducts electricity! Therefore, measuring the electrical conductivity between two re-

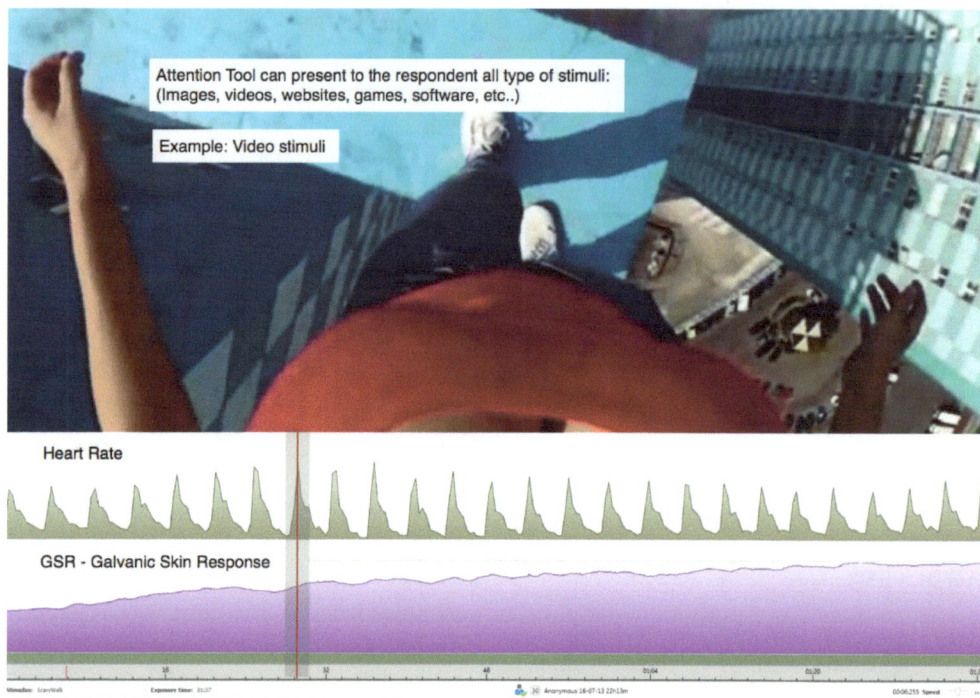

Figure Effects of arousing stimuli on palm sweating (GSR) and pulse, as shown for a single person in Attention Tool. Courtesy of iMotions (www.imotionsglobal.com)

gions that alternate in their level of perspiration has long provided a valid measure of arousal.

The analysis of sweating responses are rather simple: it consists of measuring the relative change in electrical conductivity between two electrodes – typically placed on two fingers on the same hand – where an increased conductivity after the onset of a stimulus is indicative of an increased arousal response.

That said, it is important to analyse the relative change, as there is a natural fluctuation in sweating that is not caused by arousal. Furthermore, as sweat does not vaporise instantly, any arousal response will produce a new, higher level of conductivity for seconds after the original arousal response occurred. Finally, the sweat response is known to be relatively slow and sluggish, which makes the temporal resolution suboptimal if one wants to understand emotional responses in minute detail.

Today, skin conductance, or Galvanic Skin Response (GSR) is seeing a renaissance in becoming more used in many situations where we want to understand people's emotional responses. As such, it is a relatively cost-effective and non-intrusive approach to assess people's emotional responses.

RECOGNITION OF FACIAL EXPRESSIONS

Novel methods abound, and one of the fastest increasing and most discussed methods is so-called automated facial coding. We have known for decades, even

centuries, that human facial expressions are signs of our inner emotions. While many expressions are controlled, it is also known that many facial expressions are driven by unconscious emotions. Thus, an index of facial expressions can be a potent tool for understanding consumer emotions and responses to anything from ads to gaming and other consumer behaviours.

The advantages of facial coding as a neuromarketing tool are plenty, including:

- **Price** – only requires a camera, even a web camera, and software that can reliably assess facial expressions.
- **Versatility** – due to technological availability (hardware and software), facial coding can be used to assess emotions in many different environments.
- **Scalability** – the method also allows a large scaling factor, where one can assess emotional responses on web-based platform (provided only that the participant has a web cam of a certain quality)

Today, we see many companies offering facial coding to companies, but there is still little research on the relationship between facial expressions and consumer behaviour.

Thus, we could have a validity issue for this method, and today, we see that there still remains certain issues before facial coding can be used as a reliable tool, and before we have insights into how results should be interpreted, including:

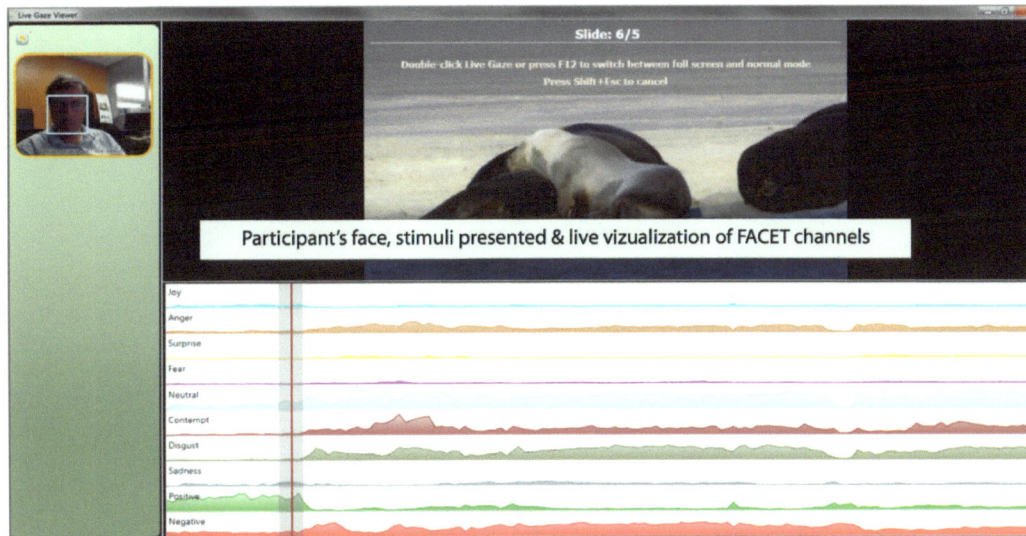

Figure Facial coding, when done properly and validated well, can provide detailed insights into consumer responses to a number of situations. This example shows facial expressions using the Emotient FACET solution running in Attention Tool™. Courtesy of iMotions (www.imotionsglobal.com)

- **Internal validity** – how reliable is facial coding, and what is the error rate?
- **Construct validity** – to what extent can facial coding reliably measure different types of emotional responses? How can it assess basic vs complex social emotions?
- **External validity** – how well does the tool assess actual emotions, and how fictitious is the testing situation? To what extent is the labelling of expressions reflecting actually occurring emotions?
- **Social effects** – facial expressions are social vehicles, and we need to see whether facial expressions are affected by the social context in which they are measured, such as whether the participant are together with other people, or whether they are conscious about being observed.
- **Introspective effects** – being aware of one's emotional responses tends to affect and dampen emotional expressions. To what extent are facial coding methods assessing expressions with or without people's informed consent?
- **Predictive power** – Are facial expressions predictive of consumer behaviour?

In a recent newsletter, Phil Barden and his colleagues at decode marketing tested one platform for facial coding. Here, they challenged the system in many ways, and in two notable studies, they made some stunning observations: first, facial expressions were found to vary, even for a static expression as when they put a doll's face in front of the camera. Second, decode marketing looked at a separate test assessing real people's facial expressions during ad watching, and commented:

"When looking at the results based on 535 people, the following becomes apparent: The dominant emotion is the neutral state, i.e., there is no obvious emotional reaction and the software had classified the facial expression as neutral. A rise in the happy emotion (= smiling) can only be seen at the end of the commercial and the valency curve rises accordingly. On the other hand, all other emotions do not seem to play a part at all. Admittedly, considering the narrative arc of the commercial, this does make sense: At the beginning of the commercial very little happens and it only becomes amusing at the end."

This study shows us that facial expressions need more validation than has currently been offered. This is not to say that facial coding is invalid, but that with the current interest seen especially in commercial neuromarketing, we need to address validity much more before it can be used in full scale.

ELECTROENCEPHALOGRAPHY (EEG)

Electroencephalography is a neurophysiological measurement of electrical activity in the brain. This recording is performed by placing electrodes either on the scalp or directly on the cortex. The resulting brainwave output is referred to as an *electroencephalogram* (EEG), originally named by the German psychiatrist Hans Berger. The work on EEG techniques was already performed by Richard Caton and Vladimir Vladimirovich Pravdich-Neminsky. EEG is today used both in clinical settings to assess brain damage, epilepsy and other brain disorders. In many jurisdictions around

Emotional reaction (valence)

Neutral

Happy

Valency

TV commercial   ; N = 535 people measured

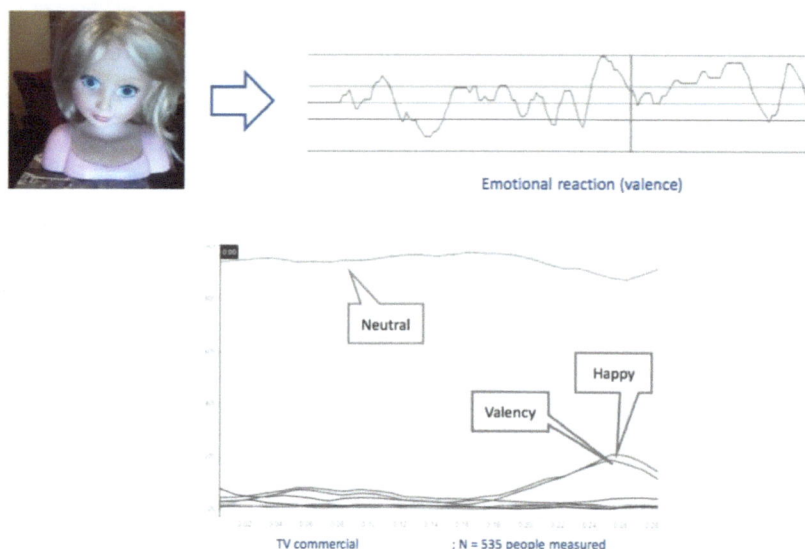

Figure In their newsletter, decode marketing made independent tests of a facial coding system that was available online. In particular, two tests were notable. In the first test (top), Phil Barden and his colleagues showed that facial expressions varied significantly in an otherwise static doll's face... In a second test (bottom), in commenting on an available online test, they showed that the emotional expressions to an ad were minuscule compared to the dominant facial expression: neutral. These and other cases demonstrate that facial coding may work, but that much more validation work needs to be done. Courtesy of decode marketing (www.decodemarketing.co.uk)

the world it is used to assess brain death. It is also a cognitive neuroscience research tool due to its superior temporal resolution, and its results are often compared to other brain recording techniques.

The EEG ranges from several to about 75 µV in the awake and healthy individual. The EEG signal as such is mostly attributable to graded postsynaptic potentials occurring in the cell body and large dendrites. The pyramidal neurons of layers 3 to 5 of the cortex are the major contributing units to the signal, and these neurons are synchronised by rhythmic discharges from thalamic nuclei. The degree of synchronisation of the underlying cortical activity is reflected in the amplitude of the EEG. Not surprisingly, most of the signal being recorded in the EEG stems from the outer cortical layers near the electrode, and the folded organisation of the cortex in humans contributes to the electrical summation of neuronal signals rather than mutual cancellation.

The EEG recorded at the scalp represents a passive conduction of currents produced by summating activity over large neuronal aggregates. Regional desynchronisation of the EEG reflects increased mutual interaction of a subset of the population engaging in 'cooperative activity' and is associated with decreases in amplitude. Thus, from the raw EEG alone it is possible – even for the untrained eye – to determine the level of synchrony. To the trained eye, it is also possible to see pathological patterns following states such as epilepsy. As can be seen in below it is possible to make

out some basic differences in healthy and pathological brain activation.

In general we can think of the EEG as two kinds of measures. First, *spontaneous activity* is the activity that goes on continuously in the living individual, as measured on the scalp or directly on the cortex. The measurement of this signal is what we call the encephalogram, which can be thought of as a measurement of electrical signals within a time window, but with no additional time factors. The amplitude of the EEG is about 100 µV when measured on the scalp, and about 1-2 mV when measured on the surface of the brain (so-called intracranial recording). The bandwidth of this signal is from under 1 Hz to about 50 Hz. Figure AX displays different kinds of spontaneous EEG activity. Today, the EEG is used extensively for clinical purposes, especially in the testing for epilepsy, but recently the combination with imaging methods with high spatial resolution, such as fMRI, has led to renewed interest in the EEG.

The second kind of EEG measure, *evoked potentials* (EP) and event-related potentials (ERP) are components of the EEG that arise in response to a stimulus (e.g. auditory, somatosensory or visual input). Such signals are usually below the noise level and not normally possible to distinguish in the raw EEG output. In order to see the effects of the stimulus one must apply a train of similar stimuli and then average the signal for all these epochs. In this way, the signals recorded at the time of a stimulus are grouped into one category.

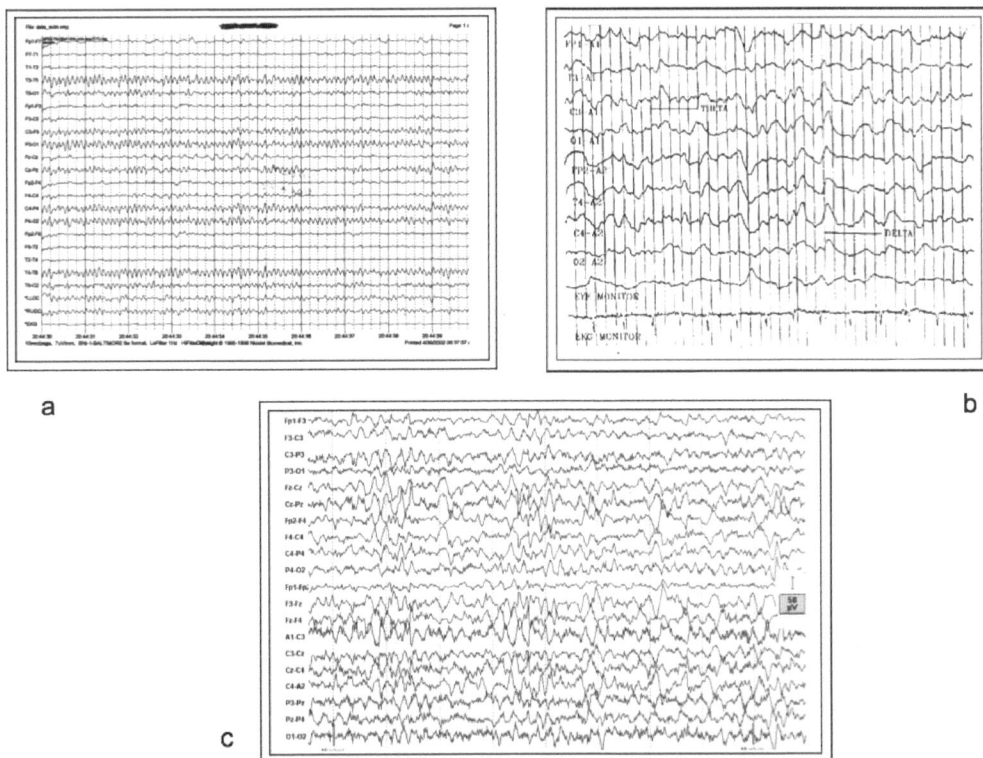

Figure The raw EEG in different states. The record of a healthy and awake individual shows different low-amplitude and high-frequency bands of activation, indicative of rapid communication between ensembles of neurons (a). When going to sleep the brain can operate at several different levels of sleep, ranging from Rapid Eye Movement (REM) sleep to stages of deep sleep. Deep sleep stages display high-amplitude and low-frequency patterns, which indicate lower cooperation between and within brain areas (b). Finally, epileptic seizures show up clearly on the EEG as both high-amplitude and relatively high-frequency patterns on all recorded channels. This demonstrates that most parts of brain are affected by the seizure (c). This pattern of activity is indicative of areas that get into a mutual loop of activation, eventually leading to chaos and non-information processing in the brain.

Averaging the signal for a time frame cancels out any spontaneous fluctuation in the EEG, and it is therefore possible to make out and display the signal intensity within a time frame common to all stimulus epochs. So, for a visual stimulus such as a word we get different waves of increase and decrease in signal intensity. Each of these components is today identified as relevant indicators of cognitive processing states. For example, EEG recordings of a subject reading a word show early increases or decreases indicative of early visual processing and the drawing of attention to the stimulus, and later we find a significant peak after approximately 400 milliseconds thought to indicate semantic processing. This component is called the P400, where P = a positive signal change.

The ERP shows how it is possible to identify specific activation patters within the brain according to the changes in signal intensities across time. Although EEG is traditionally thought of as superior in temporal resolution but with poor spatial resolution, continuous technical developments have improved the spatial resolution dramatically. Today EEG is performed by ap-

plying hundreds of electrodes on the scalp, typically by using a head cap with predefined positions. As a research tool, EP/ERP can thus provide valuable information about both the precise timing and now also the gross cortical distribution of the activity generated during mental activity.

Intracranial EEG and deep electrodes

Not all EEG recordings are done from the outside of the skull. In some cases it is also possible to record the electrical properties directly from the brain itself. There are in general two ways of doing this; by laying an intracranial grid of electrodes on the surface of a part of the brain, or by using deep electrodes directly into the brain. Both methods are invasive and are only applied in humans as part of a clinical evaluation, typically in the search for epileptic foci for surgical planning. In this clinical procedure, the surgeons need to know not only where the epileptic seizure initiates, but also where other important cognitive functions are located. If possible, surgeons avoid ablating brain

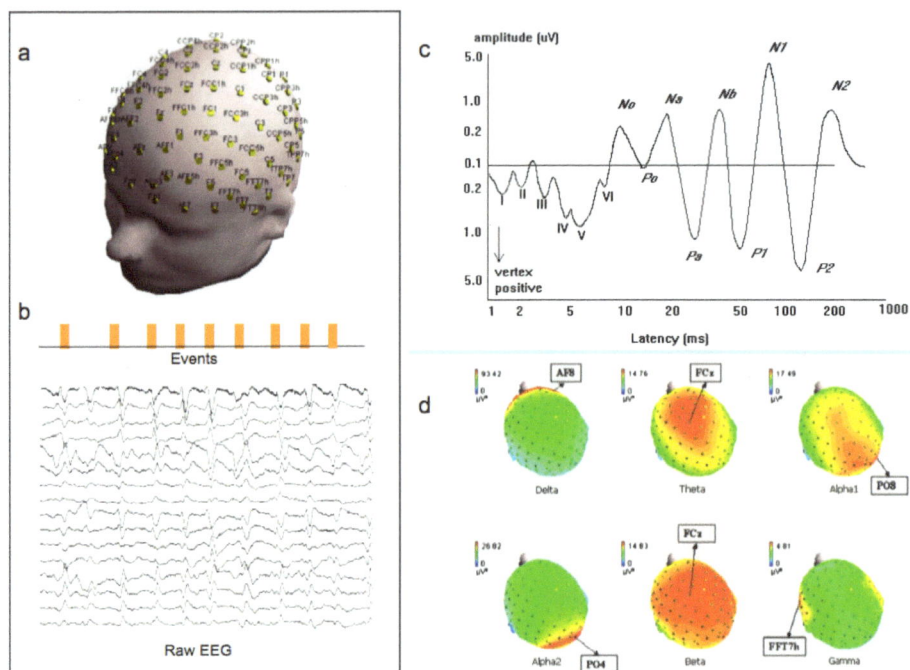

Figure The raw electroencephalogram and two ways to analyse it. A number of electrodes are placed on specific places around the scalp according to a pre-specified system (a). The readouts from each of these electrodes, or channels, can be plotted as separate curves over time (b), and make up the raw encephalogram. Events such as sound stimuli can be plotted on the same time curve, both to give an idea of trends in the raw EEG, but also to be used in the later averaging of the data. The events are shown for illustrative purposes (b). The event-related analysis of the raw EEG is made by averaging the EEG signal changes at the time interval around each stimulus event (c). By averaging the signal spontaneous activation is cancelled out, while the signal change common to all stimulus events will be visible. In this way, several components have been determined. These include early positive or negative peaks thought to involve subcortical activations (e.g. from the brainstem) and later onsets thought to involve more elaborate, cognitive processes. An example is the P400 component (P400 = positive peak around 400 milliseconds after stimulus onset, not shown here) which is thought to be an indicator of semantic processing. Finally, more recent developments have made it possible to analyse the data both in terms of the EEG bandwidths during a period and the relative spatial localisation of these. The EEG power spectrum topography (d) shows the average bandwidths at rest. As can be seen, delta activity is focal at the prefrontal site and maximal at the AF8 electrode site. Theta activity is found at the frontal midline area, maximal at the FCz electrode site. Alpha-1 activity was centred on the parietal–occipital midline area and maximal at the PO8 electrode site. Alpha-2 activity was focal at the occipital area and maximal at the PO4 electrode site. Beta activity was highly diffused across the scalp, and a maximal site at the FCz electrode site was selected. Gamma activity was focal at bilateral temporal areas and maximal at the particular electrode site.

tissue that is involved in a cognitive function such as memory or language. Therefore, the patient needs to be awake during the test, both in order to respond to the set of cognitive tasks that are applied, and in order to report any changes occurring during the test. While the electrodes are implanted, researchers are sometimes allowed for a limited time to do scientific testing on these patients.

Intracranial imaging is also done on non-human animals such as the macaque monkey. By using this animal "model" of the human cognitive system, further exploration of the neural correlates of cognitive functions are made. Important discoveries have been made using this approach, such as the findings of mirror neurons made by Gallese and colleagues.

Together, EEG presents a whole range of different recording (and stimulation) approaches that each makes significant contribution to the study of brain-mind relationships. With its superior temporal resolution the EEG provides valuable information to combine with other imaging techniques that have a high spatial resolution. We will return to the combination of such data at the end of this chapter.

STEADY STATE TOPOGRAPHY (SST)

Steady State Topography (SST) is an evoked potential methodology for observing and measuring human brain activity that was first described by Richard Silberstein and co-workers in 1990 (Silberstein, Schier et al 1990). While SST has been principally used as a cognitive neuroscience research methodology it has also found commercial application in the field of neuromarketing and consumer neuroscience in such areas as brand communication, media research and entertainment. During an SST study, participants wear a

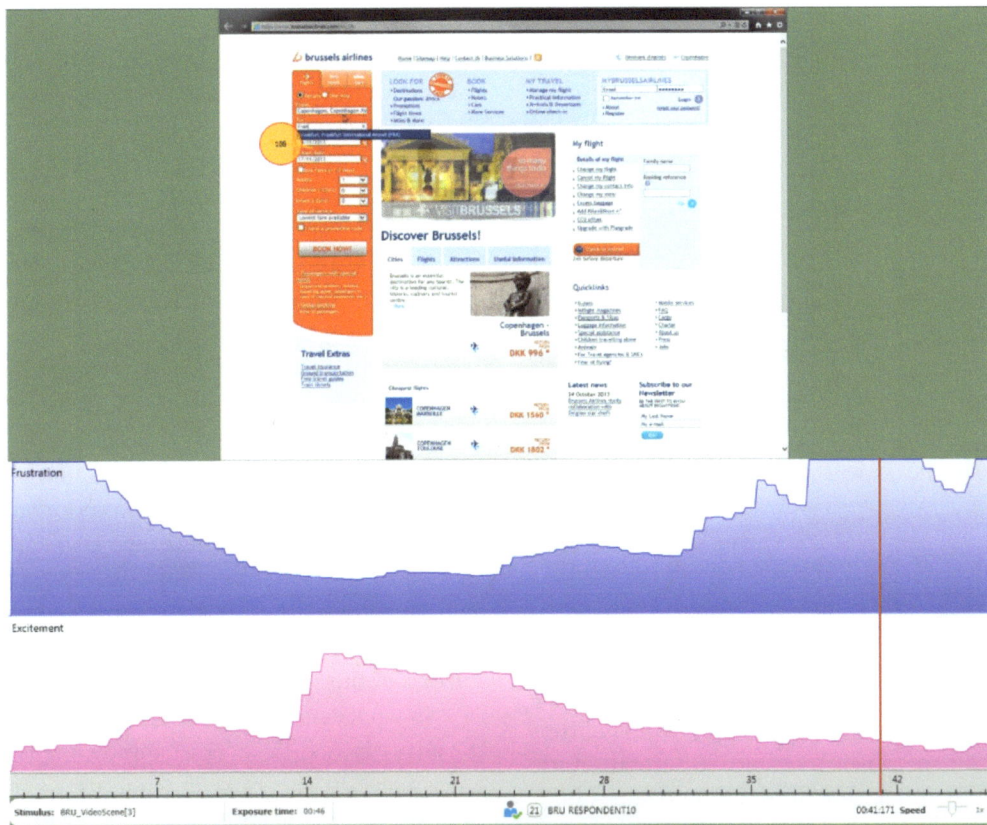

Figure   Certain EEG metrics may assess consumer relevant emotions to behaviours such as web browsing. In the above example, we can see the recording of a single person searching for a particular flight between two destinations, and not finding it (top). During the visual search (eye fixation indexed by orange circle), EEG based metrics of emotional engagement and frustration are shown (bottom). This allows researchers to understand when a particular emotion, such as frustration, becomes the dominant emotion, and shaping consumer preference and attitude. Courtesy of iMotions (www.imotionsglobal.com)

special visor while viewing a video screen or other display where advertising or entertainment is presented. The visor creates a dim flickering stimulus in the visual periphery and the brain response to flicker is measured over the scalp. This allows one to measure variations in the neural processing speed, and hence activity in different parts of the brain. That is, higher processing speeds imply higher levels of activity. As different parts of the brain specialize for different cognitive and emotional processes, the patterns of brain activity determined from SST enable researchers to infer the psychological processes taking place in response to the viewed material.

Compared to EEG and fMRI, SST possesses a number of specific advantages. These include the ability to track relatively rapid changes in brain activity (high temporal resolution) over an extended period of time, strong signal strength (high signal signal-to-noise ratio) which means it is possible to work with data based on a single trial per individual. However, compared to EEG, SST cannot be used in a brightly lit environment as the external or ambient light levels must be low.

Specific measures derived from SST that are relevant to consumer neuroscience and neuromarketing include:

- **Visual attention.** This is determined from electrodes located over the left (attention to detail and text) and right (attention to global features) visual or occipital cortex, (Silberstein, Schier et al 1990)
- **Engagement,** or sense of personal relevance is determined from electrodes located over the prefrontal cortex from scalp electrodes located near the forehead (Abraham and von Cramon 2009).
- **Approach-Withdraw, or Motivational Valence**: This measure is determined from the difference in brain activity over the left and right frontal cortex. It indicates whether one is attracted (left frontal hemisphere more active than the right) or repelled (right frontal hemisphere more active than left) by the material being viewed. (Silberstein and Nield 2012)
- **Emotional Intensity**. This measure is determined from brain activity at the right hemisphere parieto-temporal site. The measure is similar to 'arousal' and indicates the strength or intensity of an emo-

Figure   Neuro-Insight recording facility in Melbourne, Australia. Courtesy of Neuro-Insight (www.neuro-insight.com)

tion, irrespective of the specific emotion. (Kemp et al 2002)

- **Long-term memory encoding.** This measure indicates how strongly the current experience is being stored, or encoded in long-term memory. Activity in regions of the left and right lateral frontal cortex have been shown to be correlated with the strength long-term memory encoding (Silberstein 2000, Rossiter et al 2001). This measure is most important as Neuro-Insight has identified the strength of long-term memory encoding of *brand* or *key message* in an advertisement as the most consistent indicator of the likelihood of the advertisement influencing consumer behavior, that is, a measure of advertising effectiveness. The link between long-term memory encoding while viewing an advertisement and the subsequent change in consumer brand preference was demonstrated in a 2008 study (Silberstein & Nield 2008)

An example of truly effective advertising.

In 2010, Neuro-Insight was commissioned by Think-box, the premier free to air television research body in the UK, to evaluate a highly effective advertising campaign created by the advertising agency, DDB. The advertisement was for an electric shaver, but rather than using the usual approach of exhibiting the tangible benefits of the razor (multiple diamond coated blades etc, etc) the advertisement featured a female shaped robot or 'fembot' shaving a naked man in the shower.

The campaign was an outstanding success with a declining sales trend in electric shavers reversed. Neuro-Insight were asked to evaluate the advertisement so as to identify what specific features of the advertisement were the ones most powerfully encoded in long-term memory and hence responsible for the commercial ef-

fectiveness. SST data was recorded while 50 males and 50 females viewed the advertisement in a normal advertising break of a prime time television program.

SST measures of long-term memory encoding in the 50 males. The red trace illustrates the left hemisphere encoding site thought to play a crucial role in encoding the detail information while the blue trace illustrates the activity at the right hemisphere encoding site thought to play a crucial role in encoding global information or 'the big picture'. The units of activity are expressed in radian and 1 radian increase represents a speed-up of brain processing speed by approximately 12 msec. The dotted horizontal line at 0.7 represents the median level of long-term memory encoding based on the Neuro-Insight database derived from over 4000 advertisements. There are two clear peaks of long-term memory encoding. The first as the fembot approached the man in the shower and the second at the time of the appearance of explicit branding near the end of the advertisement.

So what can we see from the data that accounts for the advertisement's success? Firstly, the levels of long-term memory encoding at these times are extremely high by Neuro-Insight standards. These peak values fall in the 99.5% percentile. Secondly, the level of long term memory encoding is very high during the narrative or 'story' and especially during the branding. This means that the emotions and associations elicited during the early part of the advertisement will be linked to the brand. Thus, even though there was no explicit message about the electric shaver, the males clearly responded to the inferred 'sexual theme' of the ad and linked this to the brand. Unsurprisingly, this advertisement was less effective with females where long-term memory encoding during the final branding scene only reached the 46% percentile.

In summary, SST is a brain imaging methodology that has some unique advantages that make it especially suitable for academic and commercial consumer neuroscience studies.

Mobile EEG

EEG is the only method for assessing brain activation in mobile environments, including stores, virtual reality, outdoors and any other place where we want to allow free movement. There are at least three technological advances underlying this recent development:

- **Better, cheaper electrodes** – The use of better electrodes, conducting gels, dry electrodes (instead of gels or saline water based ones), and at a significantly lower price

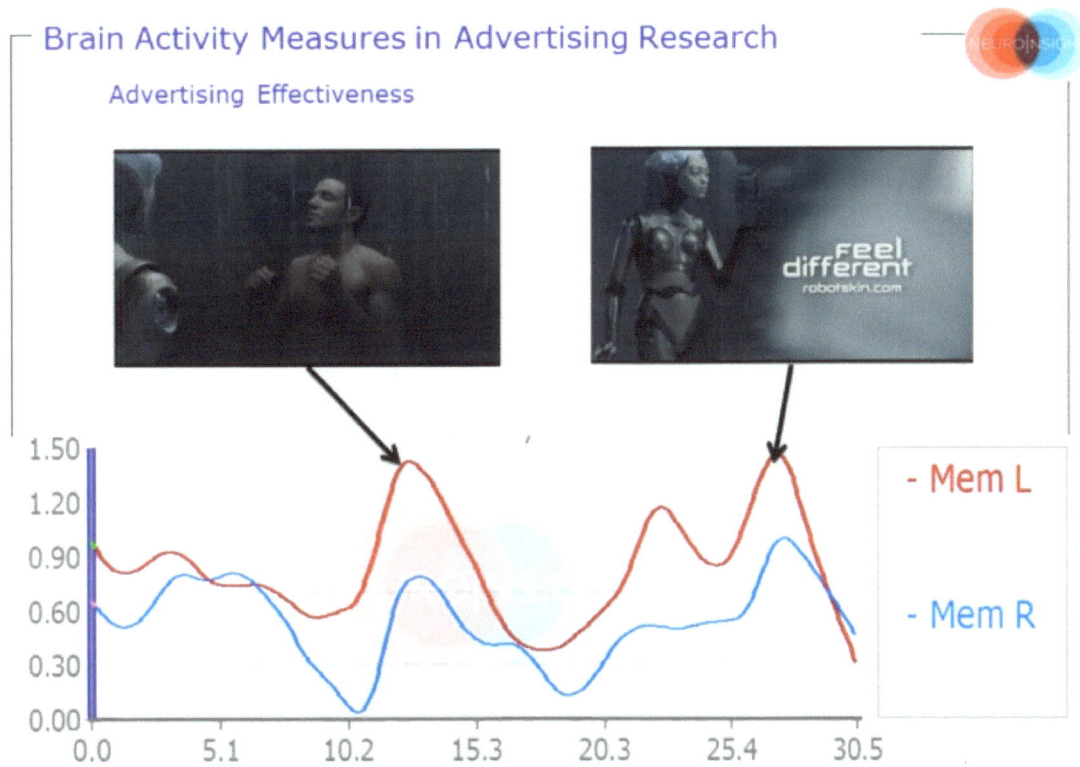

Figure Brain-based long-term memory scores during an ad. Stronger engagement of the left hemisphere (red line) relative to the right hemisphere (blue) demonstrates time points of crucial value during ad viewing. This use of EEG allows an unprecedented temporal resolution of consumer engagement and responses during ad viewing, which is soon becoming an important diagnostic tool for creatives and companies alike. Courtesy of Neuro-Insight (www.neuro-insight.com)

- **Improved noise reduction** – Methods such as Independent Component Analysis (ICA) and similar approaches allows identification and filtering of unwanted noise in the raw EEG, such as spikes due to eye blinks and motor activity (see below).
- **Better neuroscience** – Our understanding of the basic brain mechanisms underlying choice has improved tremendously, probably exponentially, during the past couple of decades. This knowledge, while valid in its own right, now guides us in search for relevant neural markers of consumer behaviour in mobile settings.

Novel ways to reduce the noise

The EEG is traditionally very noisy, even when a person is sitting still. Noise blinks, facial expressions, and motor movements are all picked up by the EEG. While traditional methods such as Independent Component Analysis (ICA) are used to pre-process the data (i.e., before the actual analysis), several novel solutions contribute to improve the noise filtering:

- **Eye-blink recording** – Very often, studies using mobile EEG are combined with mobile eye-tracking. As people blink, the eye-tracker loses the signal from the eye in a highly specific manner (with high-resolution eye-trackers it is possible to identify eye blinks, since the progression of eyelids closing can be captured by the eye-tracker). In this way, eye blinks can be precisely determined from most eye-trackers and thus be used to remove blink-related artefacts in the EEG.
- **Accelerometry** – Just as in today's smartphones, in which we can use sensors to know when a phone is tilted or shaken, one can use accelerometers to assess specific movements of the head and body.

Figure Certain EEG systems, such as the ABM X-10, allow researchers to assess brain responses while they are in different environments such as stores, amusement parks, and even perform sports activities. Courtesy of Neurons Inc (www.neuronsinc.com)

Using this solution, it is possible to equip the EEG headsets and make body sensors that record this, and use data from accelerometers to improve the noise filtering of the EEG even further.

- **Facial coding** – While facial coding is mostly used for decoding facial expressions (see the dedicated section for this below), one can use the same metrics for assessing facial movements, and correcting the EEG with these metrics. Facial movements are, after all, known to induce noise in the EEG. Therefore, an appropriate assessment of, e.g., talking, smiling or frowning is valuable for EEG pre-processing purposes also. This solution is very novel and not yet completely validated, but with the need for noise filtering being so high, it should be used to a larger extent, especially as the necessary technologies are becoming available and used in combination.

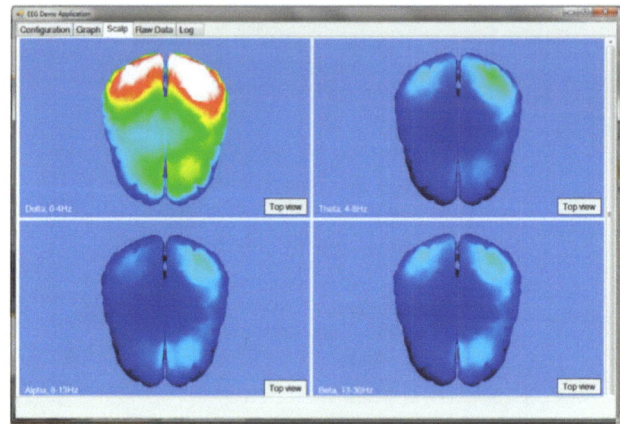

Figure   Live view of brain activation with EEG, showing the spatial distribution of four different frequency bands (delta, theta, alpha and beta) in a live view module. As can be seen, delta activation during this live view is very high particularly in the occipital region (brain shows front at bottom, back at top). Courtesy of iMotions (www.imotionsglobal.com)

Can you use the Emotiv scales for anything?

How should you do a neuromarketing test? I'm increasingly being asked whether the scales from the Emotiv EPOC Affective™ Suite system can be used to assess cognitive and emotional responses in e.g. customers. After all, it would be really appealing if we could use a full box set with responses such as Engagement, Meditation, Frustration and Excitement. I also see that some new up and coming companies use this system more or less from the box. After all, who wouldn't just pay $5,000 for a neuromarketing study rather than the more expensive studies that require whole teams and specialists?

I have used the Emotiv system myself for some of my studies. For recording EEG it works decently. But when it comes to their emotional scales, the truth is that they are, at best, just a black box with many unknowns. Quite realistically, it's more like a can of worms.

From one of my own data, it's easy to check whether the Emotiv scales are distinct. They are produced very nicely through the export function in iMotions' Attention Tool – the best scalable neuromarketing suite I know and can think of. Best of all, it allows export of time-synced raw EEG data, along with data from eye-tracking, GSR, facial coding and much more.

So how do the Emotiv scales fare? Let's take the example from "Meditation" and "Frustration". After all, we should expect that scores for "Meditation" would be quite different from "Frustration" scores, right? Well, the truth is that these scales are highly correlated:

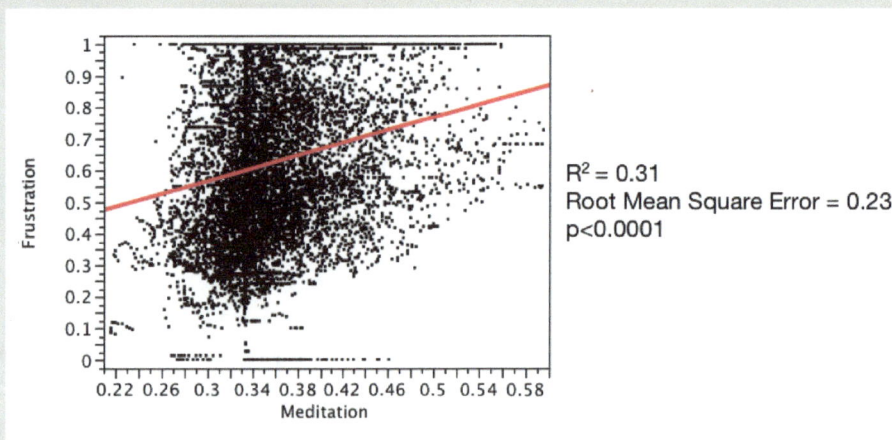

$R^2 = 0.31$
Root Mean Square Error = 0.23
$p < 0.0001$

So unless you think that meditation is really about frustration (it may be to some...), you should be skeptical towards the Emotiv scales. Some of the other scales seem to fare better, such as the Engagement score. My own studies so far suggest that the Engagement score is related to working memory load, but this is indeed still a heterogenous construct, and much too premature to draw any conclusions. I do not yet know how specific Engagement scores are to working memory...and should it then be called "Engagement"?

The basic problem remains, however, that the Emotiv system is still a black box solution, and more or less impossible to determine how the scales are made.

As you can guess, I am not a proponent of black boxing, particularly not in neuromarketing where we should be able to converge on the same solutions. Quite the contrary. I simply do not understand the need for secrecy among neuromarketing companies. The science is already out there, so why make up new scales? It opens up the possibility of cheating, snake oil production and what is less. Think about the strategic blunders that may be made based on erroneous and unscientific hand waving.

If you want to do neuromarketing studies, make sure you do it right from the beginning. There is already too much hype and BS in this industry, so let's start being self-corrective.

MAGNETOENCEPHALOGRAPHY (MEG)

Magnetoencephalography (MEG) is the measurement of the magnetic fields produced by electrical activity in the brain, usually recorded from outside the skull. It is a highly interesting tool for investigation of functional activation of and connectivity in the brain. The spatial resolution with the most advanced instruments comes down to a few millimetres and the temporal resolution is, as with EEG, down to milliseconds. This allows us to record how activation spreads from one to another region.

The principles of MEG

The physical principles of MEG are based on the old observation from H.C. Ørsted from 1820, that an electrical current in a wire will generate a surrounding circular magnetic field. Since impulses propagating in the brain are generated by electrical currents an abundant amount of small local magnetic fields will be generated. The magnetic field from a single neuron is far below detection level; however, the combined fields from a region of about 50,000 active neurons can give rise to a net magnetic field that is measurable.

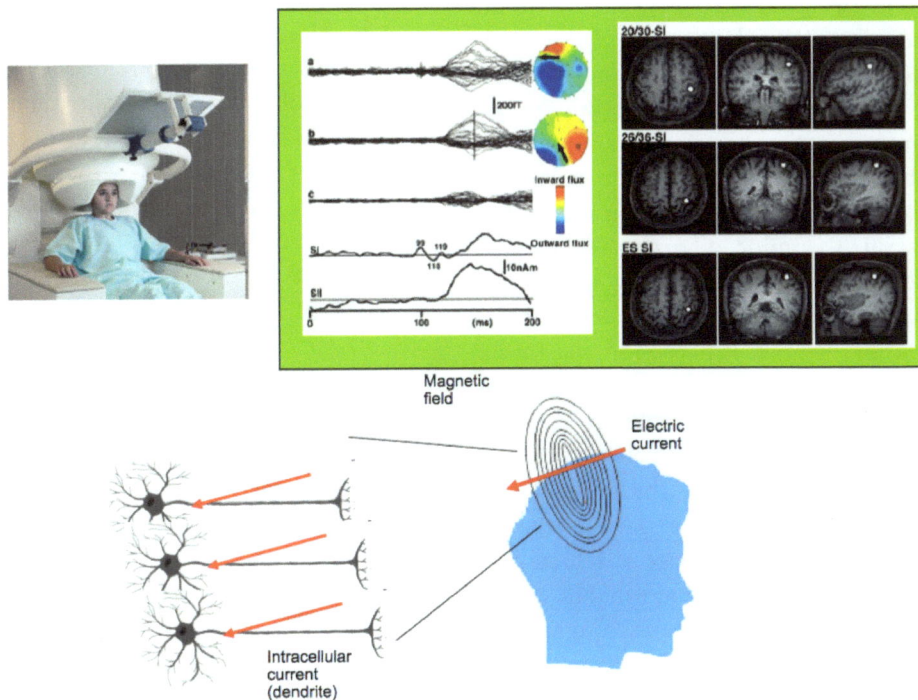

Figure Magnetoencephalography and its analyses. The subject is placed in the scanner that has a large set of shielded sensors. The signals themselves derive from the net effect of ionic currents flowing in the dendrites of neurons during synaptic transmission and in the extracellular medium as return currents (bottom). Action potentials do not produce an observable field because the currents associated with action potentials flow in opposite directions and the magnetic fields cancel out. Inset left: magnetic fields following painful (epidermal) stimulation where (a) shows the recorded data; (b) and (c) displays residual magnetic fields obtained after filtering the somatosensory processing signals from the recorded data. The bottom two lines show the time course of the source strengths during the painful stimulation. Inset right: source locations of the MEG data overlaid on MR images.

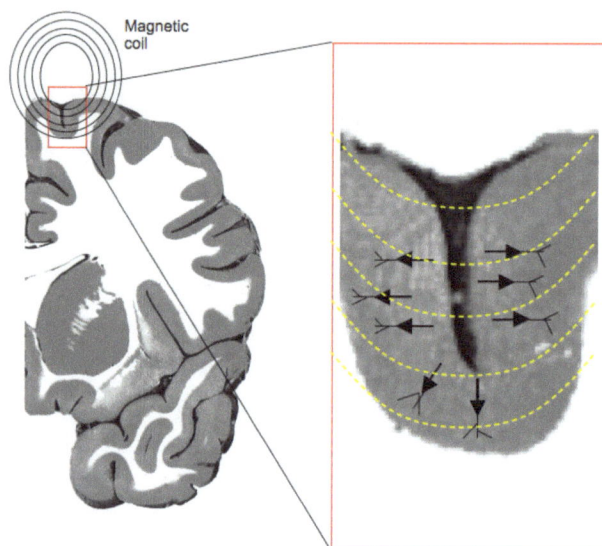

Figure The magnetic neurons in a sulcus of the brain. The magnetic gradient of the neurons at the top and bottom of a sulcus do not have an orientation that maximises their contribution to the MEG recording. The magnetic changes occurring in neurons on the sulci walls project are better measurable by the MEG apparatus.

Let us for simplicity consider a single electrical wire which will be surrounded by a circular magnetic field orthogonal to the electrical current. If it corresponds to an electrical current in the brain, an electrical dipole, parallel to the surface of the skull, then the magnetic field will go out of the skull on one side of the current and back into the skull on the other side. Changes in the electrical current in the brain and in the magnetic field will induce an electrical current in a circular lead placed parallel to the surface of the skull. The induced current will change direction if the lead is moved from one side to the other of the current in the brain. If, by contrast, an electrical current in the brain is perpendicular to the surface of the skull, then no magnetic gradients are produced outside the skull and an external lead will remain silent.

The electromagnetic signals in the brain derive from the net effect of ionic currents in the dendrites of neurons and in the extra-cellular space as return currents during impulse propagation and synaptic transmission. Action potentials do not produce significant fields as the currents associated with action potentials flow in opposite directions cancelling out the magnetic fields.

From the above considerations on the direction of the electrical dipoles and generated magnetic fields it appears that it is the neurons located in the wall of the sulci of the cortex with orientations parallel to the surface of the head that project measurable portions of their magnetic fields outside of the head. Thus, neurons on the top and in the bottom of the sulci will have an orientation which yields magnetic fields with minimal gradients outside the skull and will not be recordable. Still, deviation of the radial direction of the convex source by only 10 to 20 degrees can be enough to give a detectable signal. Therefore, it seem likely that especially the convex sources near the skull and thus near the recording apparatus contribute significantly to the MEG signals.

MEG recording

The magnetic signal emitted from the working brain is extremely low, of a few femtoteslas (1 fT = 10-15 T). Therefore, extremely sensitive and expensive devices such as the Superconducting QUantum Interference Device (SQUID) are used. The SQUID is an ultra-sensitive detector of magnetic flux. It acts as a current-to-voltage converter that provides the system with sufficient sensitive to detect neuromagnetic signals. In order to record these week magnetic fields shielding from external magnetic signals, including the Earth's magnetic field, is necessary. An appropriate magnetically shielded room can be constructed from so-called mu-metal, which is effective at reducing high-frequency noise, while noise cancellation algorithms reduce low-frequency common mode signals. With proper shielding, the SQUID acts as a low-noise, high-gain current-to-voltage converter that provides the system with sufficient sensitive to detect neuromagnetic signals of only a few femto Tesla in magnitude.

The first detection of magnetic rhythm from the brain dates nearly 40 years back and used an induction coil magnetometer in a magnetically shielded room. Modern systems have now up to about 300 SQUID channels placed around the head, and have a noise level of around 5 to 7 femtoteslas. This has to be compared to an overall magnetic field of the brain of around 100 to 1000 femtoteslas.

Data analysis

The primary technical difficulty with MEG is that the problem of inferring changes in the brain from magnetic measurements outside the head (the "inverse problem") does not in general have a unique solution. The problem of finding the best solution is itself the subject of intensive research today. Adequate solutions can be derived using models involving prior knowledge of brain activity and the characteristics of the head, as well as localisation algorithms. It is believed by some researchers in the field that more complex but realistic source and head models increase the quality of a solution (for more details see here3).

Relation to other brain recording modalities

MEG has been in development since the 1970s but has been greatly aided by recent advances in computing algorithms and hardware, and promises good spatial resolution and extremely high temporal resolution (better than 1 ms); since MEG takes its measurements directly from the activity of the neurons themselves its temporal resolution is comparable with that of intracranial electrodes. MEG's strengths complement those of other brain activity measurement techniques such as EEG, PET, and fMRI whose strengths, in turn, complement MEG. Other important strengths to note about MEG are that the biosignals it measures are not distorted by the body as in EEG (unless magnetic metal implants are present) and that it is completely non-invasive, as opposed to PET.

In research, the primary use of MEG is the measurement of time courses of activity as such time courses cannot be measured using fMRI. Due to various technical and methodological difficulties in localisation of sources using MEG, its use in creating functional maps of human cortex plays a secondary role, as verification of any proposed maps would require verification using other techniques before they would be widely accepted in the brain mapping community.

FUNCTIONAL MAGNETIC RESONANCE IMAGING (FMRI)

Functional Magnetic Resonance Imaging (fMRI) is by far the most popularised version of all neuroimaging measures. While the method actually covers multiple "functional" imaging methods using the MRI machine, most accounts cover what is labelled the BOLD method (Blood Oxygenation Level Dependent), which we will describe below.

While fMRI is a powerful measure with high spatial resolution, it also has several drawbacks. Here, we address both the basic building blocks of fMRI as a method, as well as the limits of the method. Before embarking on fMRI as sun, we need to understand what MRI is, and this opens up for presenting MRI as a large toolbox in itself for assessing brain structure and function.

Coupling of brain activity to flow and metabolism

The physiological basis of the neuroimaging signal is the coupling between regional cerebral activation and blood flow and further the uncoupling between flow and oxygen consumption during activation. Put simple, an increase in activity in the cerebral tissue leads to a blood flow increase that can be measured. The increase in flow markedly exceeds the increase in oxygen consumption so that the concentration of deoxyhaemoglobin decreases. This local decrease in deoxyhaemoglobin gives rise to the fMRI signal. We will return to the blood-level dependent technique later.

Theoretical calculations suggest that a large share of the metabolic energy is spent on action-potential propagation along axon collaterals. However, empirical studies using simultaneous electrophysiological and hemodynamic recordings by Lauritzen and co-workers (Lauritzen 2001) and by Logothetis and co-workers (Logothetis 2001) have demonstrated that the increase in flow and BOLD-signal reflects the increased synaptic activity and local field potentials in the dendrites, rather than a higher firing activity in the postsynaptic neurons. Thus, a release of stimulating as well as of inhibiting neurotransmitters will result in an increased metabolic turn-over that increases blood flow and the BOLD-signal.

Magnetic Resonance Imaging (MRI)

MRI is an imaging technique used primarily in medical settings to produce high quality images of the inside of the human body. It is based on the principles of nuclear magnetic resonance, which is a spectroscopic technique used by scientists to obtain information about the chemical and physical properties of molecules. As such, MRI started out as a so-called tomographic imaging technique – a method for obtaining pictures of the interior of the body. Today, MRI has advanced far beyond this and now represents a battery of different approaches that can measure the structure, function, connectivity, and chemistry of any part of the body.

MRI is based on the absorption and emission of energy in the radio frequency range of the electromagnetic spectrum. The human body is mostly made of fat and water – body tissues that have many hydrogen atoms. As such, the human body consists of about 65% hydrogen atoms. These hydrogen nuclei form the very basis for the signal in MRI. Each voxel in the human body contains one or more tissues. Zooming in on the

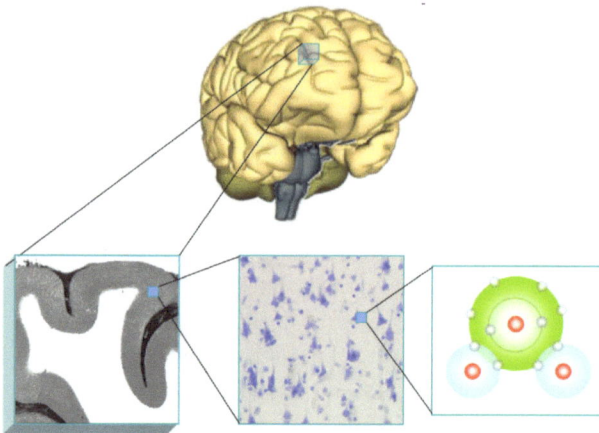

Figure A voxel of the brain. The voxel is a representation of a volume in three-dimensional space. In the brain, the resolution of the scanner determines how small the voxels can be. Parameters such as higher scanner field strengths increase the spatial resolution, and hence the ability to represent separate structures in the brain. The brain voxel extracts the signal from one part of the brain, where the local molecular environment influences the magnetic response. The voxel chosen here is much larger than usual for MRI scans, for illustrative purposes.

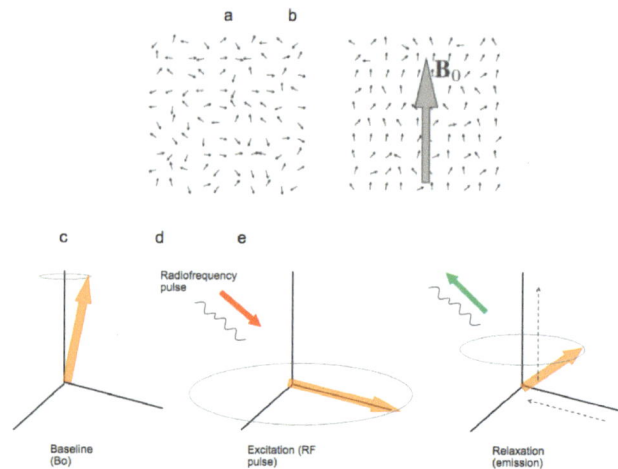

Figure The signal that makes up the MRI. Outside the scanner the atoms are oriented at random in the brain (a). When a subject is put into the scanner, the atoms align to the magnetic field of the scanner (B0). However, the alignment is not perfect, since neighbouring atoms influence each other (b). At such baseline, the atom spins along the y axis, i.e. the B0 field (c). When a radio frequency (RF) pulse is applied the spin of the atoms is influenced, and "pushed" down (d). This is a state of disequilibrium, and during equilibration towards the B0 field the atom releases energy that it received from the RF pulse. The local milieu of the atom – i.e. whether it is in grey matter, white matter, bone or cerebrospinal fluid – determines the speed of this relaxation. This is the basis of contrast in the MR image, and thus what makes it possible to visualise the different tissues of the body.

voxel reveals cells, and within each cell there are water molecules. Each water molecule consists of one oxygen atom and two hydrogen atoms. If we zoom into one of the hydrogen atoms past the electron cloud we see a nucleus comprised of a single proton.

In a magnetic field such as the MR scanner, the magnetic orientation of each hydrogen atom becomes aligned to the magnetic field and spins around this orientation. If a brief electromagnetic (radio frequency) pulse is applied, it temporarily distorts the atoms alignment to the magnetic field. When radio frequency pulse ends, the atoms start to re-align to the magnetic field, a process called *relaxation*. It is during this phase that the atom loses its own energy by omitting its own energy, providing information about their environment. The relaxation occurs in two dimensions: Time-1 and Time-2. The realignment with the magnetic field is termed *longitudinal relaxation* and the time in milliseconds required for a certain percentage of the tissue nuclei to realign is termed "Time 1" or T1. This is the basis of so-called T1-weighted imaging, which produces the most well-known structural images in MRI. T2-weighted imaging relies upon local dephasing of spins following the application of a transverse energy pulse; this *transverse relaxation time* is termed "Time 2" or T2.

The T1 and T2 constants provide the basis for most medical imaging. In different parts of the body, such as the brain, different tissues alter the speed in which T1 and T2 relaxation occurs. The three most typical tissues of the brain are grey matter (GM), white

matter (WM) and cerebrospinal fluid (CSF). The influence of these tissues produces different signal intensities – contrast – that makes it easy to distinguish between them. By varying different parameters during scanning, such as the rate and amplitude of the radio frequency pulse, or the time from excitation to recording, it is possible to highlight different properties of the tissues and their differences.

Functional MRI (fMRI)

While T1 and T2-weigthed images are superior at imaging the structure of the brain, MRI also offers ways to measure different functions of the brain. In general, there are two main approaches: BOLD fMRI and perfusion MRI. While the BOLD approach relies on a complex series of events that couple brain activation to vascular changes and the relative level of regional oxygenated blood, perfusion MRI measures the cerebral blood flow (CBF) or cerebral blood volume (CBV).

Blood Oxygen Level Dependent (BOLD) fMRI is the most used and well known way to assess brain activation with MRI. Brain activation changes the relative concentration of oxygenated and deoxygenated haemoglobin – blood with or without oxygen, respectively – in the local blood supply. While oxyge-

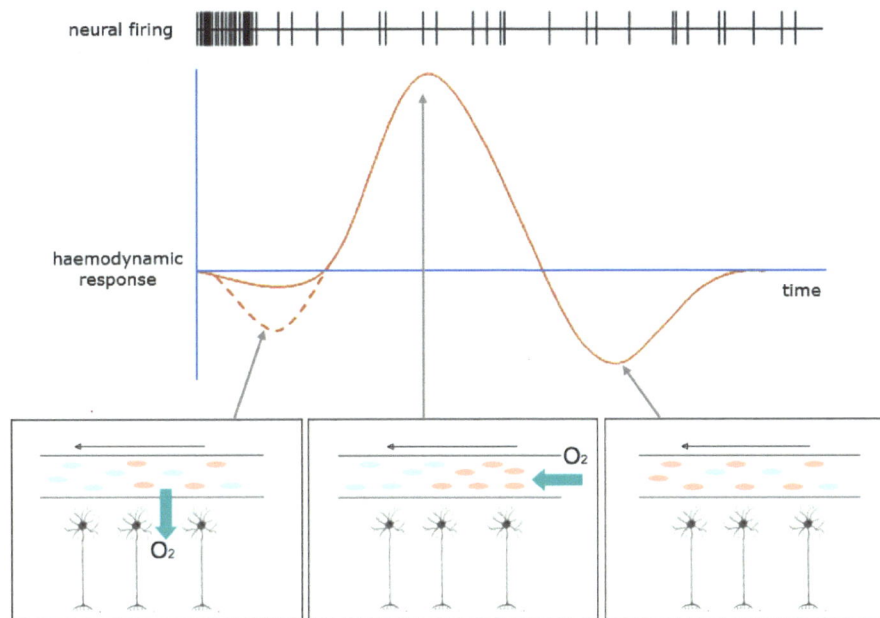

Figure The Blood Oxygenated Level Dependent signal simplified in four steps. Step 1: increased neural activation leads to an increase in the consumption of oxygen from the blood, leading to a lower level of oxygenated blood, and more deoxygenated blood, leading to a drop in the BOLD signal. Step 2: the vascular response to the increase in oxygen consumption leads to a dramatic increase in new, oxygenated blood at the same time as the oxygen consumption drops due to decreased levels of neuronal activation. Step 3: a normalising of flow and deoxy/oxyhemoglobin levels (not shown). Step 4: a post-stimulus undershoot caused by the slow recovery of blood volume.

nated blood is *diamagnetic* and does not change the MRI signal, deoxygenated blood is *paramagnetic* and leads to a drop in the MRI signal. If there is more deoxygenated blood in a region it therefore leads to a drop in the BOLD signal, and more oxygenated blood in the region leads to a higher signal. The BOLD response can be thought of as the combination of four processes:

- **An initial decrease** (dip) in signal caused by a combination of a negative metabolic and non-metabolic BOLD effect. In other words, when groups of neurons fire they consume more oxygen. When this happens, the local level of oxygenated blood drops, and there is relatively more deoxygenated blood in that area. In addition, there is also a dilation of the blood vessels, which further increases the negative BOLD effect.
- **A sustained signal increase** or positive BOLD effect due to an increased blood flow and a corresponding shift in the deoxy/oxy haemoglobin ratio. When the neurons go back to a lower level of activation, their increased consumption of oxygenated blood drops. At the same time, influx of new oxygenated blood is increased due to the previous demand. As the blood oxygenation level increases, the signal continues to increase. This drop in demand of oxygenated blood, combined with a delayed supply of oxygenated blood, leads to a dramatic 'overshoot' of the relative amount of oxygen-

ated blood. This abundance leads to the main effect of the BOLD signal.

- **A sustained signal decrease** which is induced by the return to normal flow and normal deoxy/oxy haemoglobin ratios.
- **A post-stimulus undershoot** caused by a slow recovery in cerebral blood volume.

The 'initial dip' is thought to be the measure closest related to the neural activation, since it relates to the first drop in signal intensity due to consumption of oxygenated blood. However, the signal changes at this stage are so small that they are mostly detectable by the use of extra strong MRI scanners, which are both expensive and not yet generally approved for human testing. It is therefore the second phase – the sustained signal increase – that is used in most BOLD fMRI studies. This signal is an indirect measure of neural activation, as it is the result of a delayed vascular overshoot of oxygenated blood, as a response of demand for oxygenated blood in a region. Although the BOLD response is an indirect and delayed measure of neural activation, it has been shown to have a time resolution at the millisecond scale. In addition, the method has a very high spatial resolution at below 1 mm resolution. With recent technical advances the resolution has become even smaller than this, bringing MRI to the sub-millimetre scale.

There are different ways in which a study can be conducted using fMRI. In general, there are two main

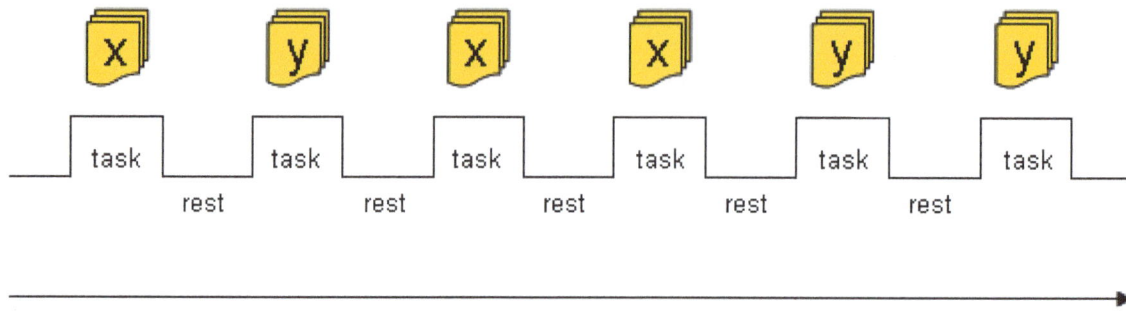

Figure The block design. Intervals of separate trials (X and Y) are separated by periods of rest. Note that the order of X and Y trials are random. In such a design, X trials could be "reading letters" while Y trials could be "reading numbers".

design categories – block designs and event-related designs. Following PET studies, the *block design* was the most used design in the earliest days of fMRI. Typically, a block design consists of a presentation of stimuli as blocks containing many stimuli of the same type. For example, a blocked design may be used for a sustained attention task, where the subject is instructed to press the button every time he or she sees an X on the screen among other letters. Typically blocks are separated by equally long periods of rest, although one may design blocked experiments without rest. A block design could be used if we were interested in seeing the difference between encoding of the identity vs. the position of objects.

In an *event-related design* stimuli are presented in a random or pseudo-randomised order. Here, individual stimuli of various condition types are presented in randomised order, with variable stimulus onsets. This approach provides a means to look at different trial-relevant changes such as correct vs. error responses. An event-related paradigm therefore lies closer to a traditional psychological experiment, and it allows post-hoc analyses with trial sorting (accuracy, performance, response time etc.). This design is more efficient, because the built-in randomisation ensures that preparatory or anticipatory effects (which are common in blocks designs) do not confound event-related responses. An example of an event-related paradigm can be seen below:

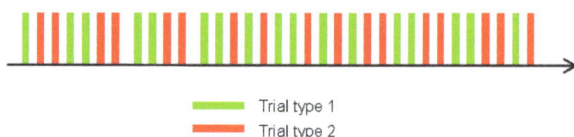

Trial type 1
Trial type 2

Figure Unlike the block design the event-related design focuses on individual trials. Hence, the order of stimuli is much more random, and is good for use in psychological designs where we do not want the subject to be able to predict the order of stimuli. An example paradigm could be to present stimuli at very briefly durations and ask for each stimulus presentation whether the subject has seen the stimulus. Responses will vary according to the duration of the stimulus, as well as endogenous processes (e.g. inattention). The subjects are not able to predict their next response (or stimulus experience).

When do you choose to use either a blocked or event-related design? In general, one can say that if you want to study state-related processes (as in "encoding state" vs. "retrieval state") then a block-design would be the best choice. However, if you want to study item-related processes (as in correct vs. incorrect responses) an event-related design is optimal. In certain experimental setups it is possible to combine block and event-related designs, making it possible both to track overall differences in cognitive states, as well as being sensitive to differences from instance to instance.

When considering the experiment design, another important issue concerns the states that are compared. There are several different contrasts one can choose to make. In some cases, especially in older neuroimaging studies, a comparison is made between a specific task and a rest period. Today, it is more usual to compare two or more active states. In general, we can distinguish between within-subject comparisons and between-subject comparisons. In *within-subject designs* a comparison can be made between two different processing states, between different contents during a given state, or as effects of time. *Between-subjects designs* allow for studying group effects in a number of ways, including comparison of patients and healthy subjects, between different groups of patients, or between two groups who have been manipulated (e.g. trained) differently. A list of some possible designs is presented in Table 1. In this way, it is imperative to pay attention to what comparisons are being made in a given experiment.

When studying the effects of one condition, one must always contrast that to some other condition to achieve any meaningful data. Let us demonstrate this with a study on the perception of facial expressions. In a standard emotion task in neuroimaging, subjects are asked to look at images of different faces. Half of the faces are female, and the subject's task is to report whether the face is a male or a female. At the same time, the faces also vary according to what kind of

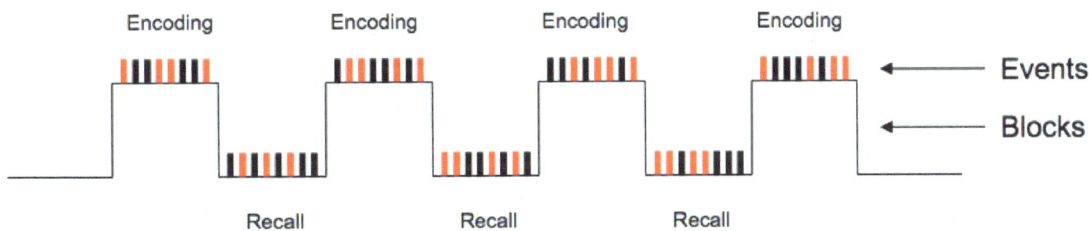

Figure   Combined block and event-related designs allow for more sophisticated analyses of both states and events.

emotion is expressed. The faces can either be neutral or aversive (e.g. sad or frightened). During the experiment, our main focus is on the difference between seeing aversive and neutral faces. During the analysis of the data we first determine when the aversive and the neutral faces were shown. We then determine the signal intensities for each of these periods and look at the mean signal intensity for neutral and aversive faces, respectively. The mean signal and distribution in the brain for each condition is shown below. If we are interested in identifying areas that are involved in the processing of aversive faces, we take the mean activity during aversive stimuli, and subtract the mean activity

of all neutral stimuli. In this way, all activations that are common for both conditions, including primary visual processing and rudimentary face perception is cancelled out. What we end up is where in the brain we see specific rise (or fall) in activation when the subjects see aversive faces.

BOLD fMRI is probably the most used measure of brain activation today. It has demonstrated its utility in every aspect of cognitive neuroscience, and is continuously being developed as a research tool. With increasing technological advances the method will lead to a better understanding of the temporal and spatial workings of the mind and brain.

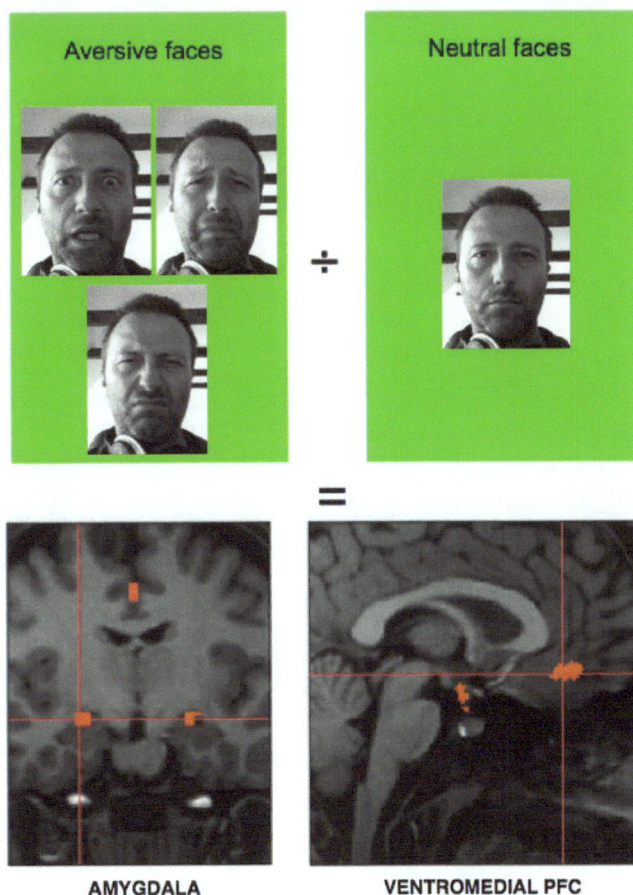

Figure   Looking at aversive faces activates emotion areas of the brain. In a series of different face stimuli, some faces are aversive (e.g. frightened, sad and repulsion) and others are neutral. If we subtract the activity associated with neutral faces from aversive faces, we get neural signal that is selective for looking at aversive faces. Here, bilateral amygdala and orbitofrontal activation can be seen in an individual subject using this contrast.

Validity check

If there is any neuroscience method that has gotten the attention of media, researchers and the public, it has to be fMRI. While this is very good, there is a clear tendency that people overly simplify the actual methods behind fMRI studies, the premises and restrictions this method imposes, the extreme data processing needed, and the significant caution one has to display when interpreting results.

One such study example was recently demonstrated by Bennett and colleagues (Bennett 2009), who showed that by doing the fMRI analysis on particular ways, they would make "findings" of brain activation in a dead salmon! Moreover, the observed brain activation would be tied to "interspecies communication." Put more simple: if you do your fMRI analyses wrong, you may find anything!

fMRI is still a method that requires an enormous amount of expertise. In fact, most MRI labs have a full team of engineers, physicists, MDs, IT experts and much more, that more or less are required just to have a scanner running, and getting quality data out of the scanners. When doing fMRI studies, the methods used require people that have a deep understanding of both the brain, the scanner, magnetic resonance physics, and multiple other domains. In this way, fMRI results cannot be taken at face value, but always requires a firm understanding of the methods that have been employed to get to those results. When done properly, fMRI is an extremely powerful method that allows true insights into consumers' minds. When done wrong, such results will be fanciful and colourful images that may well capture the recipients' interest and awe, but will mislead them in terms of actual insights.

Individual differences

Other factors are also of importance when speaking of neuroimaging. In particular, there is a substantial

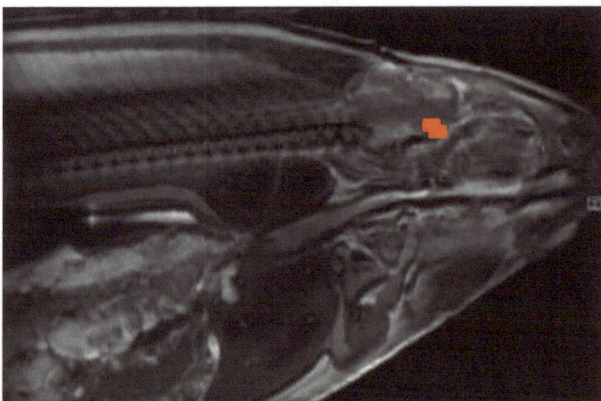

Figure   Brain activation in a dead salmon (reconstruction)

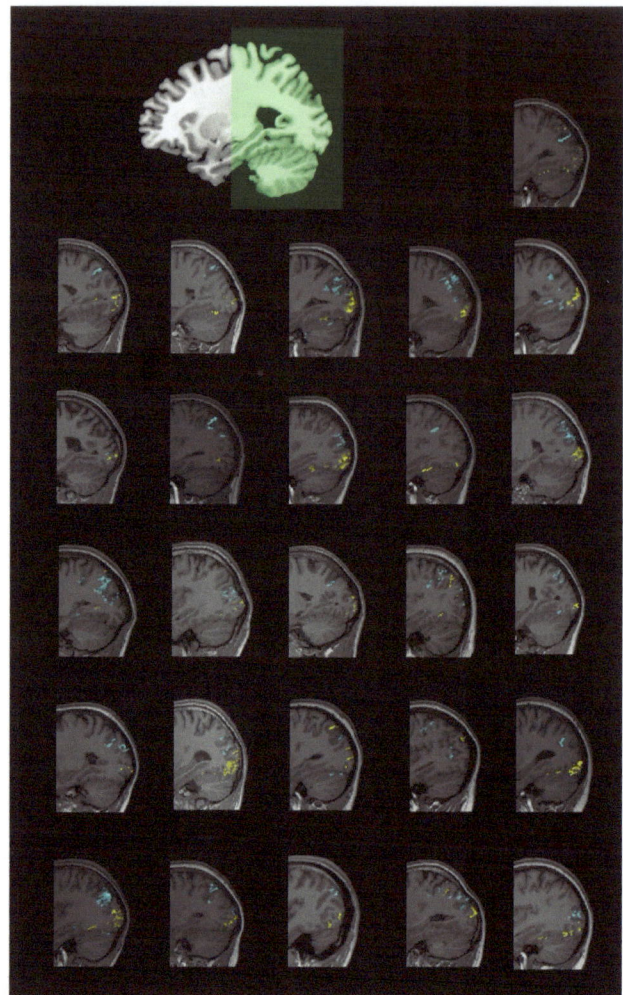

Figure   Individual differences in fMRI results. In a study of the encoding of objects, brain activation was found to be different for when subjects encoded the identity (ventral stream, yellow) or the spatial location of objects (dorsal stream, cyan). While the overall effect was found across the cohort, there was also a substantial individual variance in the extent and size of this brain activation for the 26 participants that were part of the study. The images show the posterior parts of the brains of each individual (section highlighted on top brain in green). From (Ramsøy 2009)

averaging of the signal, meaning that individual differences are washed out, and what remains are the activation that is common to all participants of a study. As there are substantial differences in neuroanatomy, blood flow, and in brain responses, one can say that vital information may be lost in the across-group analyses. This was illustrated in one of my fMRI papers, where we found substantial variation in how the brain was activated during intentional encoding of the identity or spatial location of objects.

Arterial Spin Labeling (ASL)

By altering the magnetic properties of the blood flooding into the brain, an MRI measurezment can also be made

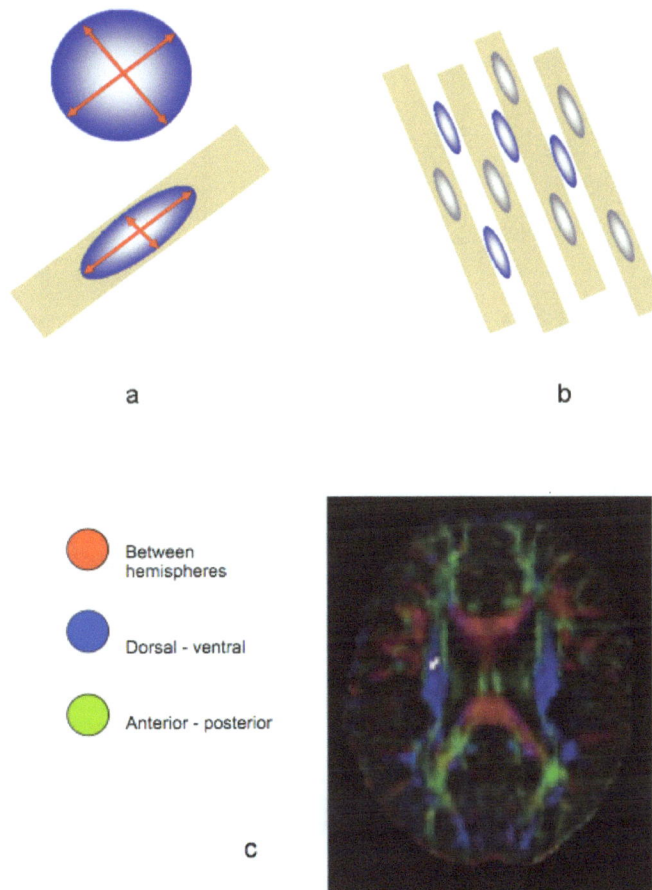

Figure Diffusion of water is dependent upon the local environment of the molecule. In the freed and unrestricted medium (i.e. a glass of water) water can diffuse freely (a). The diffusion is isotropic; it has the potential to move in all directions. If the water molecule is physically restricted it can no longer move freely in any direction. This diffusion is anisotropic; it cannot move in any direction. In a medium of fibres such as the brain's white matter (schematically shown in b) water molecules are highly restricted by the axonal fibres. In this way, it is possible to visualise the fibres tracts of the brain and furthermore to estimate the integrity (homogeneity) of white matter in a given region. Such visualisation produces the typical coloured DTI brain image (c) that displays different trajectory trends in regional white matter.

sensitive to the blood perfusion in the brain. A technique called Arterial Spin Labelling uses this idea by "labelling" the blood in the carotid artery by applying a brief radio frequency pulse. This alters the magnetic properties of this part of the blood. As this blood continues to flood into the brain, it makes it possible to measure the relative change in magnetic susceptibility of a region of the brain, or the entire brain. A so-called perfusion weighted image can be performed by subtracting a baseline brain image with no magnetised blood from a brain image with magnetised blood. In this way, one can measure the blood perfusion in a region of the brain, and it is also possible to compare perfusion images between groups.

Other specialised MRI sequences

Diffusion Tensor Imaging (DTI) and tractography

MR can also be used to measure the movement of water molecules over time. In a free and unconstrained (isotropic) environment water will diffuse equally in all directions, also known as Brownian movement. If you measure the diffusion in this medium, the resulting image will be a sphere. If water is put into a more constrained (anisotropic) environment, however, it cannot diffuse freely but can only move along the structures that it is physically limited by. For example in a glass of water, the water in the middle can move freely in all directions. The movement potential of one water molecule is equal in all directions. However, if you put in a drinking straw into the water, the movement potential of the water molecules within that tube is dramatically limited. If you now measure the water diffusion within this tube, it is no longer circular or equipotent, but an oblong sphere.

In biological matter such as the brain, the water diffusion is significantly limited. However, there is a systematic difference between grey and white matter. While grey matter has little inner structure in the sense of limiting water diffusion,

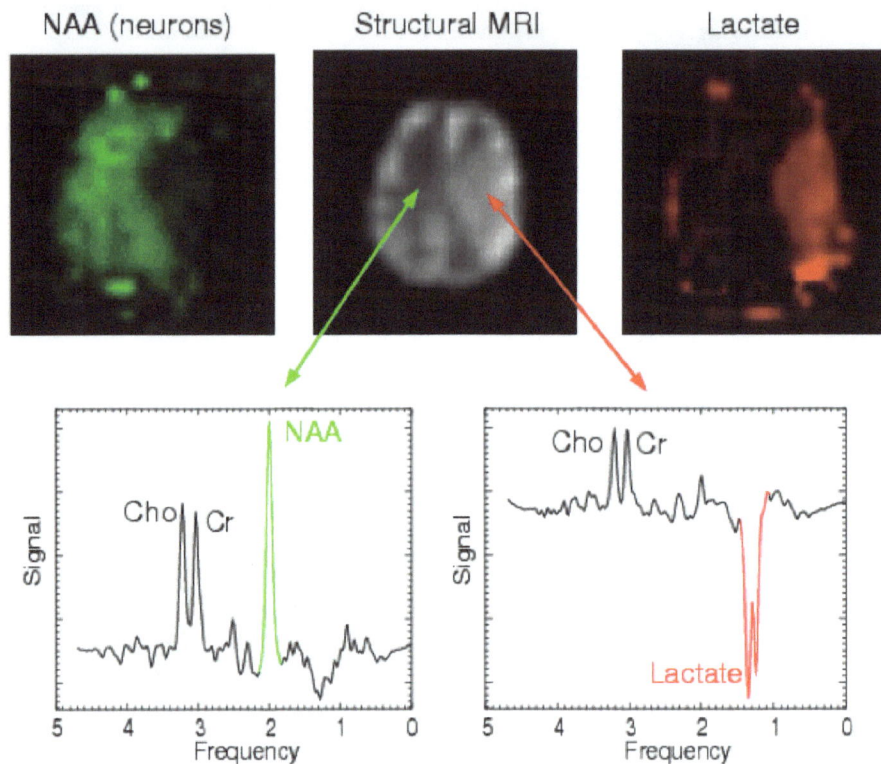

Figure Spectroscopic MRI detects chemical differences in brain pathology. On the left hand side (green) the healthy part of the brain shows normal NAA levels (green peak), while on the left hand side this NAA level has diminished, while lactate has a significantly altered change (red drop)

white matter consists of fibres that constrain the diffusion in some directions more than others. Just like putting one or more straws into the glass of water hinders water from moving through the straw walls, neuronal fibres constrain water diffusion across, but not along, the fibre direction. Connections between brain areas occur as bundles of fibres (axons). By using diffusion tensor imaging it is possible to measure the relative direction and coherence of these white matter tracts. Thus, the DTI can be used to measure white matter changes in neurological disease, but it is also possible to determine how a selected area is connected to other brain areas by following the fibres from the selected region, a method called *fibre tracking*.

MR Spectroscopy

Magnetic Resonance Spectroscopy (MRS) is closely related to Magnetic Resonance Imaging. Both techniques use the magnetic properties of atomic nuclei to get information about a biological sample. However, there is a crucial difference: While MRI measures the spatial distribution of magnetisation, MRS measures the amount of signal from each chemical environment (chemical or molecular distribution). In other words, MRI allows one to study a particular region within an organism or sample, but gives relatively little information about the chemical or physical nature of that region. MR spectroscopy, which is an NMR method, provides a wealth of chemical information about the same region. The frequency distribution is called a spectrum, and is analogous to the optical spectra of substances which are responsible for their visible colours. In this way, MR spectroscopy can measure the chemical composition in a given region of the brain. MRI and MRS can be combined to provide the spatial distribution of chemical compounds. This figure shows the spectrum from a region of the brain, and the corresponding chemical distribution of this region. If you compare this healthy area with an area from a diseased brain region, you can see some dramatic changes. First, from the healthy hemisphere it is possible to see that the N-acetylaspartate (NAA) signal is the highest of all molecular compounds. The NAA is a marker of mature axonal white matter in the living brain. By contrast, this is precisely the same signal that drops dramatically if we measure the same area in the lesioned hemisphere. Even without such dramatic effects, MRS can be used to study degeneration of brain tissue in diseases such as multiple sclerosis, stroke and Alzheimer's disease, and even the effects of substance abuse.

MRI – a tool for the future

With its superior spatial resolution and multiple uses, the MR scanner is an indispensable tool in cognitive neurosciences. The advances in MRI come from many directions, but in general we can speak of two categories of advancements: technical and analytical tools. *Technical advances* include the production of scanners operating at higher field strengths. Increased field strengths enhance the scanner's ability to record signals. So, by exchanging a 1.5 Tesla (1.5T) scanner with a 3 T scanner we get a higher spatial resolution in both the structural images as well as the functional images. While 3T or 4T is the current high-field standard in scanning subjects and patients, scanners are already available at higher field strengths such as 7 T and 11 T. Scanning the brain at 7T or higher field strengths has now demonstrated the possibility to make out the different layers of the cortex. This is important not only because we get a higher resolution for studies already performed, but since it will also generate whole new ways to study the brain, and our ideas about its workings.

Advancements in field strengths are complemented by other areas such as improvements in the apparatus that generate the magnetic pulse, or the receiver that records the signal. Such ongoing improvements are likely to make significant contributions to the possibility to scan the brain with increasingly higher resolution, and sometimes invent novel ways to acquire the data. While these advances promise a better resolution they are also associated with specific problems. For example, the higher field strength that produce better signal to noise ratio in most parts of the brain, lead to greater loss of signal in other areas. Since the BOLD signal (see previously) relies on the relative amount of oxygenated blood in an area, areas with oxygen that are not part of the brain influence the signal in that area. The medial temporal lobe and ventral prefrontal cortex both lie close to the nasal air cavities. As a result the BOLD signal in these brain regions is corrupted by the non-brain areas filled with oxygen. When moving from field strengths of 1.5T to 3T this became a problem that needed to be addressed. Acquiring fMRI activation data from the hippocampus, amygdala and orbitofrontal cortex were distorted by a loss of signal at 3T. To overcome this problem, studies now use specialised sequences that minimise the artefacts in these areas. However, when moving from the standard 3T or 4T to 7T or higher field strengths these problems re-emerge. Only new advancements in noise reduction methods can solve these problems.

The other general area of advancement is *software improvements*. In a few years the field of neuroimaging has seen a significant expansion in the number of ways to analyse data. Many of these improvements are not isolated to MRI alone, but are relevant to most or all neuroimaging approaches. While the early days of fMRI neuroimaging analysed trial blocks, later advances such as the event-related paradigm has led to new analytical tools. One such is the Dynamic Causal Modelling (DCM) analysis of neural activation patterns. In general DCM is a way to analyse how activations across the brain occur at the same time or are caused. There are two main ways to analyse such data. An analysis of the *functional connectivity* focuses on correlations between brain areas. For example, one can ask whether there is a contingent relationship between activation in the hippocampus and the dorsolateral prefrontal cortex (dlPFC) during an encoding task. Here, one can get an estimate of how correlated two (or more) areas of the brain are, in other words the relative activation strength between two or more areas. The second kind of analysis is to look at the *effective connectivity* between areas. Here, one tries to move beyond mere correlation and estimate the relative cause and effect relationship between areas. In our example of hippocampus and dlPFC we move beyond asking how correlated the regions are and now how one area is causing the other to activate. Today, both approaches are being developed continuously and are increasingly more being used in the analysis of neuroimaging data. However, since they rely on specific experimental design and data sampling, only paradigms specifically tailored to the analysis can be used.

Optical imaging

Optical imaging is a quite recent addition to the brain imaging toolbox. A laser source of near-infrared light is positioned on the scalp. A bundle of optical fibres are used as detectors and placed a few centimetres away from the light source. The detectors record how the path of light from the laser source is altered, either through absorption or scattering, as it traverses brain tissue. This information is used in two ways. First, it can measure the concentration of chemicals in the brain by measuring the absorption of light in an area. Second, it can provide information about more physiological properties of the brain that is associated to the level of neuronal firing. This is done by measuring the scattering of light, which is an indicator of swelling of glia cells and neurons In this way, optical imaging provides a simultaneous measure of both the source and time course of neural activation within an area of the brain.

In a study using optical imaging, Sato and colleagues (Sato 2005) studied the activity of somatosensory cortex in the pre-surgical planning stage of patients with brain tumours or epilepsy. As with previous findings in the human and non-human animal literature, The researchers found that the organisation of the somatosensory cortex consisted of neighbouring response areas for e.g. the fingers, and that the areas demonstrated a certain amount of functional overlap. It is also noteworthy to consider the relation between these findings and the study of brain plasticity in phantom limb sensation presented earlier in the section about MEG scanning. This study also demonstrates that optical imaging has a potential use in pre-surgical planning.

Multi-modal brain imaging

Neuroscience today rests on a number of different imaging techniques (modalities), and we have only mentioned the most prominent here. Each approach continues to provide novel findings that contribute to our understanding of the brain and our cognitive apparatus. Each imaging technique has its strengths and weaknesses, especially in terms of their relative spatial and temporal resolution. An obvious solution is therefore to compare findings from studies using these different imaging approaches. However, since each method is different from the rest, we will never get identical results. Instead, what we get is added information about our area of interest. For example, if our study focuses on subjects reading a text aloud, we can compare the results from EEG and fMRI. In this way, we can get a better understanding of both the localisation of areas that are responsible for text reading (fMRI), and at the same time get results about the millisecond to millisecond changes in activation levels during the task (EEG).

Imaging modalities can be compared in many ways. In addition to the comparison of the results in different studies an often used approach is to combine imaging techniques more directly. A PET study will most often co-register its findings to a standard MRI or CT structural brain. The same is actually done with fMRI images: the activation images are co-registered to the structural scans. This process gives us the opportunity to have a better ability to see where in the brain our changes are found. The activation images in themselves bear too little information about the localization of the activations to be meaningful.

Simultaneous imaging from different sources

While most comparisons are done on images that have been recorded separately, for example the PET scan and the structural MRI scan, it is also possible to record and co-register results from recordings that have been done simultaneously. For example, Dang-Vu and colleagues (2005) used combined EEG and PET to study the contributions of brain areas in different levels of sleep. The EEG and PET recordings were done simultaneously while the subjects were sleeping. The researchers then identified the unique stages of sleep from the EEG patterns, and focused on delta activation during non-REM (NREM) sleep and compared this to REM sleep and normal wakefulness. The researchers found a negative correlation between delta power and regional cerebral blood flow in the ventromedial prefrontal cortex, the basal forebrain, the striatum, the anterior insula and the precuneus. These findings thus hint about areas that vary in delta activation according to our level of awareness.

Similarly, Laufs and colleagues (Laufs 2003) combined EEG and fMRI measurements in the study of subjects at rest. The study of "resting state" must be seen as controversial, due to the fact that neither the mind nor the brain can be seen as being at "rest" at any time. Laufs and colleagues assessed the different patterns of activity occurring within phases in which subjects were "at rest", and found that the resting phase consisted of "intertwined yet dissociable dynamic brain processes" (p. 11056) in the EEG. In other words, the neuroimaging studies imply that the brains of subjects at rest are doing different things at different times within the rest period. Furthermore, the researchers were able to make out separate neural networks underlying EEG bands such as the alpha and beta band activity.

It should be noted that there is a substantive difference between combining EEG with PET and with fMRI. While the PET scanner does not induce artefacts that are insurmountable for the EEG, the MR scanner produces artefacts in the EEG so substantial that attempts until only recently have been unsuccessful at such a combination. One approach has been to use read the EEG from people between fMRI runs, and in this way assess the state of awareness, e.g. whether they are sleeping or awake. Through proper noise filtering it has now become possible to filter out the scanner artefacts and make the EEG data useable even from within the fMRI scanner. In this way, it has become possible to combine the high temporal resolution from the EEG with the high spatial resolution in the MRI. Still, the filtering of artefacts is seen as problematic – we may see a significant loss of true signal, while some residual artefacts (even from the filter applied) can occur. Combined measures such as the EEG/fMRI must therefore still be interpreted with caution.

Imaging genetics

Within the area of multimodal brain imaging new important steps are taken to combine our understanding of how genes make up your mind. Genes act at the molecular level in the body and thus acts as the very building blocks of neurons. In this way, individual differences in the genetic makeup can influence the workings of the brain in significant ways.

A natural variation in the promoter region of the serotonin transporter gene (called 5-HTTLPR) has been linked to alterations in serotonin transcription as well as serotonin uptake. Individuals who have two copies of the long (l) variant have a higher concentration of the serotonin transporter mRNA and therefore have greater serotonin uptake in comparison to individuals who have two versions of the short (s) variant. These subjects have a relatively lower concentration of the transporter and as an effect relatively greater synaptic serotonin levels. Hence, the genetic makeup you might have in this area influences how much or little serotonin you have available in the brain. Just looking at your genetic makeup with respect to the serotonin transporter gene, it is now possible to predict the level of active serotonin in any given person.

By applying such a combination of neuroimaging and genotyping, Hariri and colleagues (Hariri 2002) demonstrated that the type of genetic makeup had implications for the response of the amygdala to pictures of facial expressions. Subjects with the short genetic version demonstrated elevated levels of amygdala activation, presumably caused by a higher concentration of serotonin in these subjects. This genetically driven difference in amygdala excitability might contribute to the increased fear and anxiety typically associated with the short genetic version. As such, the study by Hariri and colleagues is a powerful demonstration about what can drive the individual differences in how the brain works. Today, many researchers hold the view that this area of "imaging genetics" holds the promise to give significant new insights into the workings of the brain and mind

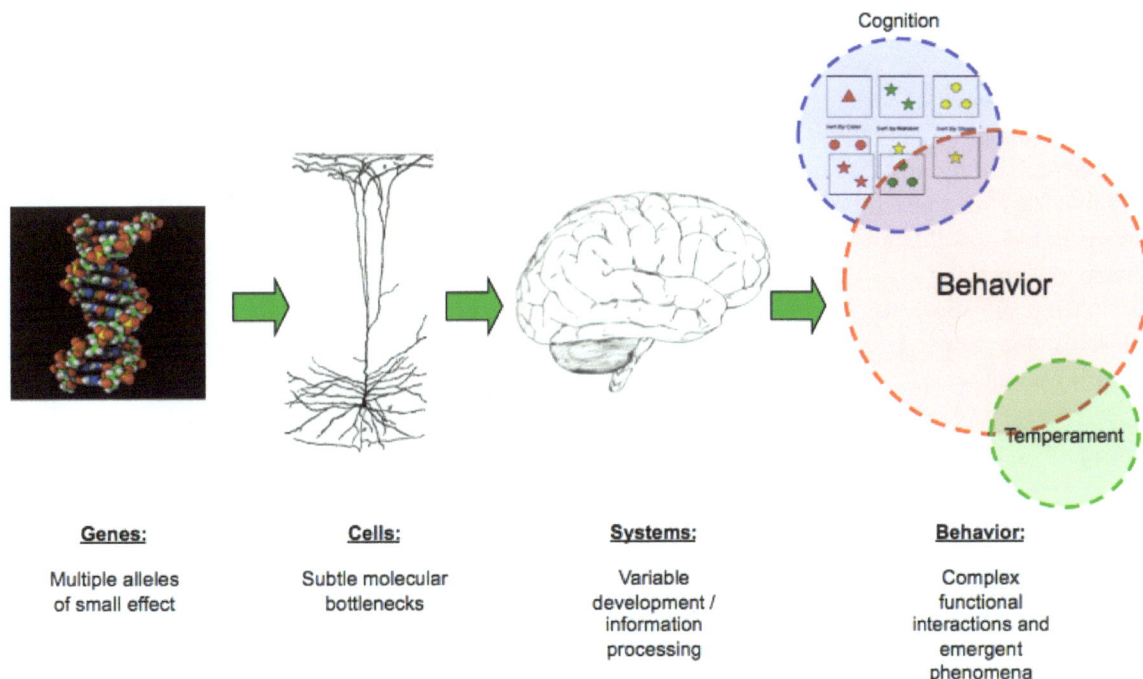

Genes:	Cells:	Systems:	Behavior:
Multiple alleles of small effect	Subtle molecular bottlenecks	Variable development / information processing	Complex functional interactions and emergent phenomena

Figure The effect of genetic variation in serotonin function on the response of the human amygdala. Using BOLD fMRI researchers have demonstrated that short allele carriers of the serotonin transporter gene exhibited greater activation of the amygdala to threatening stimuli than long allele carriers. This is indicative of the short allelic carriers having greater synaptic serotonin levels, and it might contribute to an increased fear and anxiety in this group.

CHAPTER 4

Senses & Perception

Walking down a street, you are exposed to a multitude of sensory experience. The sight of the street and the store windows; the delicious odour from a bakery; the sound of people talking on the street; and the tactile feeling of walking down the uneven street. All senses contribute to your experience of the moment, and can in different ways affect the way in which we experience, prefer and remember these events.

Much today is spoken and written about the way in which different senses can affect consumer behaviour. Does it help playing French music to increase the sales of French wine? To what extent can odour affect your preference of certain foods? How do you best design a jingle?

Before embarking on these questions, it is necessary to first have an understanding of how our basic senses work, and how they are represented in the brain. Furthermore, knowing about so-called "convergence zones" in the brain allows us to better understand how senses can merge to create coherent experiences, and even affect each other.

SEEING

Humans are visual creatures: the by far most dominant sense we have is vision, both in terms of how we orient ourselves and in the sheer real estate that the brain devotes to vision. But the process of seeing is so sophisticated that it has taken us centuries to figure out just the basics of how it works. Not surprisingly, it is also the sense by which we by far have the best understanding of.

The eye

Your eyes are a pair of extremely powerful photosensors. Actually, your eyes are collections of millions of photosensors. Rods and cones, as they are called, capture light that travel through the lens and hits the back wall of the eye.

While cones are sensitive to multiple wavelengths, they have a lower sensitivity to contrasts; rods, on the other hand, have a high sensitivity for contrast and movement, but a low detection of colours and finer details.

The distribution of the rods and cones is uneven on the back of the eye. Most of the area is most densely covered by rods, a small area called the *fovea* is more densely backed with cones. Within this region, the eye can make out much more details

The visual field has a vast span, but only a small fraction of this visual field is part of our visual focus wherein we have a strong visual acuity (see figure below). These two properties of the eyes allow us to both detect things in the periphery (with only low acuity but high sensitivity to contrast and movement) and focus on a few selected items with strong acuity (with high acuity and colour sensitivity but low sensitivity to contrast and movement).

This means that we can only see things clearly that make out about 2 degrees of our surroundings. The rest of our **visual field** (an adult normally sees 200 degrees of the visual environment) are good at detecting movement and edges.

The brain – an interpreter

Another important feature of vision is that it is filled with errors. Nevertheless, you often do not notice these flaws. One simple "poor man's" illusion is shown below – follow the instructions below the image.

Contrary to many popular notions, the eyes are not well designed light detection machines. One such notable

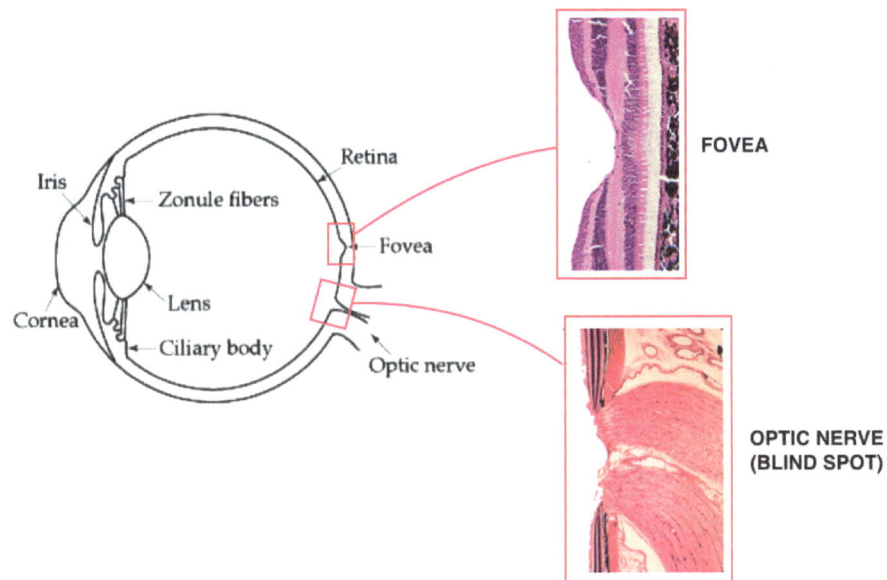

Figure Cross-section of the eye, showing different areas of relevance to seeing. Thez fovea (inset, top) has a high concentration of cones, and is seen as a fault in the cellular layers. The optic nerve exits at the back end of each eye, leaving this region without any light sensitive cells (inset, bottom).

flaw is the so-called blind spot. Each of your eyes have this problem. When all cones and rods gather their information, they project this information backwards towards the brain. Crucially, they do so through a bundle called the optic nerve, and at this very spot of the eye, there are no light sensory receptors – no rods or cones. This means that you are virtually blind within a small region of each of your eyes. If do the exercise below correctly, you will know that you have a blind spot on your eye.

Why have you never noticed this blind spot before? This hints at one of the most powerful properties of the brain. It is a feature that transcends all levels of what the brain does – from the early sensory system to complex social behaviour: the brain tries to *interpret* the world around it. That is, it does not work like a passive receiver, like a camera. Rather, it tries to work out what is "out there" – when information is lacking, it tries to fill in the gaps. Therefore, the blind spot does

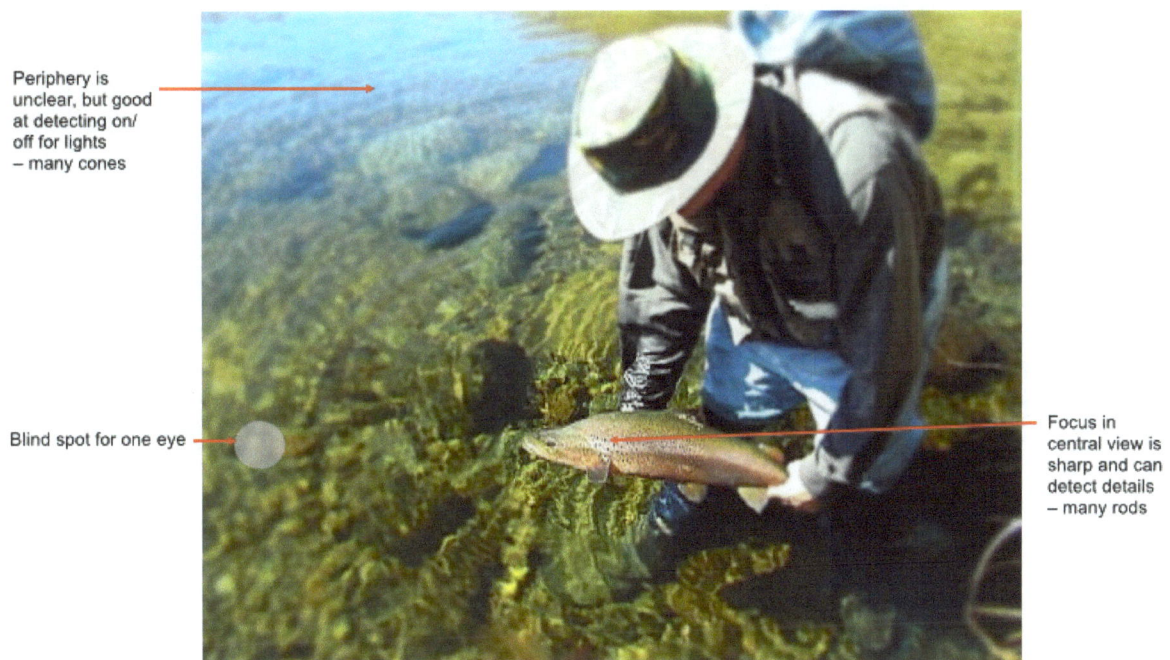

Periphery is unclear, but good at detecting on/off for lights – many cones

Blind spot for one eye

Focus in central view is sharp and can detect details – many rods

Figure Summarising figure of the visual field.

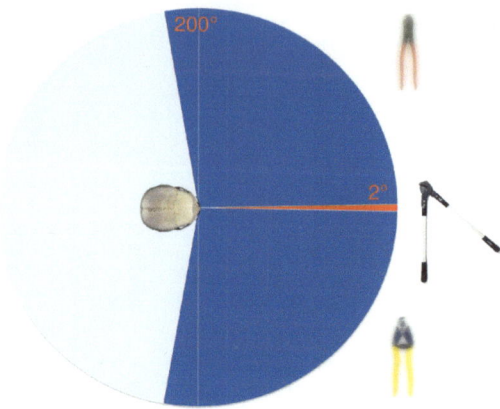

Figure   The visual field showing as the entire visual field (blue) and the central focus area (red). Courtesy of Khalid Nassri

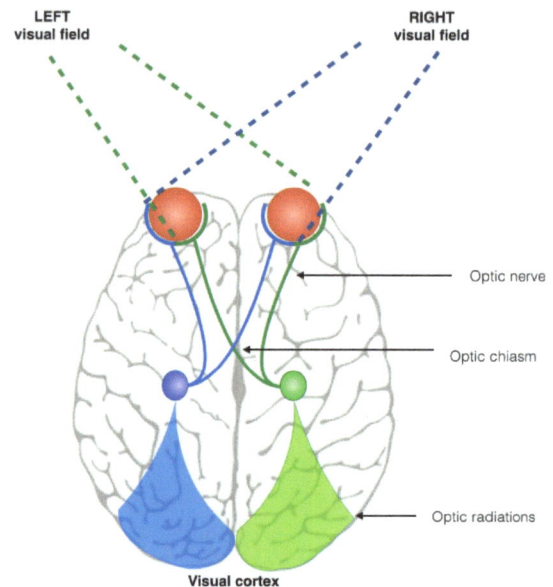

Figure   The stream of visual information from the eyes, through the thalamus and finally to the primary visual cortex. What occurs in the left visual field enters the right side of each eye, and is transferred to the right side of the brain (coloured green), and vice versa for the right visual field going to the left side of the brain (coloured blue).

not appear to you as a blind spot: your brain has already interpreted your world for you so that it appears coherent and consistent.

We say that the brain is an interpreter, and even a "forecaster." It tries the best it can to automatically make sense of the world around it. It is a great function that assists us in everything from avoiding falling, to acting in social situations. However, as we will see, this basic function of the brain can also lead to some problems.

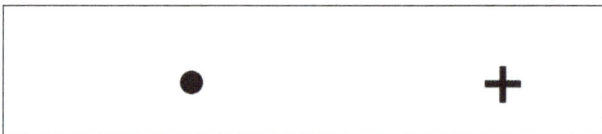

Figure   Hold the book/tablet at about an arm's length and hold a hand over your RIGHT eye. Look at the cross. While doing this, move the book/tablet slowly closer to your face. At some time the dot will disappear. Why does this happen?

Visual pathways

Visual information is projected from the eyes back through the optic nerve, until it reaches the thalamus. Here, the optic nerves from each eye cross each other, an X-shaped feature called the *optic chiasma*. Beneath this chiasma there are areas (the suprachiasmatic nucleus) that pick up fluctuations in lighting related to the day-nigh cycle.

The cross is not complete, but functional: only light that is processed on the left side of the visual field, which hits the right side of the retina, is projected to the left side of the thalamus. In this way, there is a lateralisation within the body, ensuring a gross spatial representation of what goes on in the left and right parts of the visual field. This means that whatever goes on to the left of you will reach the right side of your retina,

and is projected to the right side of the brain. That is, things to your left is processed on the right side of the brain. We will see this crossing for most sensory modalities.

Thalamus and vision

Within the thalamus, specific layers of the so-called lateral geniculate nucleus (LGN) respond differently to the same functional differentiation seen in the eyes (see the LGN figure here). In the magnocellular layers of the LGN cells are responding to the activation of the rods, and are thus implicated in perception of movement and small differences in brightness. The parvocellular regions of the LGN respond to the activation of the cones, and thus are implicated in the perception of colour and form. A more recently classified types of cells – the koninocellular (or inter laminar) cells – respond only to light in the blue wavelength.

As we will see in later chapters, the thalamus is a conglomerate of different nuclei that hold other functions of early detection, fast responses to danger, and other important features.

Early cortical vision

From the thalamus, the crossed visual signals are further projected back to the primary visual cortex

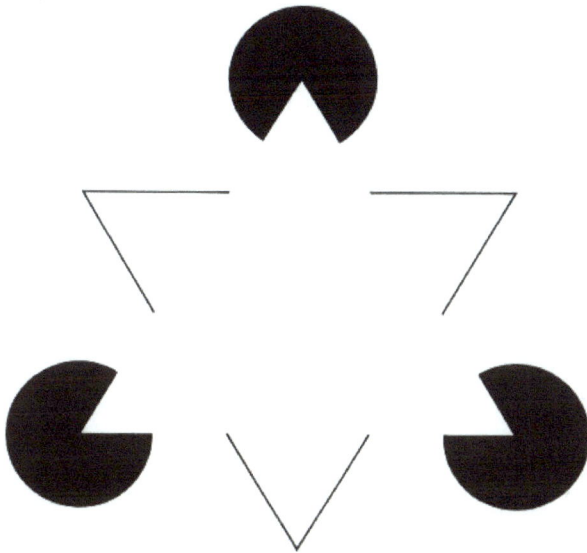

Figure The Kaniza Triangle is a perfect illustration of the filling-in phenomenon in primary visual cortex. Most people see a white triangle floating above a line drawn triangle. In fact there are no triangles, but your brain interprets and "makes sense" out of the incomplete information in the picture, and assumes two triangles.

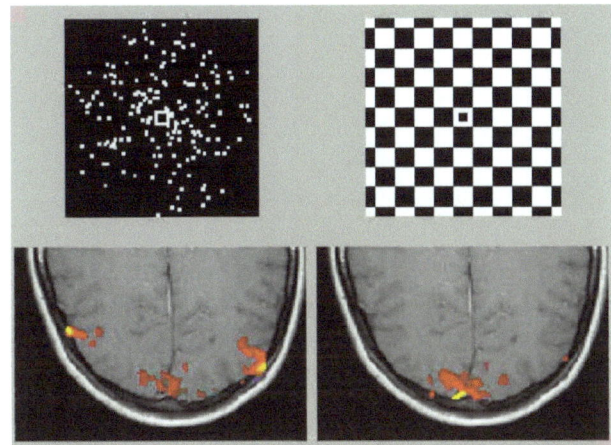

Figure Functional activation of the brain using fMRI. When showing a flickering checkerboard (top right) we see an enhanced activation of the primary visual cortex (bottom right). When we show dots moving in an "in and out" zooming movement (top left), we both observe engagement of the primary visual cortex but also in two additional areas on both the left and right side of the brain. These are the functional visual areas we call V5. The brains are shown as seen from above (horizontal view), with activation shown at the back (caudal) parts of the brain (courtesy of Olaf Paulson 2014).

through the optic radiations. As we saw, things to your left is processed on the right side of your brain. Indeed, this representation of the outer world on your inner "brain-scape" is very specific. If you see a particular pixel on a screen, that pixel is spatially represented in your brain. If you then present another pixel slightly to the right to this first pixel, the spatial representation of this pixel in your brain will be at a relative distance and angle that matches the real world. We call it that the visual system is *retinotopic*, meaning that there is a topographical mapping between the "real" world and the way the brain processes this information.

Early vision seems to serve many functions beyond mere retinotopy. One crucial function seems to detect borders and shapes. Another is to "fill in" where information is lacking, just as we saw in the blind spot. Below, you can see an example of the Kaniza Triangle illusion where this "filling-in" is obvious.

Specialised visual units

A basic feature of the brain is that it has specialised modules, and early vision is one of the prime examples. At the back of your brain, the brain has specialised units for processing colours, movement and other features.

This structure to function relationship is so specific that if you were able to knock out such a single module you would also find that this person would indeed lack that capacity. Take for example the case of a

woman described by famous neurologist Oliver Sacks (Sacks 1998) who had a suffered a very specific lesion to a region known as V5. As a consequence, she was unable to process movement! That is, she saw the world in a series of still images, lasting a few seconds at a time. Her ability to pour a cup of coffee, cross the street or even follow a conversation (since we all lip-read when we talk to each other) was rendered impossible. While such lesions are rare, they are a strong testament to the organisation of our brains, and how they build up our embodied minds.

We can see that people with lesions to specific regions of the visual cortex become blind within specific areas of their visual field. Others lose the ability to see or recognise colour (not the same as retina-driven colour blindness).

Similarly, when showing images with and without colours while doing an fMRI scanning, it is possible to turn the colour coding regions (often called V4) on and off. The same applies for movement, where we can turn area V5 on and off by applying static or moving stimuli (see figure).

From sensation to visual cognition

The many specialised units for processing different aspects of the visual world may have a good function, but only as specialised units. How can we recognise objects and know their position? Here, a crucial

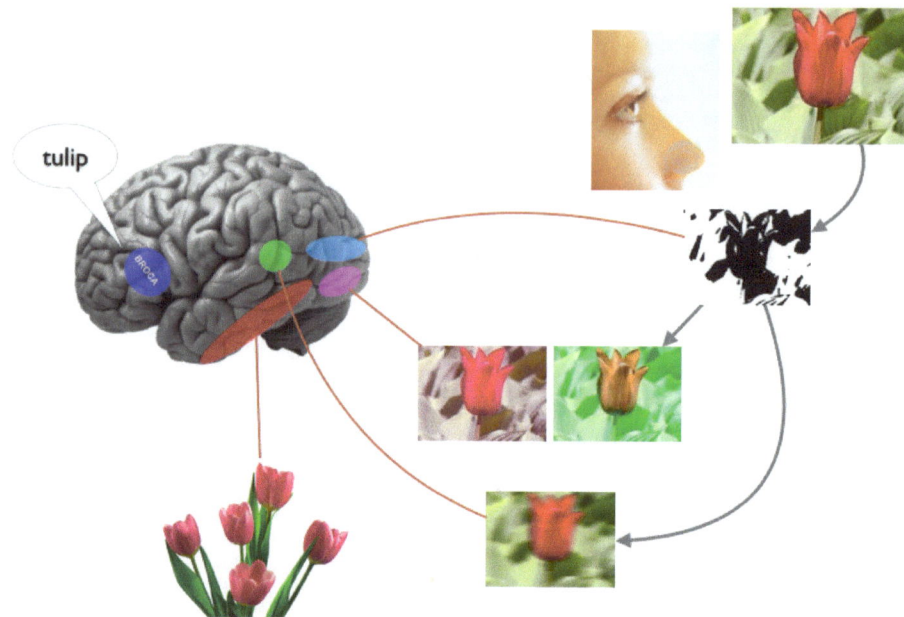

Figure From sensation to cognition (to naming). When we see a flower, the visual signals are first fragmented into specialised processing (colour, movement, contrast etc). Then, along the inferior temporal cortex (red) these specialised processes are joined and synthesised to provide a more coherent picture and concept of the item. Object recognition occurs. In order to name an item, we employ a different region of the left frontal lobe – traditionally called Broca's area. Brain region indications are only tentative for pedagogical purposes.

component of the brain comes to play. From the visual cortex we can see that parts of the lower (inferior) temporal lobe serves as a region in which this divergent information processing is converging. That is, colour, contrast, movement and other features that have been processed independently until this point, are now sending their specialised information to the same region of the brain.

This suggests that the ventral (or the "inferior") temporal cortex plays a role in integrating these different processing streams, and this again allows for more complex visual processing, such as object processing. We can say that we are moving from visual sensation to visual cognition. We are moving from a detailed processing of visual information to a more "holistic" processing. In historical German psychology, this kind of processing was called Gestalt – when we are viewing an entity as a coherent whole, rather than the sum of its parts. A cup is, well, a cup, rather than the sum of all the different parts that make it up. Similarly, a face is perceived as a whole thing, and not so readily as the sum of eyes, ears, nose, mouth, skin, and so on.

Visual cognition is a key function in how we can survive as living species, and as consumers its a foundational operation that allows us to recognise objects, brands and other entities that are the very building blocks of our actions as consumers. What if you were not able to recognise the Coke brand? What if you did not see a pair of jeans as a product?

In fact, there are particular cases in which this ability is lost. If you suffer a lesion to the bottom part (ventral, or "inferior") of the temporal lobe, chances are that you will become unable to recognise objects, a phenomenon called *visual agnosia*. People suffering from this will not recognise a hammer as a hammer, or a bottle as a bottle. They have lost the ability to recognise objects. Now imagine a person with visual agnosia wanting to buy a water melon. The task will be insurmountable, unless he or she was able to read their way to the right place. Objects appear as objects, but with no identity. Particular lesions to the temporal lobe lead to a loss of the ability to recognise faces, or even see a face as a coherent thing – something called *prosopagnosia*.

Thus, the role of visual cognition in consumer behaviour cannot be overstated. The first phase in recognising and buying a product is the ability to recognise something as a product at all! The same applies to brands and other relevant consumer items.

SINGLE NEURONS, JENNIFER ANISTON AND YOUR GRANDMOTHER

Did you know that there are single cells within your own head that respond ONLY to one particular item? About a decade ago, researchers at CalTech (Quiroga 2005) recorded the activity of single neurons (with deep electrodes inserted into the medial temporal lobe region) while epilepsy surgery patients watched images of people, objects and places. Here, they found one cell that responded only to actress Jennifer Aniston! Another cell only responded to images of the Sydney Opera House, and yet another cell responded only to images of actress Hale Berry.

These findings suggest that there is a highly specific coding of single cells on to objects, people and places. But how specific is it? Do you have one brain cell devoted to one single percept? Do you have one single neuron that only responds to your grandmother? Will you lose the ability to recognise your grandmother if that cell perishes?

Rather than thinking that there is only one single cell devoted to one item, we should think that the cell we record is connected to a larger network that is responsible for coding your memory of specific people, places or items. So if that single cell is lost, nothing happens. You will still have the network, which will typically have hundreds of thousands of cells (at least), and a few brain cells lost won't matter.

Still, the finding demonstrates how neatly our knowledge about our surrounding world is encoded in our brains. We have specific responses to each and every thing and person we know well. By the same line of reasoning, we can be certain that the more a person is acquainted with a brand, the more it will be "hardwired" into our brains and the more specific it will generate a response. By this token, companies can indeed create a strong and lasting impression into consumers' brains.

Two streams of vision

Strange phenomena exist in the realms of brain injury. As noted, a patient with a lesion to the inferior temporal cortex may lose the ability to recognise objects. What is really odd is that they are still able to show what an object is used for! For example, such a patient may be unable to recognise a hammer as a hammer, but when given the tool in his hand, he will easily demonstrate what the hammer is used for.

This demonstrates a disconnection between knowing WHAT and knowing HOW. Similarly, the same patient knows exactly where in his visual field the hammer is, and won't make any mistakes about grasping out for it. So we also have a disconnection between knowing WHAT and knowing WHERE.

What seems a strike of luck to researchers looking into this phenomenon is that we can also find examples of exactly the opposite phenomenon. Some patients with lesions to their parietal lobe show a specific opposite of the patients that have temporal lobe lesion. These patients demonstrate an intact ability to recognise and name objects and faces. Show the person a hammer, and he will say "that's a hammer." But give him the hammer and ask him to show how to use it, and he will first maybe reach out in the wrong direction, and then tumble around with the hammer, with the risk of injuring you or himself. These patients, with lesions to the parietal lobe but intact inferior temporal lobe, know WHAT it is but not WHERE it is or HOW to use it.

The WHAT vs WHERE/HOW distinction attests to two different streams that our brains are processing information. Indeed, the **Two-Streams Hypothesis** by Milner and Goodale (Milner 2008; Milner 2006; Goodale 1992) is one of the prominent theories of brain function. Just as we have seen in the examples above, this model states that the *dorsal stream* (going via the parietal lobe) is devoted to spatial and action oriented perception, while the *ventral stream* is devoted to pro-

FACTOR	VENTRAL STREAM	DORSAL STREAM
Function	Recognition/identification	Visually guided behaviour
Sensitivity	High spatial frequencies - details	High temporal frequencies - motion
Memory	Long term stored representations	Only very short-term storage
Speed	Relatively slow	Relatively fast
Consciousness	Typically high	Typically low
Frame of reference	Allocentric or object-centered	Egocentric or viewer-centered
Visual input	Mainly foveal or parafoveal	Across retina
Monocular vision	Generally reasonably small effects	Often large effects e.g. motion parallax

cessing of things like object identity. The table below demonstrates a list of different properties related to this distinction.

The WHAT vs WHERE/HOW distinction is important for understanding consumer behaviour and communication effects. Although there are multiple different possibilities, let us focus on two distinct ideas:

- **PRODUCTS:** Our relationship with products need to be seen as at least having two distinctions: what we perceive it to be, and how we use it. In many products, we may well recognise the item as a product, but we can nevertheless be hopefully misunderstanding of how to use it. Bridging this gap – from knowing WHAT to knowing HOW – can often prove crucial to whether consumers find the product interesting and are interested in buying it.
- **BRANDS:** Building a brand is most often connected to the WHAT stream: we know brands like Coca-Cola, Volvo, and Nike by their name, logos and their most typical products. But in which way do we have a HOW or WHERE related to those brands? Coke, for one, has a product that is highly specific with regard to HOW (drink it), but also WHERE (positioning in store, where store is etc). Other brands, such as LEGO, are also scoring high on both HOW, WHERE and WHAT. Yet, other brands can score low on either of these metrics. Whether this is deliberate or incidental is a matter of marketing strategy.

Further convergence of visual information

What happens with visual information after the split into ventral and dorsal streams? After all, we don't experience objects, products and faces as disassociated positions and identities. We experience the world around us as spatially and temporally coherent. We see an object in space and time at the same instant.

One of the places that the dorsal and ventral streams converge, is in the medial temporal lobe. First, in the ventral-anterior parts of the temporal lobe, we see that object processing occurs quite anterior along the so-called fusiform gyrus and the perirhinal cortex. Faces are thought by many researchers to be processed in a specific part of the fusiform gyrus (the so-called Fusiform Face Area, FFA), but other researchers suggest that this region is not specific to faces alone. Similarly, scenery and spatial locations – such as when you are in the living room of your home – are thought to be processed in the more posterior positioned region called parahippocampal cortex, and it has been suggested that this also has a specific region which

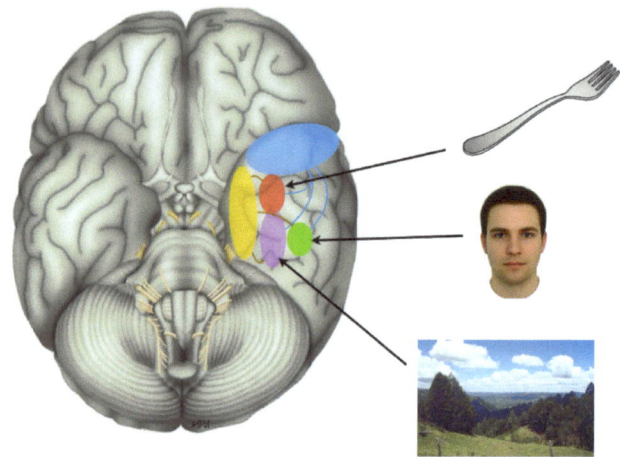

Figure Specific regions of the lower/inferior temporal lobe are suggested to hold highly specific coding of highly complex features such as objects (red), faces (green) and places/scenes (purple). These regions are thought to project forward to the frontal part of the temporal lobe (blue), as well as the medial temporal lobe (yellow), providing the neural basis of cognitive convergence. Please note that the regions are only tentatively defined, and not anatomically correct.

processes places (the Parahippocampal Place Area, PPA) see (Ramsøy 2009) for an overview.

The specialised units then project forward to regions that do even more convergent processing. These regions, such as the hippocampus and the temporopolar cortex are regions that receive information from many of these specialised units. As we will see, this information does not need to come from specialised *visual* processes, but is something that is *cross-modal*, i.e., the information is collected from the many different senses. That is why we are now turning to the other senses.

Before doing so, consider this: what is the general trend that you can extract from this view of visual processing? If there is one general trend, it is that information moves from highly specialised and segregated units to increasingly conjoint processes. A face is nothing more than a coherent collection of specific features: the angles of eyes, the colour of the skin, the roundness and colours of eyes, and so forth. The perception of an object is therefore a *synthesis* of elements into something more than the sum of its parts.

In the very same way, we can see that as faces, places and objects are sent as specialised kinds of information from their units to the convergence zones, to become conjoint and whole entities. This perceived conjoint and space-time coherency is what we can call an *episode*. Your experience of being here, in this moment, reading this book, in the particular surroundings you are in, is exactly such a coherent experience. It relies on both the specialised units AND their joining together in a process with which the end product seems unproblematic to you. You are only presented with the end result, but the specialised units of your brain are working hard to provide you with a consistent image of the world. As we will see throughout this book, this helps us in many ways, but it can also lead us astray.

HEARING

Sounds are as diverse as any input your body can receive. From the unpleasant noise of a baby's scream, to the soothing effect of music, the emotional richness from a long awaited text message beep on your phone, to the familiar noise of crackers you eat. Sounds are all around us, and although constantly present, we cannot say that our hearing takes up the same amount of "brain space" as vision.

Indeed, if we consider the mere real estate in the brain that is devoted to seeing, as opposed to hearing, vision takes up considerably more space. The primary visual cortex is many times larger than the primary auditory cortex. And from there, vision seems to dominate neural processing.

However, the distinction between vision and hearing is quite pronounced. For example, in visual images, we have an easier time to remember simpler images rather than complex ones. The more complex an image is, the more likely it is that we will miss particular kinds of information. In studies by Demany and colleagues, this has been found not to be the case for audition. Indeed, our sense of hearing does not make it easier to learn and remember simple sounds relative to complex sounds (Demany 2008) and does not depend on selective attention, as vision does (Demany 2010). This suggests that hearing and vision operate quite differently.

Brain processing of sound (an ultra-brief tour)

Sounds are compressions of air that hit the ear – and if the frequency of the airwaves is within the range that the ear responds to, specific cells within the ear reverberate to this frequency, leading to neural signalling that projects from the ear towards the brainstem. Here, just as the visual system has the Lateral Geniculate Nucleus of the thalamus, the hearing system has its own thalamic neural core, the Medial Geniculate Nucleus, which is one of the earliest neural processing of sounds. This nucleus then forwards those signals to the primary auditory cortex, which is located at the superior (top) part of the posterior (back) temporal lobe.

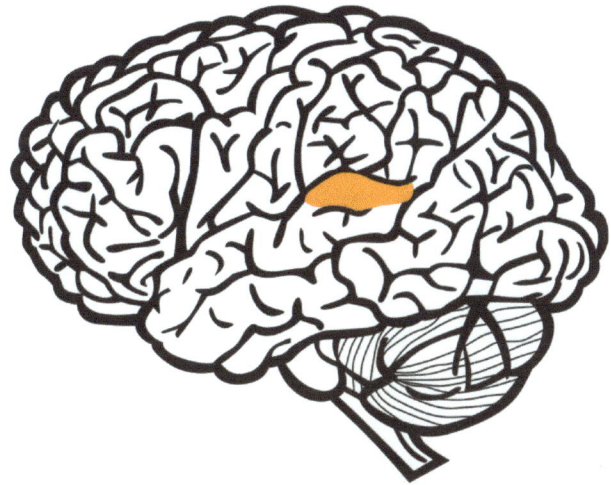

Figure The primary auditory cortex lies on top of the posterior temporal lobe, and includes regions such as the Planum Temporale (orange).

Sounds that are processed in the primary auditory cortex are then processed further by the temporal lobe, as well as the frontal lobe. In the temporal lobe, the *identity* of the sound is processed, including the recognition and processing of sounds or sounds from objects (just as with the visual system), while our ability to remember auditory information for a few seconds are processed in the prefrontal cortex.

Just in the same way that visual objects are recognised, auditory "objects" can be recognised, too. Whether a sound comes from a particular animal, has a specific meaning, or otherwise the context of the sound (are you in an echo-like cave, or in the cityscape?), is also processed in the temporal lobe.

One particular region of the temporal lobe was discovered by Carl Wernicke (1848 – 1905), where he observed a class of patients that demonstrated an inability to understand language, but with relatively spared speaking abilities. These patients turned out to most often have a lesion to the left temporal lobe.

Taken together, the temporal lobe is important for the initial processing and categorisation of sounds. As the temporal lobe is close and densely connected to regions related to emotional response and memory (amygdala and hippocampus, respectively) it is not far fetched to assume that sound can have a strong impact on our emotional and cognitive response to the world around us.

Sound branding

Sounds are part and parcel of the way in which we live in the world today, and as consumers this is particularly obvious. As we will see, sounds are indeed more

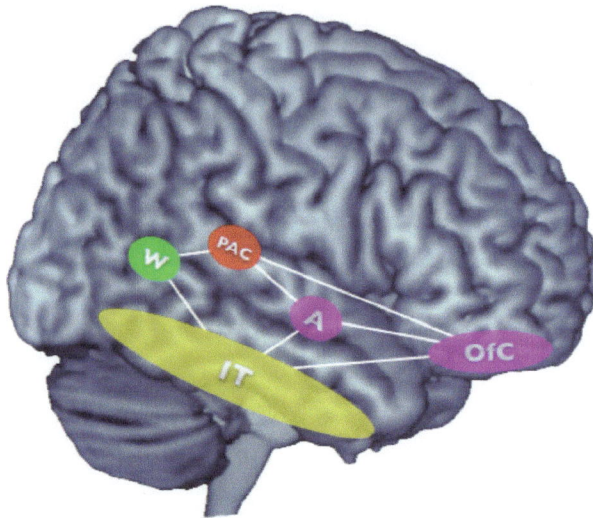

Figure The interpretation of sound. Sounds enter the primary auditory cortex (PAC) from the brainstem, and projects to other regions within the temporal lobe. If there is language to be understood, Wernicke's region (W) comes into play. The left region is more related to syntax and semantics, but right sided lesions to Wernicke's region can lead to failure to understand intonation and emotions in language. The decoding of meaning and emotional content depends on the inferior temporal lobe (IT) as well as the amygdala (A), while the hedonic pleasure of sounds such as music is associated with increased engagement of the medial orbitofrontal cortex (OfC).

readily used – planned or unplanned – by companies in their branding efforts. Here, we will take a few examples: jingles and product sounds.

JINGLES

If there is anything we recognise beyond what we see, it is the short and snappy sounds from brands. The snappy and fresh sound of Intel; the building expectation associated with the sound of THX in cinemas, or the direct brand associations we get from of phrases such as "Just do it!" or "I'm loving' it", which automatically makes you think of Nike and McDonald's, respectively. Brands own sounds in many ways, and jingles is one way of using the auditory sense to brand themselves.

You can make the same exercise right now – or you might have generated some automatically already. Make a list of different brand sounds and jingles, and discuss with colleagues or fellow students. What made it to the list? Why do you think some were more easy to recall than others?

As we will see, the use of jingles are tightly connected to brands in several ways, and they can even lead to expected outcomes. In the chapter about learning and memory, we will see how sounds can imbue value to an otherwise neutral event, something we will learn as "Pavlovian learning".

PRODUCT SOUNDS

Yes, products have sounds. If you didn't think too much about it, please answer some of the following questions:

- What is the sound of eating potato chips? Or how about opening the chips bag?
- How does it sound to close the door of a Mercedes 500c, as opposed to a Fiat 500?
- Can you recall the clicking sound when opening or closing your smartphone?

Products have sounds. In many cases, these sound are carefully designed, but surprisingly often they are the incidental result of the product design process.

In a study I did for LEGO Duplo, participants (mothers of children aged 3-5 years old) were tested with a number of tasks. One of these tasks were to listen to different sounds, while their pupil dilation response was recorded. As we will see in the chapter on emotions, pupil dilation can be a measure of emotional arousal, indicating whether something is deemed relevant by an organism. We tested a number of sounds, and then looked at the different responses. Compared to white noise, we found that the sound og LEGO Duplo bricks produced the strongest emotional response. In comparison, a jingle did not make much impact. Finally, children's laughter produced lower emotional response. This may initially seem surprising, but studies have demonstrated that laughter may indeed lead to a *lower* arousal, and that laughter can be stress-releasing. Thus, one can say that depending on the aim of the use

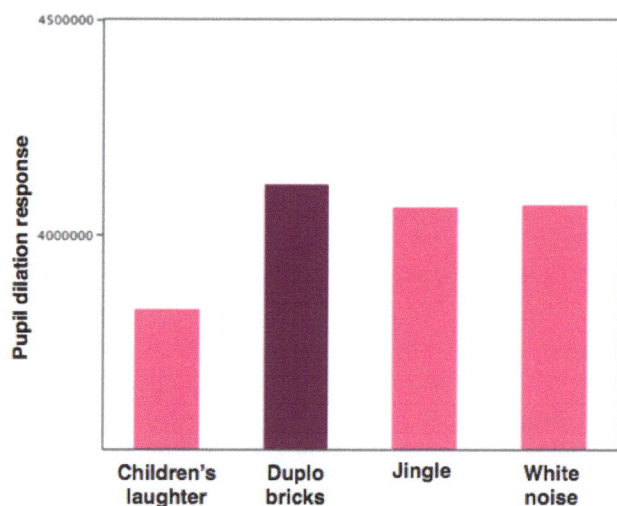

Figure In a study for LEGO Duplo, participants were exposed to different sounds. The largest emotional response was caused by the sound of crashing LEGO Duplo bricks. While the suggested jingle did not show any difference to white noise, the sound of children's laughter had a lower pupil response.

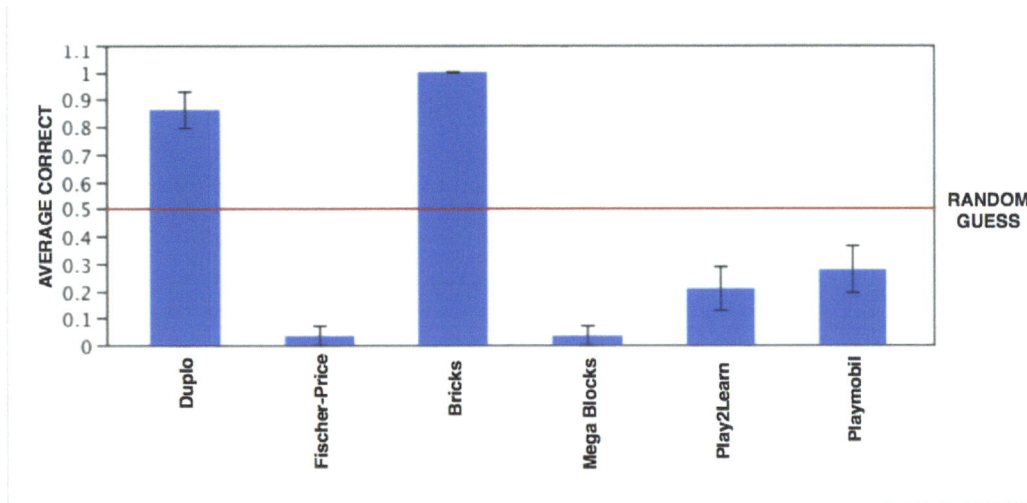

Figure Guessing from shaking the product. When customers tried to guess the product after holding and shaking it, they were above chance levels whenever they shook bricks and LEGO Duplo packages. Whenever they were tested with other products, they were below chance levels of guessing. This suggests that for this particular group, a particular product sound was closely linked only to the LEGO Duplo brand. Please note that chance level guessing in some instances was 50%, but across the options, chance level guessing was in fact 1/6, or 16.7%. Still, this does not change much to the interpretation of the study, in which LEGO Duplo was by far the most often recognised product sound.

of such a sound, marketers at LEGO should consider bricks when they want to engage the consumer, and children's laughter when they want to have the consumer to be more relaxed.

In a related study, participants were asked to guess which packaging sound they were listening to. Here, it was found that the participants were better than chance at guessing LEGO Duplo, as well as the more generic category "bricks", but that all other comparable brands had a much lower than chance level at guessing correctly. However, it was also found that customers were more likely to guess that the sound was from LEGO Duplo, even though they were exposed to sounds from other products. This means that for the tested population, LEGO Duplo had a stronghold on the sound of anything "brick-like". Indeed, it is a brand asset that many brand managers would give their right hand for obtaining.

We will dive deeper into the effects of sounds on attention, preference, memory, and other responses in the coming chapters.

In-store music and wine choice

Does playing music inside stores have any effect on consumer choice? In a oft cited study by North, Hargreaves and McKendrick (North 1997), consumers were found to be affected by the type of music that was played inside stores. In particular, playing of French music in the wine section was associated win increases in purchase of French wine, while playing of German

music led to increases in purchase of German wine. Notably, the effect of German music on purchase of German wine persisted even in the face of a more general preference for French wine.

In a similar study, Areni and Kim (Areni 1993) found that there were also effects of playing classical and pop music on wine purchase. Classical music made consumers spend more money on wine, and the purchases were not driven by purchasing more bottles of wine, but rather buying more expensive wines.

SMELLING AND TASTING

It is well established that smelling and tasting are two sides of the same coin. It is hard to taste much when there is no odour. Think about when you are having a cold, and things do not taste like much. Or just hold your nose next time you are eating something. Taste is indeed, to a large extent, in the nose.

There is a famous quote from the poet Marcel Proust, in his book "Remembrance of Things Past" (Volume 1), where he carefully carves out the effect of an odour that automatically and unexpectedly leads to a vivid recollection of a past memory:

> "But when from a long-distant past nothing sub-sists, after the people are dead, after the things are broken and scattered, taste and smell alone, more fragile but more enduring, more unsubstantial, more persistent, more faithful, remain poised a long time, like souls, remembering, waiting, hoping, amid the ruins of all the rest; and bear unflinchingly,

in the tiny and almost impalpable drop of their essence, the vast structure of recollection."

Odours and tastes can lead to specific memories. Some may have particular odours associated with friends, family members or other loved ones, and these memories can be automatically triggered by a brief exposure of the same odour. What is the neural fabric of such effects?

Brain bases of smell – from nose to brain

In your nose there are specific chemical sensors that pick up on the food you are exposed to, as well as other chemicals in your surroundings. Such chemicals move through the back of the throat and the nose reach to reach olfactory nerve endings in the roof of the nose. The molecules bind to these nerve endings, which then signal the olfactory bulb to send smell messages to a few critical regions of the brain:

- The **olfactory bulb** – a multi-layered cellular structure positioned at the bottom of humans' brains, which is thought to serve (at least) four different functions: 1) odour discrimination; 2) increased sensitivity of odour detection; 3) filtering of background odours to enhance the transmission of a few select odours; and 4) allowing modification from higher brain areas involved in arousal and attention of the detection or the discrimination of odours

- The **amygdala** – a structure now well known to be involved in emotional responses – both positive and negative – which thus makes it act as a *relevance detector* (as we will see more about in the chapter about emotions and feelings). More recently, a structure called the **habenula** has also been discovered to process odour information, and play an important role in emotional responses to odours.

- The **entorhinal cortex** of the (medial) temporal lobe and the **hippocampus** – structures we will see are engaged in certain forms of learning and memory, and thus providing the very essence of Proust's vivid recollection of past memories from the mere whiff of an odour

Brain bases of taste – from tongue to brain

Taste cells are located on the tongue and throat, and there is an uneven distribution of such cells that provide what we call *taste buds* (on the tongue they are called *papillae*). At the tip of such cells there are specific receiving endings that bind with chemicals in the mouth, which are thus affecting the response of the cells.

Signals from the tongue are then projected back to specific brainstem nuclei and carried further to the thalamus. Also, as with the olfactory system, medial temporal lobe structures and orbitofrontal cortex receives the information downstream, thus being the neural foundation for the merging of the two senses.

Joining forces in the name of taste

As can be seen in the previous figure, smell and taste are joined in the brain – the sensory information is said to *converge* to provide a coherent experience of, say, a taste. When you eat, you are in fact experiencing a conjoint process: you are tasting AND smelling at the same time. Often, smell comes before tasting (as in smelling delicious and freshly baked buns and having to wait for them to cool down). This can be illustrated with the following figure:

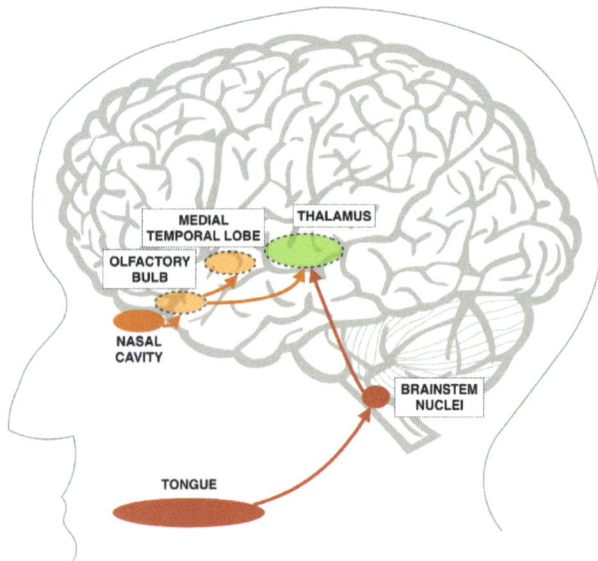

Figure Simplified model of basic smelling and tasting brain processes. The tongue translates chemical signals and project them to brainstem nuclei, which again project to thalamic nuclei. Olfactory processing operates similarly by an initial translation of chemical signals in the nasal cavity leading to responses that are processed by the olfactory bulb of the brain, which then projects to the same thalamic nuclei, as well as medial temporal lobe structures such as the entorhinal cortex, hippocampus and amygdala.

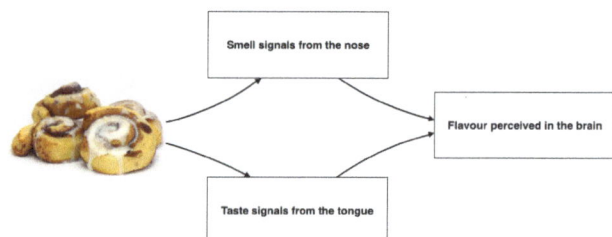

Figure The joining of smell and taste to provide the full flavour. Taste and odour are initially treated separately in the brain, to then become joined later in the neural processing of the more comprehensive taste experience.

As we will see later in this chapter, sensory convergence is not unique to taste and odour. It is, just like we have seen with the visual sense, a basic function how the brain first distributes processing to specialised units, then integrates and joins these processes to provide a more coherent and congruent representation of the world. The joining to taste and odour is meaningful to the extent that it can be used to guide behaviour: if something has an awful odour, it is much less likely that you will go on to eat it. Our ability to link odour to (expected) taste appears so seamless, but is in fact the result of this converging experience.

When is a smell too much?

Have you ever tried to have too much of the perfume Chanel No. 5? Or any similar experience with an odour? Basically, we could ask the question whether an odour can be "too much". Anecdotal evidence suggests that when an odour becomes "to much" It can go from pleasant to unpleasant, and in a recent study we wanted to test this experimentally. With my Master Thesis student Lars Frederiksen and Professor Per Møller at the Department of Food Science at the University of Copenhagen, we exposed participants to different concentrations of flower odour, and asked them to report their degree of liking for this odour.

What we found was a non-linear effect: up until a certain concentration, the relationship was positive: participants showed an increased liking to increasing concentrations. However, after a certain point, this effect was reversed: stronger concentrations were now related to *lower* preference ratings. This suggests that there is a crucial optimal concentration for an odour in how it is processed and preferred by consumers.

Interestingly, when we did further exploratory analyses, we found that there was a gender difference in this effect. Compared to men, women showed an earlier peak and a more rapid decline in their preference with higher concentrations after this. This suggests that beyond the general effect, knowing your target audience and their responses to odours can be crucial for the success of your product.

Effects of odour on preference

In the same study, we also ran a test to study whether an odour could have a similar effect on items that were presented simultaneously with the odour. In other words: can the emotional effect of odour concentration affect something that is presented simultaneously? By providing odours with a certain intensity at the

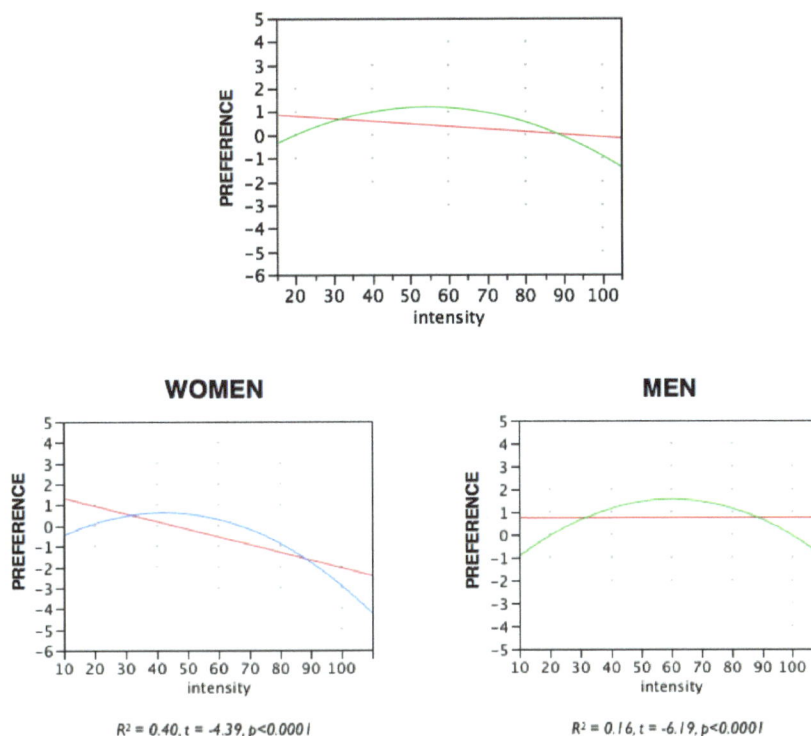

$R^2 = 0.40, t = -4.39, p < 0.0001$ $R^2 = 0.16, t = -6.19, p < 0.0001$

Figure Effects of chemical intensity on odour preference. Top: On average, the intensity of a chemical compound showed a non-linear effect on preference, with a peak effect that switched the relationship between intensity and preference. Bottom: We also found gender differences in which women demonstrated an earlier peak in their maximum preference, compared to men. Curved lines indicate the non-linear effects, and red straight lines indicate the linear regression line.

same time as we showed novel brand logos and abstract art, we found a significant and similar effect: up until a certain odour intensity, brand logos and art was preferred more with higher concentrations; but above a certain threshold, any higher concentration was associated with a *plummeting* of preference for the brand logo and abstract art.

This study tells us a crucial story: what we smell cannot be seen as something static and non-emotional or cognitive. Rather, odours affect our emotions, moods and even how we respond to otherwise unrelated items such as art, brands and products. Knowing these effects can prove essential to brand managers and product developers alike. With the methods provided here, we have a set of tools to *measure* those effects, which allows us to run new studies to curb the effects of odour and taste on product and brand experience.

The underlying fabric of this effect is quite obvious: odours and taste is directly linked to the emotional and memory-related (mnemonic) brain structures such as the hippocampus. By comparison, the visual system does not have the same direct route (or at least the same strength of connections) between the sensory and the emotional/mnemonic systems. Indeed, the effect of odour on memory reflected upon by Marcel Proust can be explained by the wiring of the brain.

Smell to enhance brand memory?

Does it make sense to use scents in stores to enhance consumer responses? And how specific does that odour need to be? Several companies around the world claims that a brand or a product needs to have a very specific match between the brand and the odor. However, few studies have explored this experimentally.

In a study by Morrin and Ratneshwar (Morrin 2003) the researchers set out exactly to determine whether odours could indeed enhance brand memory, and whether the odours had to be relevant to the brand for this to occur. What the authors found was that ambient scents – weak odours presented in a room during the testing session – indeed had a positive impact on memory for the brand related information they learned. Crucially, the authors found two further facts: 1) the odour did *not* have to be congruent or relevant to the product; and 2) the effect was by far largest when odours were present during the *learning* stage, rather than only during the recall stage.

Together, this study shows us that odours *can* indeed affect consumer memory, but contrary to what many companies are claiming, there seems to be no

reason to design a very specific odour. Also, the fact that odours had the largest effect during learning, and not recall, makes strong recommendations as to when companies should put their efforts in combining brand related items with odours.

BODY SENSE

Just as with any other sense, our sense of our body is the result of sensory organs sending information to the brain. Body senses are heterogenous: we are here speaking of both external senses such as those that allow you to delicately feel something at your fingertips, deep pressure sensors of your skin, the hair follicles on your arm allowing you to feel a fresh breeze, temperature sensors that allow you not to be burned, and many more. Furthermore, we are also broadly talking about the more "inner" senses of the gustatory system, such as the system that allows you to feel a stomach ache.

Body and brain

The crucial thing is that for the body sense, the same mechanism (with few exceptions) is very similar to what we have already seen in the other senses. A sensory organ receives a stimulus that it is highly tuned to pick up (e.g., the soft touch of another person's hand on your forearm). This signal is translated into electrochemical signals that are transmitted back to the spinal chord and up towards the brain. If the event is harmful, such as the burning of a hand on a hot stove, this will lead to a spinal chord response that will subtract the hand. Thus, it is not you that consciously are responding to the hot hand and *then* withdraw your hand. Rather, your hand is *already* withdrawn when you experience the burn.

In the somatosensory cortex, just as the visual system, there is a close 1:1 mapping between a body part and a particular region of the brain. Indeed, the mapping is so close that you can manipulate a particular region of the somatosensory strip, and see dramatic changes on people's sensory experiences. Stimulation of the "hand" region will leave people feeling a tingle or stroke across their hand. People with lesions to a particular area of the strip – e.g., the face – will lose the ability to sense anything from this region.

Just as with the visual system, the laterality of the somatosensory system is crossed. This means that what you sense on the right hand of your body is treated by the *left* side of your brain. The lateralisation of function has been suggested to hold many functions, allowing for adaptation of the body sense to differen-

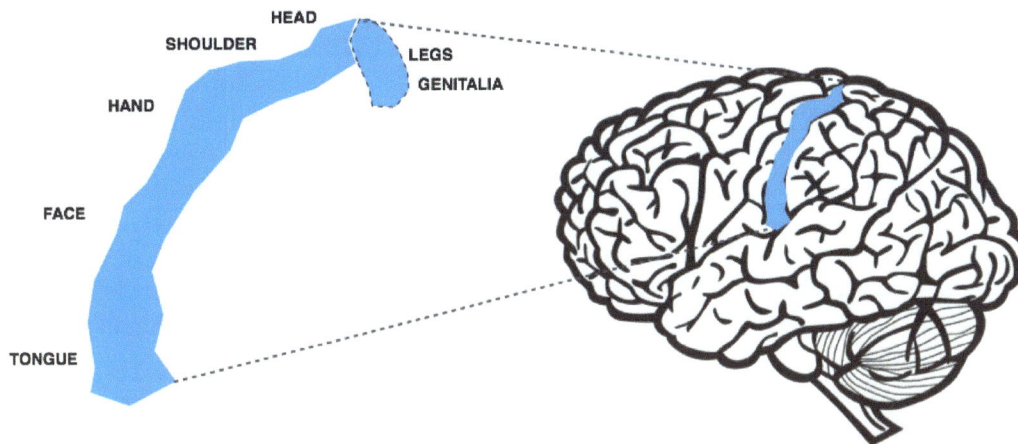

Figure Somatosensory mapping. Within the sensory strip, there is a neat ordering and representation of each body part. The strip extends into the medial surface of each somatosensory cortex (dotted line) and contains representations of the legs and genitalia.

tiate between the two halves of your body (and brain). For example, you may have noticed that your sense of touch is better in your right hand than left hand (if you are right handed). This is suggested to be the result of a finer tuning of the somatosensory response to finer details in your dominant hand, relative to the non-dominant hand.

Somatic responses and product preference

Certain cells in your body system are responding to pressure and thus also "weight" (although some of the perception of weight is also registered as the compensation that your motor system needs to enforce to hold an item). But in a recent study my research group and I did (unpublished data) we demonstrated that certain somatosensory effects can affect consumers' product preference. In two related studies, we studied the effects of sensory stimulation on preference for hammer drills.

In study 1, we studied how the thickness of the neck of the hammer drill could have an effect on preference. By measuring the thickness of the neck on hammer drills and relating this to the preference for each drill, we could explain more than 30% of the variation of the preference. This suggests that at least some of the variation in people's preference for hammer drills could be explained by variations in neck thickness, and this implies that thickness is a variable that product designed should take into the product design process.

Furthermore, when combining this study with exploring how much people focused on the neck (by using mobile eye-tracking), we found that thicker necks were associated with longer fixation times to the neck region during product viewing, and that this drove much of the subsequent product rating and choice.

In study 2, we explored the effects of product weight on product preference. In other studies, it has been shown that product weight – such as the weight of a B&O TV remote control – can affect the quality evaluation and preference. Heavier products seem to lead to higher preference – the added metal in the B&O remote control has nothing to do with its product performance; nevertheless, it still leads you to think of it as a better product! In this study, we weighed all hammer drills and had participants hold the

Figure Hammer drill preference. Two properties of a hammer drill that drive consumer preference is the perceived **weight** of the drill, and the **thickness of the neck** of the drill. Both properties show a positive relationship to preference, in that higher weight and thicker neck is related to higher perceived quality and product preference.

product before rating their product preference and willingness to pay (WTP). Here, we found a close relationship between product weight and preference and WTP, in that heavier products were more likely to be preferred. This explained about 40% of the variance in product preference.

Taken together, the two studies suggest that particular aspects of the sensory input can affect the way in which we perceive, evaluate and choose products. While neck thickness and weight may show such effects, some of this variation may be related to more general features of the product, such as the mere size and robustness of the product. However, these findings suggest that specific elements can be tailored to provide an optimal sensory experience that boosts customer interest and preference.

PRODUCTS, BRANDS AND SENSORY LOAD

Sensory experiences abound when we talk products and branded experiences. From walking down a shopping street to eating at a French restaurant, and riding a roller coaster, our senses are exposed differentially. We can say that different products, brands and services are *loading* your senses differently.

Jinsop Lee (see http://www.jinsop.com/), is a famous designer that has thought long and creatively about how we can understand how our senses are engaged by design items. In his TED talk, he introduces this as a way to "design for all five senses", where he explains a new way in which we can both chart and influence how consumers' sensory systems are exposed during the interaction with products and services.

Here, we will focus on the assessment aspect, by adopting his 5-senses chart. In reality, the chart is really simple. You ask a bunch of people to rate how they experience a particular product, by making 0-10 scale judgments for each sense.

How much does this product affect your vision; your hearing; your touch; your sense of smell; and your taste? The chart – which we from now will call the Sensory Load Chart (or SLC) – will look something like this:

When looking at different experiences and products, it is obvious that there are plenty of differences and individual profiles associated with different product or service experiences. Eating at McDonald's loads the taste system, but is relatively low on sound. Nintendo Wii does not have any smell or taste associated with it, but are rated as high on touch – much due to the new controlled it came up with. Going to the Opera is experienced as high on sound (of course) but also perhaps surprisingly high on sight. Going to the fair is seen as an extreme event to your senses that basically plays on all senses.

Figure The Sensory Load Chart. On a scale from 0-10, you should rate how much a particular product experience, service experience or even brand taps your sensory system. The more you think it is using, e.g., your sight, the higher the score. Do this for all senses.

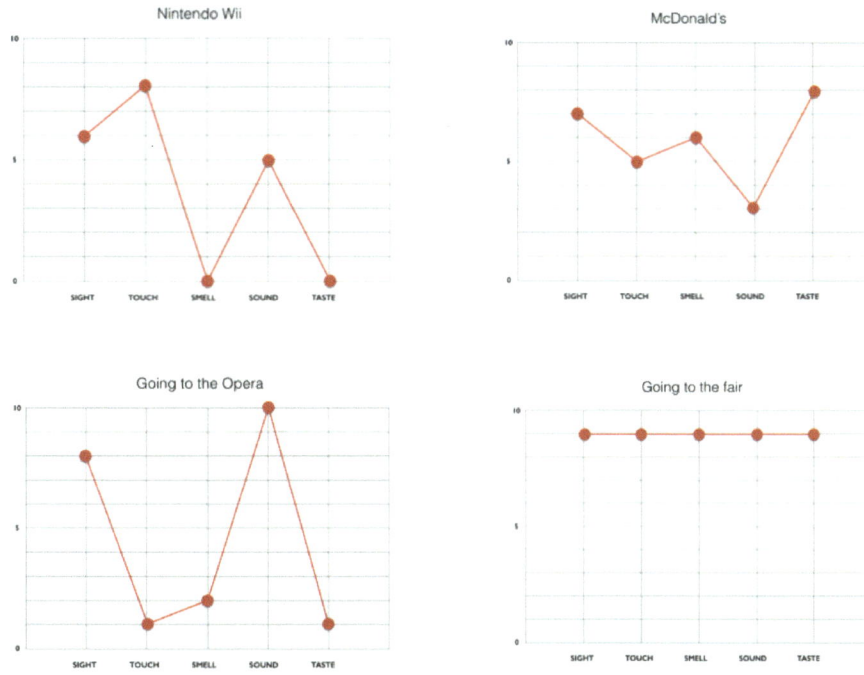

Figure   Different products and experiences tax the sensory system differently, as assessed by the Sensory Load Chart. The experiences of playing Nintendo Wii, eating at McDonald's, Going to the Opera and visiting the fair all load your senses differently. The last item is exaggerated on purpose. Make your own sensory chart of different brands and products. What does these charts tell you?

The graphs below highlights this, and demonstrates how well the SLC is at visualising the different load of the senses.

Just as we can use the SLC to assess effects of individual products or services, it is possible to ask consumers about their sensory associations to particular brands. Here, it is also obvious how different brands have different sensory profiles. Here, you can also make your own charts for different companies.

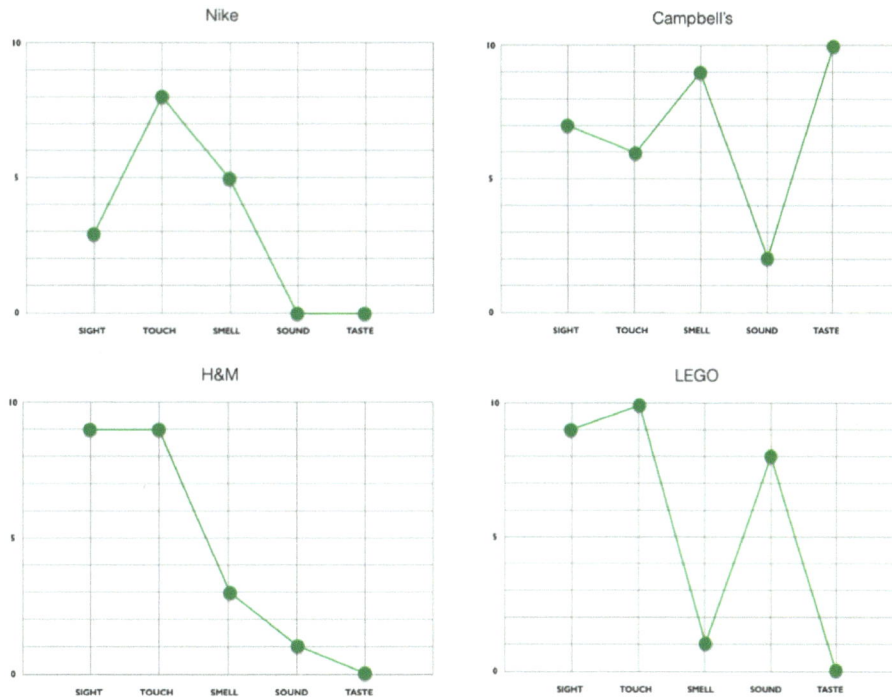

Figure   Sensory profiles for different brands. Not surprisingly, brands differ substantially in how they are perceived to load the different senses.

Tweaks and topics to the Sensory Load Chart

When you have become used to using the SLC, it is time to focus on additional topics. Here, I will present three selected topics that can have broader interest.

1) Variance and certainty

When looking at the SLC it is possible to look at the variance in how people rate the different items. If the variance is very low, it suggests that people tend to agree on a specific score. For example, if the mean score is about 9 on touch for H&M, and the standard deviation is 0.3, then chances are that most people tend to think of H&M as taxing the somatosensory system quite much. On the other hand, if the mean score for Nike on smell is 5, and the standard deviation is 3.5, then this suggests that there is a very high variance in how people think about Nike and odour. Maybe there are some hidden clues in there? Maybe a sub-category of the consumers you asked are runners and have a very particular association between Nike and odours?

2) Groups and segments

It is very likely that there will be group differences within the consumer segment you are studying. If you are testing the SLC on H&M, maybe there is a huge difference in the SLC profiles for men and women? Maybe one group perceives LEGO to have more sound than another group? Such differences may prove crucial in our understanding of consumers' perception of brands, and a pointer towards what drives key differences between groups of consumers.

3) Planned and perceived branding

It is one thing that brand managers and designers strive to make consumers perceive a brand in a particular way. However, what if their ideas were not matching the reports from consumers? Such deflections and differences pose a challenge to brand managers, and should be treated with care. What if brand managers of Nike were surprised to see that many customers associated Nike with odour, when they themselves rated Nike brand as having a low odour load on the SLC? This suggests that there is some hidden process at stake that differentiates the ideal and planned brand communication with the perceived brand qualities.

Often, however, the differences are not enormous, but rather small. Still, even a significant difference (as tested with statistical comparisons between brand managers' and consumers' SLC responses) can be meaningful to discuss. Why do brand managers at Campbell's score their brand to be 9 on odour when customers only rate them at 6? What causes this discrepancy? Maybe further analyses and testing reveals that customers indeed do not associate Campbell's products with any particular odour?

The SLC allows us to explore both the expected and planned, as well as the perceived sensory load that brands, products, services and other consumer items have. When used correctly, it can be a powerful and cost-effective tool for brand managers and product designers alike in understanding how consumers relate to their brands or products. In any case, use of the SLC can provide many highly relevant – and sometimes heated – debates about how companies are thinking about the use of different senses.

SENSES MERGE!

A basic feature of how the brain processes information is that it joins, or *merges*, this information. That is, processing of odour, taste, haptic cues, sight and sound, are all joined together in some regions of the brain. We can here mention three specific regions, and note how they will be treated in later chapters.

Convergence and memory

When sensory information converge in the medial temporal lobe, it will be closely linked to memory. Indeed, some accounts suggest that the convergence of information is the brain substance of processing of

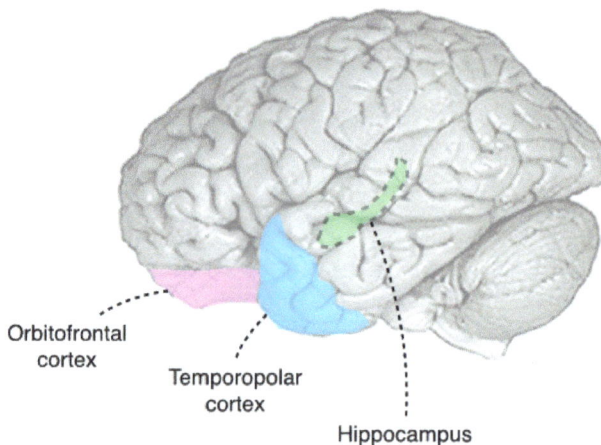

Orbitofrontal cortex

Temporopolar cortex

Hippocampus

Figure Convergence zones of the brain. The temporopolar cortex (blue) receives input from many diverse regions of the brain but does not fully integrate these signals. The hippocampus (green) and the surrounding medial temporal lobe integrates sensory, emotional and cognitive processes to produce coherent experiences and memories of the world. The orbitofrontal cortex (purple) integrates sensory and emotional signals to produce a hedonic experience of the world.

"episodes". That is, your experience of being here and now, in this moment, is the result of this convergent process in the brain. Until this point, sensory experiences are detached and not associated. But with the convergence of information on to the hippocampus through the inferior temporal lobe, the medial temporal lobe and amygdala, all provide the building blocks to create a coherent experience of the world.

Our experience of products rely very much on this particular aspect, and some authors have suggested that a company's ability to combine several senses simultaneously is indeed related to the success of that company in building a strong brand, or *brand equity*.

As we will see in the chapter about learning and memory, the building of associations – both within and across senses – is a crucial building block for any company who wants to succeed in winning their customers' memory and emotions.

Convergence and liking

A second convergence zone in the brain can be found in the orbitofrontal cortex (or OfC) – the space just above your eyes. Here, signals also converge and are closely linked to hedonic experiences. That is, whether you like the taste of sugar, the sound of music, or the view of a piece of art, this region is engaged. The more you like it, the more parts of this region is engaged. As we will also see in the chapters about emotions and feelings, and wanting and liking, there is a relationship between how "basic" or "abstract" a reward is, and how posterior vs anterior in this part of the brain we see activation.

One notable thing about this is therefore that our sensory experience of the world around is closely connected with some kind of "evaluation" – there is rarely if ever an experience of the world without some kind of emotional flavour to it. The sensory convergence in the OfC – possibly together with other regions such as the amygdala and ventral striatum – allow for a rapid evaluation of events, both based on the properties of the events themselves, but also based on past experiences.

AGAINST NAÏVE REALISM

If there is one thing you can take from this chapter on our sensory system, it should be that WYSINWYG..... or "What You See Is Not What You Get". How you perceive the world is NOT how the world actually looks, or put differently: your experience of the world is a mere *representation* and *interpretation* of how the world looks. If you think that your experience of the world is *exactly how the world is* then you are fooling yourself. You are indulging in what is called *naïve realism*, the belief that the world is just as you see it. On the contrary, the world contains far more content than our limited sensory system and brain can capture and comprehend.

Think of how bats can hear sounds at much higher (ultrasonic) pitches than we do, or how elephants and whales hear rumbles from their conspecifics across far distances, something that is impossible to detect with the human ear. Some snakes can see infrared, allowing them to capture prey in the dark by homing in on their body warmth signature. Indeed, as humans, our sensory experience of the world around us is limited!

Our sensory system is fallible and limited, and the brain has to go extremes to fill in the gaps. Often it works fine, but sometimes our perception of the world around us is flawed. Illusions pinpoint this forcefully, just as the demonstration of the blind spot suggests how non intrusively this works in an everyday world for all of us.

As we will see throughout this book, our experience of the world is highly influenced by numerous factors. Some are due to filtering and noise, others due to filling in phenomena, and yet others are due to plain and simple self-delusion. But crucially, we are constantly deluded by our perception of the world, believing that "this is it", while so many other hidden factors are either not sensed, filtered, or perceived consciously. As we are turning to the topics of attention and consciousness, we will see that our idea of the world is merely the tip of the iceberg.

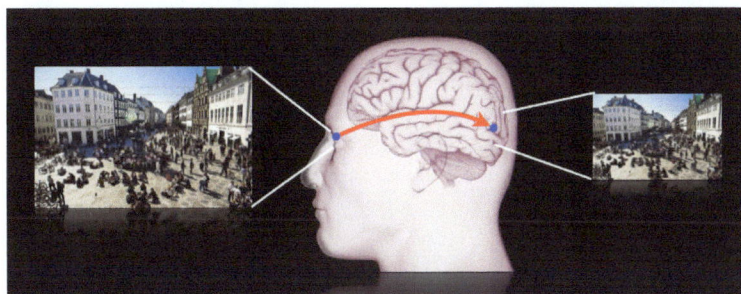

Figure   The lure of naïve realism. Do we indeed experience the world "as it is", or are we merely seeing an interpretation of the world around us?

Dwindling brain specialisation with increasing age

What happens in our brains as we get older? We probably all know what happens at the behavioural level. Most notable is the changes in memory, and the ability to couple information together. Remembering a name, or mixing names on people is a frequent effect. Forgetting what happened when and who did what are well-known memory problems of ageing.

But what causes these problems? And are they caused at the stage of learning or retention? Or may the changes even occur during preparation or rehearsal?

In a study now published in Neurobiology of Aging, we studied the effects of healthy ageing on how the brain processes different kinds of visual information. Based on prior work showing that visual attention towards objects predominantly recruited regions of the medial temporal lobe (MTL), compared to attention towards positions, we tested whether this specialisation would wither with increasing age.

Basically, we tested the level of brain specialisation by comparing the BOLD fMRI signal directly between object processing and position processing. We looked at each MTL structure individually by analysing the results in each individual brain (native space) rather than relying on spatial normalisation of brains, which is known to induce random and systematic distortions in MTL structures.

Running the test with functional MRI, we found that several regions showed a change in specialisation. During encoding, the right amygdala and parahippocampal cortex, and tentatively other surrounding MTL regions, showed such decreases in specialisation.

During preparation and rehearsal, no changes reached significance.

However, during the stage of recognition, more or less the entire MTL region demonstrated detrimental changes with age. That is, with increasing age, those regions that tend to show a strong response to object processing compared to spatial processing, now dwindle in this effect. At higher ages, such as 75+, the ability of the brain to differentiate between object and spatial content is gone in many crucial MTL structures.

This suggests that at least one important change with increasing age is its ability to differentiate between different kinds of content. If your brain is unable to selectively focus on one kind of information (and possibly inhibit processing of other aspects of the information), then neither learning or memory can operate successfully.

One important feature of this study is that it provides a new means to study age-related disorders such as Alzheimer's Disease. It is well known that this disorder initiates in the MTL region, most likely the trans-entorhinal region, and it does so long before the clinical symptoms of Alzheimers or even its predecessor, Mild Cognitive Impairment. Hence, the search for ways to assess changes in this region has become a growing field of interest. One possibility could be to employ the methods developed in this project to assess early functional and morphological changes in the MTL region, and possibly improve early detection of Alzheimer's and related disorders.

Another interesting option would be to explore to what extent healthy ageing is related to changes in everyday functions, such as shopping behaviour, learning and remembering movie contents and other complex kinds of information.

CHAPTER 5

Attention & Consciousness

Every day, your senses are bombarded with information. Although your outer senses process all the input they receive that is within their sensory domain, there is no way that you can consciously perceive all this information. From the movement of a single hair on your arm caused by air compression, to the response of a single cone due to a photon of light, and to the response of a single olfactory node to a chemical compound, our sensation of the world is the result of an intense filtering, abstracting and summing of events in our sensory system. Our conscious experience of the outer world is a mere fraction of what our senses are exposed to.

Indeed, it is said that we are exposed to an estimated 11 million bits of information each second through all our senses. As a contrast to this enormous number, we humans are claimed to only be capable of processing only around 50 bits of that information (Wilson 2009). By the time you are trying to hold on to information for a few seconds – such as a phone number – this number drops to 5 to 9 bits of information! This suggests that tons of information in some way gets lost from the outer senses to our deeply felt conscious experience.

But there is reason and system to this seemingly meaningless loss of information. Instead of treating each sensory input as equal, our nervous system is wired to assign *relevance* and value to information from the get-go. Thus, when a stimulus reaches your senses, how it is treated and the likelihood that it will produce effects higher up in the processing system, depends on a number of factors. Is the stimulus novel or old news? Is it intense or vague? Does it fit what we are searching for already?

Even a century ago, famous psychologist William James wrote with full insight about the ease at which we talk about attention, yet are poor at understanding what it really is:

> "Every one knows what attention is. It is the taking possession by the mind, in clear and vivid form, of one out of what seem several simultaneously possible objects or trains of thought. Focalization, concentration, of consciousness are of its essence. It implies withdrawal from some things in order to deal effectively with others, and is a condition which has a real opposite in the confused, dazed, scatterbrained state."
>
> *(James 1890/1950, p.403)*

With this quote we can carve out some main insights about what attention is:

- A focusing of the mind
- Selection of some things on the cost of (many) other objects or trains of thoughts
- Selected things become more clear and vivid
- Attended items or thoughts are processed more effectively

In this chapter we will go through some of the basics of attention. We will learn about (at least) two different mechanisms of attention – the automatic *bottom-up* attention and the controlled *top-down* attention. These mechanisms are well known to be independent attentional mechanisms (Pinto 2013). We will learn about the limits of attention, and how this can create yet new "blind spots" in our minds. We will treat the difference between attention and consciousness, and the energy budget of consciousness and the brain. Then, we will look at how our consciousness is "filling in the gaps" and creating meaning where there is not necessary any. Finally, we will discuss the aspect of subliminal perception, and whether there is an optimal threshold for attention and consciousness in affecting consumer behaviour.

BOTTOM-UP ATTENTION

Sensory input can make their way to the mind by sheer force: by being extremely bright, noisy or otherwise unexpected. We call such effects **bottom-up**, as in when our lower (bottom) senses are stimulated to such an extent that they produce effects higher up in the processing system. In this way, properties of the stimulus itself can attract attention.

Bottom-up attention can be the result of events that occur either outside yourself (exteroception) or inside yourself (interoception). One example of exteroception is when there is a sudden noise, which makes your body jump and you feeling the sense of surprise and possibly slight fear. In fact, your body responding to the noise is the end of a rapid yet long sequence of events, occurring within a few milliseconds! The noise enters your ears, affects the auditory system in a particular way (sharp rise and large amplitude in the response of the sensory cells in your inner ear), which then is picked up by specific collections of neurons in the brainstem and thalamus. Due to the nature of the signal your brain automatically responds to this and directs your attention to it within a fraction of a second. At this time, you are becoming aware of the event, but already, your brain has detected, processed and labelled the event for you: "this is important! Something's going on! Check it out!"

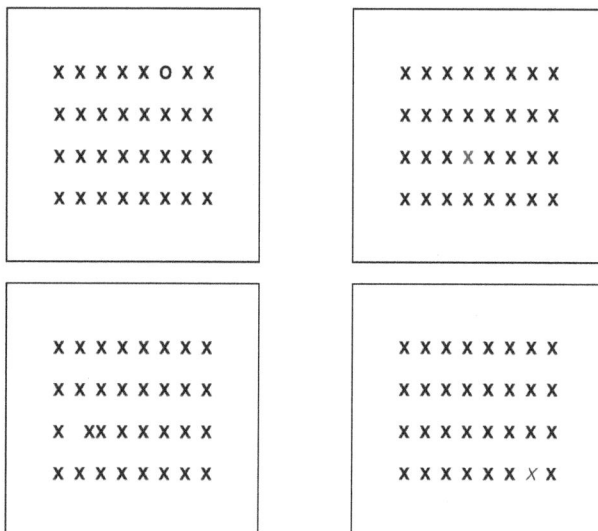

X X X X X O X X
X X X X X X X X
X X X X X X X X
X X X X X X X X

X X X X X X X X
X X X X X X X X
X X X X X X X X
X X X X X X X X

X X X X X X X X
X X X X X X X X
X XX X X X X X
X X X X X X X X

X X X X X X X X
X X X X X X X X
X X X X X X X X X
X X X X X X X X

Figure Bottom-up drivers. Stimulus properties can automatically drive visual attention, such as shape, colour and density. These effects make stimuli "pop out" relative to their surroundings. Importantly, you cannot "un-see" these effects – they will always affect your attention.

Similarly, for the visual system, we all know how annoying a misbehaving pixel can be on a computer screen. You detect it right away ant it keeps nagging your attention even when you are trying to avoid thinking about it. Below are some figures illustrating how things "pop out" just by the nature of their stimulus properties.

For the somatic sense, a mere brush on your arm will attract your attention, and a tooth ache is a classic example of how an interoceptive sensory signal from the body can force itself directly into your conscious mind.

Bottom-up attention is, as we will see, a highly powerful driver of consumer attention and even choice, and it is highly relevant in most if not all of our behaviours. Fortunately, bottom-up attention is something we understand well and are able to assess and affect.

So what is bottom-up attention in essence? Three vital characteristics can be mentioned:

- **Automatic** – it happens by itself as a response to particular events
- **Fast** – the response is immediate
- **Non-volutional** – it does not require you to wilfully focus your mind

Brain mechanisms of bottom-up attention

There are in fact many mechanisms underlying bottom-up attention, but it may be best described if we look at the visual system. Since we have such as thorough understanding of vision and the brain foundations for this, we will focus mostly on this system.

As described in the previous chapter, your eyes have cells that respond to things like movement, density, changes and contrast. In doing so, your eyes are providing crucial building blocks for your attentional system. If something moves in your visual periphery, your eyes will automatically capture this and translate that into signals that transcend all the way up to the brain. If a region stands out because it is more dense, has a different colour or a different shape, your visual system detects this early on; most likely as a consequence of responses in both the retina, the thalamus and the primary visual cortex. Indeed, recent studies have demonstrated that V1, the earliest point in the visual cortex that receives signals from the eyes, are responding only to bottom-up processes.

Information that is visually salient is therefore detected early on in the visual system, and cascaded of this information goes through the primary visual cortex and affects the attention it receives later on by other regions, such as the parietal and frontal cortex.

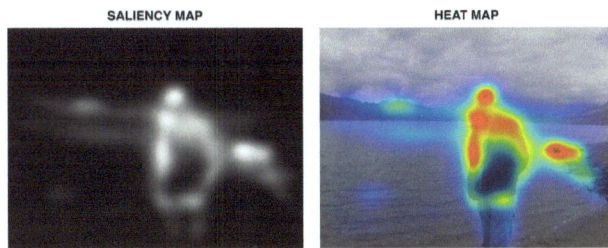

SALIENCY MAP   HEAT MAP

Figure Visual saliency is driven by predictable factors such as density, contrast, angles, colours and movement. These features are so well understood that it is possible to create mathematical models that mimic these effects, which again allows automatic analyses of images for the most salient parts. First, these effects are extracted as a "saliency map" (left), which highlights the most salient regions. One way to visualise these effects is to make the saliency scores as colour-coded scores that can be visualised as a "heat map" (right). The example analysis is made using NeuroVision™, and demonstrates that the person stands out relative to the background. Courtesy of Neurons Inc (www.neuronsinc.com)

Indeed, these processes have been suggested to create a so-called *saliency map* in which different features, such as density, angles, contrast, and colour composition, that can be modelled mathematically, allowing for automatic analysis and prediction of what the eyes will automatically fixate on (Itti 2000). That is, the mathematical models, such as NeuroVision™, are now becoming so good that they can be used to predict where people are going to automatically look, with an accuracy about 85%.

Recent studies suggest that there are in fact two parallel routes through which salient signals can be processed. While the route described here is important, new evidence suggests that another route bypasses the primary sensory cortices and go directly

to what has been labeled the *saliency network*, which consists of the insular cortex (IC) and anterior cingulate cortex (ACC). This network receives information through the route described above, but it has now been show that they also receive salient signals directly from the thalamus (Liang 2013). This happens for all senses – visual, auditory, somatic alike.

This finding suggests two things:

1. Saliency is so important to the brain (and the organism's survival) that it has evolved multiple routes that process the saliency of a signal.
2. Salient processing can occur either relatively fast or relatively slow, and the fast process is "immediate" and without the need for more detailed processing by the sensory cortices

Can bottom-up drive consumer choice?

A central question is whether visual saliency actually has some effect on consumer choice. Can things like the immediate and early responses to contrast, density, and colours really affect consumer attention and choice? In a vital paper by Milosavljevic and colleagues (Milosavljevic 2012) the researchers were testing whether bottom-up processes could indeed drive or affect consumer choice. By altering the relative brightness and contrast of selected products, they then tested whether this would influence product choice. Participants were then shown the images for different exposure times, ranging from 70 to 500 milliseconds. Here, the researchers found that the effects of visual saliency dominated the end choice during the brief exposures and decision times, while stated preference dominated choice for longer exposure and decision times.

In a second study of the same paper, the authors produced conditions in which the participants' mental effort, or "cognitive load" was higher (we will talk about cognitive load in the chapter about learning and memory). Here, they found that during higher cognitive load, the visual saliency effect was longer lasting and stronger.

Together, this means that visual saliency – which drives bottom-up attention – can be directly affecting what we end up choosing, especially under conditions of short exposure, rapid decisions and high cognitive load. We can think of several situations where this applies, everything from web browsing to the spontaneous purchases we make so often inside stores.

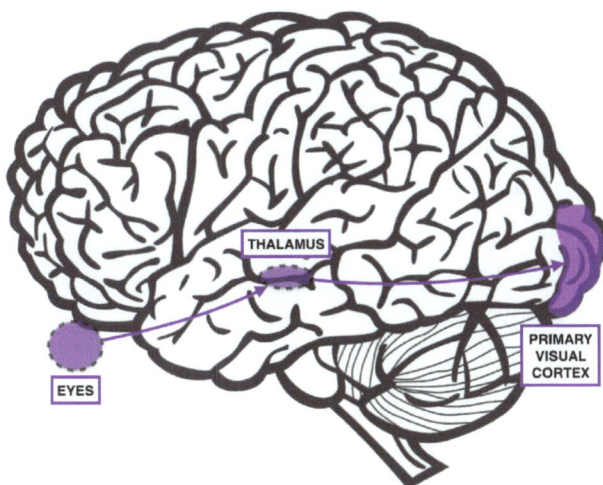

Figure Bottom-up processes are initiated in the eyes, and prompt an early response of the Lateral Geniculate Nucleus of the thalamus, which then projects to the primary visual cortex. The primary visual cortex extends into the medial/inner parts of the brain (not shown in figure).

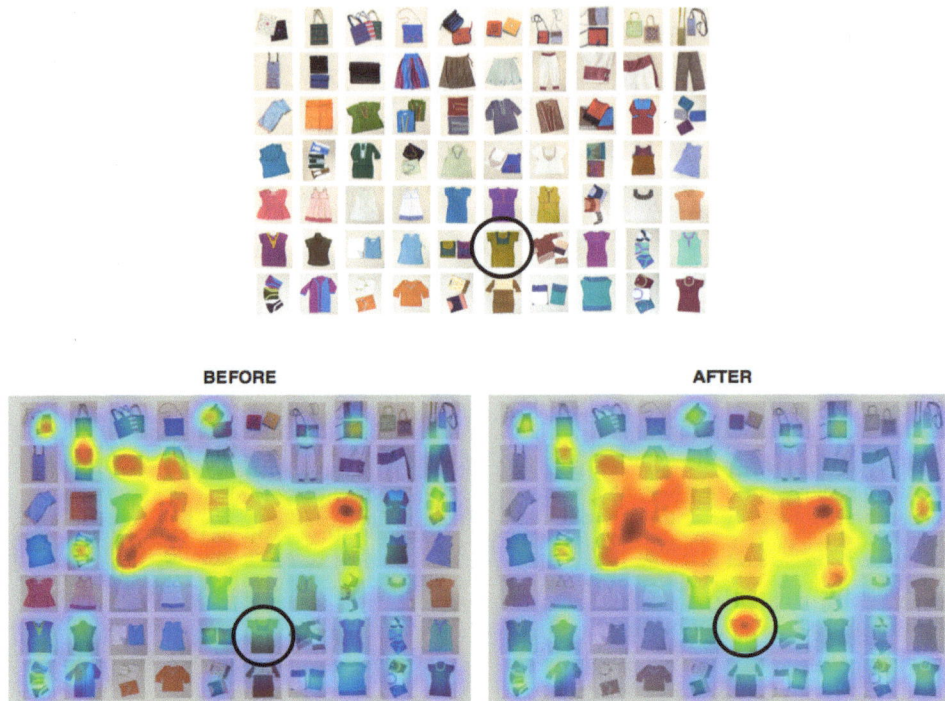

Figure   Selective boosting of visual saliency. By simply editing a selected item (black circle) by increasing brightness and contrast it is possible to increase the visual saliency and the likelihood that that particular item will be automatically seen. Before the tweaking (left) there was no saliency on the item of interest. After the simple photoshopping of the selected item, it became substantially more visually salient (right). Analysis done with NeuroVision™. Courtesy of Neurons Inc (www.neuronsinc.com)

Bottom-up attention during browsing

Prominent everyday examples of bottom-up attention comes from what you have probably done multiple times already today: browsing a webpage, looking through a magazine, or strolling down a street. In all those conditions, you are not looking for anything in particular, but merely browsing the content with an open, laid back mind. The same applies many times while you are watching commercials: you are rarely *actively* engaging in the ad (although marketers like to think you are...). In these conditions, your sensory system is highly likely to pick up on salient features in your environment.

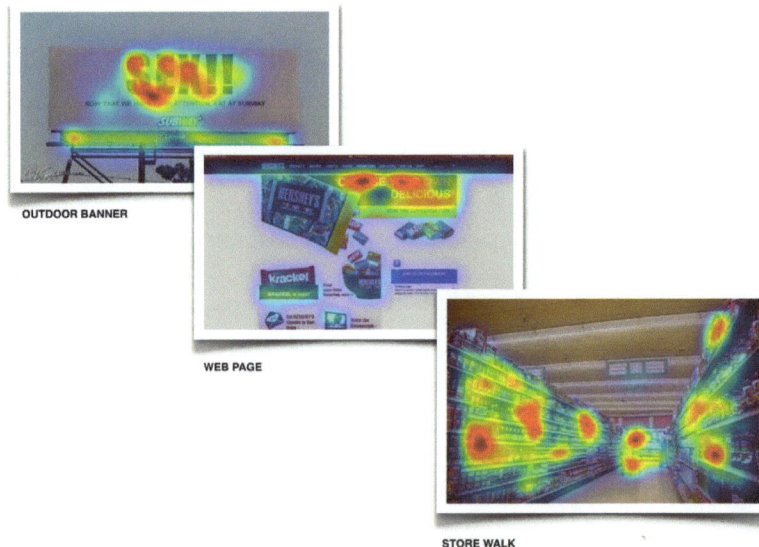

Figure   Examples of browsing examples with highly salient features in outdoor situations, web browsing and in-store walk. The NeuroVision™ heat map demonstrates which features are most likely to be noticed due to their visual saliency. Courtesy of Neurons Inc (www.neuronsinc.com)

How can we assess bottom-up attention?

In general, there are two ways to assess bottom-up attention – eye-tracking and computational neuroscience. As we have already seen above, NeuroVision™ is an online based computational neuroscience model of bottom-up attention and visual saliency. For either videos or images, NeuroVision provides a way to analyse contents for what is most likely to drive automatic attention, just as the example above shows.

The second way of assessing bottom-up attention is through eye-tracking. The immediate visual fixations from eye-tracking are indicative of the automatic drivers of visual attention. Using eye-tracking software, you can look at how long it takes from a stimulus is presented and until people look at it – the so-called *Time to First Fixation* (TFF) and *Fixation Order* (FO) metrics. The lower the TFF number and the earlier one fixates at an item (FO number is low), the more salient one can assume that a particular feature is. By comparing different features in a browsing environment, it is possible to learn what is driving visual attention, and what is more likely to be ignored. Similarly, it is possible to improve the visual saliency of items and use the TFF and FO scores as performance metrics to optimise performance.

There are drawbacks of using eye-tracking that makes computational neuroscience models more appealing for assessing visual saliency:

- **Eye-tracking** most often reflects **both** bottom-up and top-down mechanisms, and it is very hard to tease apart these processes from eye-tracking data. Computational neuroscience models can assess only bottom-up attention.
- **Eye-tracking** is more time consuming and expensive, as each study needs a test of a certain amount of participants. Computational neuroscience is rapid (analysis takes seconds to minutes) and inexpensive (a few dollars per image).

- **Eye-tracking** risks the chance of learning between trials, which will affect the results. Computational neuroscience has no learning and can test hundreds to thousands of images with no learning effect.

For these reasons, we can assume that computational neuroscience models such as NeuroVision outperforms eye-tracking in the assessment of bottom-up attention and visual saliency. However, when there is a need to understand other drivers of visual attention, computational neuroscience models are suboptimal and inaccurate. As we will see, top-down attention and other attentional drivers (e.g., faces and words) are best understood by using eye-tracking and other tools.

TOP-DOWN ATTENTION

While bottom-up attention is what we can call *exogenous attention*, i.e., attention that is generated through what is going on in the external world, top-down attention is an example of *endogenous attention*, which basically means attention that is driven by inner processes. Therefore, top-down attention can be understood as the internally generated and driven focusing of one's attention towards certain items or thoughts.

Imagine that you are reading a highly boring report (or simply find this book extremely boring), yet you absolutely *have to* read it. This leads to you force yourself to read the report, and the whole process of mobilising your mental energy to keep reading is a classic example of top-down attention. Conversely, the highly attractive and colourful magazine lying on the couch besides you, or the buzzing of your smartphone to an incoming text message, is the counterexample in bottom-up attention.

In comparison with bottom-up attention, what then is top-down attention? To match the three characteristics of bottom-up attention, the matching characteristics for top down-attention are:

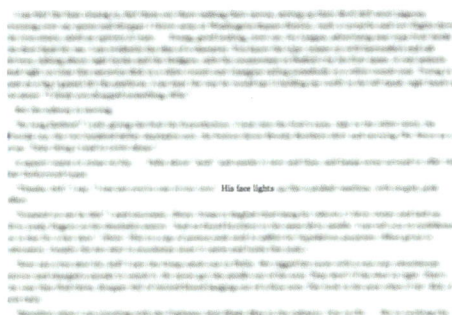

Figure   The effects of attention is an increased focus on some aspects at the cost of other aspects. In top-down attention this is typically seen for attention to the text we are currently reading (left) or the serial search for a particular person (right).

- **Controlled** – attention is not automatic but requires active selection and mobilisation
- **Slow** – the response needs build-up
- **Volitional** – it requires that a wilful focusing of your mind

Thus, we can assume that top-down attention is a process that is very different from bottom-up attention, and therefore also relies on a different set of brain structures. A we will see, this is absolutely the case.

The brain bases of top-down attention

When looking at the brain regions involved in top-down attention, we are looking for regions that are engaged when a person is **actively and consciously motivated** to attend to a specific part of the informational chaos that he or she is exposed to. In such search, we end up with a network of "higher" cortical regions, which are able to modulate the operation of "lower" cortical regions.

When you are searching for a particular item such as a red apple, you are looking for something like an apple, and something red. In doing so, one can see that parts of your frontal and parietal cortex becomes engaged. In particular, these structures have direct and indirect connections to the primary sensory areas of the cortex and they can boost or inhibit circuits within those regions. This allows the frontal and parietal cortex to "fine tune" the sensory cortices to specific kinds of input, such as a particular colour (e.g., red).

When you change your attention from one sense to another, you can easily see this on the readout from brain scanners. While the fronto-parietal system is en-

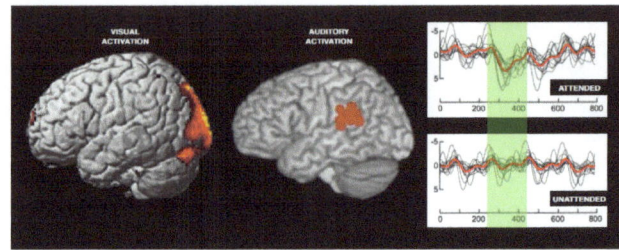

Figure Illustration of the difference between attended and unattended stimuli. In the fMRI (left and middle) this can be seen as a change in both the spatial distribution and signal amplitude of the response within the visual and auditory cortices, respectively. In the EEG this can be seen in aggregate responses (so-called event-related responses, or ERPs) as a negative deflection in brain responses. This is in fact an indication of increased activation.

gaged in both situations, changing your attention from seeing to hearing changes the brain responses of the visual and auditory cortices, respectively. In an fMRI scanning, you see this as a change in the signal extent and amplitude of the brain response, and on the EEG you will see this as larger amplitudes in the responses.

When top-down and bottom-up interact

Indeed, although attention is based on those two different systems, we should really think of them as dynamically interacting. There are clear cases when top-down attention controls what your attention will focus on, but what happens when a stimulus breaks through these barriers and forces itself – via bottom-up attention – to take attention's centre stage? What happens when you purposefully search for something on the web, and a sudden noise outside makes you look away from the screen and even wholly forget what you were doing?

Bottom-up attention has the function that it allows salient stimuli to become the attended objects, however hard you may be trying to focus at a particular task. In this way, we can see that incoming sensory stimuli can trigger an engagement of the frontal and parietal cortex, which then can lead to a top-down focusing of attention.

Can top-down attention force itself upon bottom-up attention? Sure, but not in the same way. Top-down attention is the active and slow mobilisation of mental processes that can focus attention at some things at the cost of other things. Thus, top-down attention relies on selectively *activating* and *inhibiting* selected processes in the sensory system.

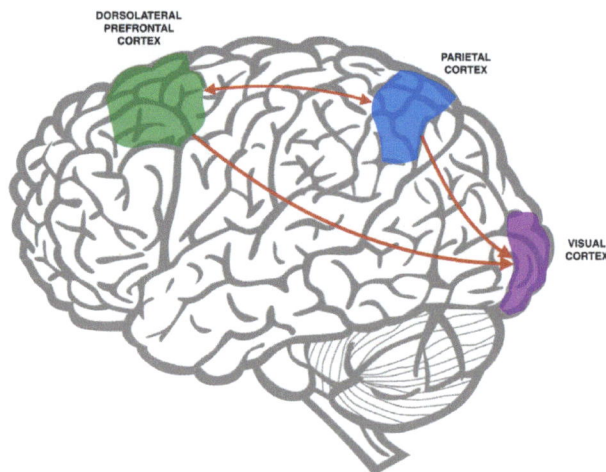

Figure Structures implicated in top-down attention. Regions of the frontal cortex (green) and the parietal cortex (blue) can modulate the activation of primary sensory cortices such as the visual cortex (purple), and through this pre-select information that becomes available to this region, such as a particular colour.

Top-down and consumer choice

One good example of top-down attention in consumer choice is a behaviour that you probably do at least once per week: shopping for groceries. You probably have some kind of shopping list of things you are going to purchase when you get to the store. So when you are looking for bread, you are actively engaging in searching for the section where bread is, and upon reaching this section you are actively searching between the different options you have. Maybe you are even reading on the back side of the packaging to find out the contents of the product? All in all, you are engaging in a focused, mobilised and effortful use of your attention to search for particular kinds of information.

How can we assess top-down attention?

Top-down attention can best be assessed by several means, but as top-down attention is closely linked to our conscious brain activity and verbal system, we should not exclude plain and straightforward self reports. If you want to know what people purposefully have been focusing on, why not just asking them?

However, in some situations it is not possible to ask people what they are actively focusing at. In other instances you want to compare top-down attention with bottom-up processes. Here, other measures such as eye-tracking can assist. In a study for a major toy brand, we asked participants to first hear a noise and then select which toy/children's brand they associated the sound with. The selection was done by looking intensely at one of four options that were presented briefly after the sound was played.

Here, we made an interesting observation: the eye-tracking responses during the first second rarely matched the responses the aggregate and more focused attentional response. In other words, we found that people's eye-fixation during the first second after

the options were presented, were significantly different than their final choice. This is a classic example of bottom-up responses (1 second) and top-down responses (aggregate response).

OTHER ATTENTIONAL DRIVERS

Do not read the following words .... **PINK ELEPHANT!**

It's hard not to read a word like "pink elephant" that is in the centre of your visual attention, even if you try not to do it. Merely having the eyes glancing over a word, it is automatically read, and reading is such an over-automatised process that it is virtually impossible for us not to read the words. You cannot un-read a word, or hinder yourself from reading it.

The same happens for faces – we are really prone to respond to faces and spend more time looking at them. Some studies suggest that more symmetrical and beautiful faces are also receiving more attention.

Are faces and text bottom-up or top-down processes? In fact, we can say they are neither. Both processes need more cognitive processing (along the ventral stream) and are therefore not to be considered "bottom-up" as such. Conversely, faces and text "pop out" to people's attention, and do not require that one engages in the slow and mobilising top-down focusing of one's attention. So one may well argue that faces and text is a different kind. Maybe we are best left by categorising them as overlearned cognitive processes. They are processes with which we have been so often exposed that our brains are simply tuned to respond to them with auto pilot responses.

Further evidence to support this comes from two different sources. First, children who have not yet become fluent in reading, do not show an attentional bias for words. Second, people suffering from social dys-

Figure   Effects of immediate and focused attention on relating a sound to toy brands. As can be seen, eye fixations during the first second of brand presentation were different from what people ended looking most at. Courtesy of Neurons Inc (www.neuronsinc.com)

Figure   Text and faces attract attention. In a passive viewing task, participants automatically look at text and faces in the text, at the cost of other things such as brand logos, products, scene items and other materials. Faces that turn towards the viewer, or contain social situations (right), are focused much more on than when people's faces are turned away (left). Reproduced with permission from iMotions (www.imotionsglobal.com).

function, such as autism, do not show the same automatic focus towards faces. Together this demonstrates that we may have a "semi-automatic" attentional focusing on things we have been overly exposed to, and here text and faces are excellent examples.

Effects of single exposures on attention

One of the crucial questions to any marketer is whether consumers change their attention and emotional responses after being exposed to an ad. After all, an ad is designed to change the way we perceive a brand, a product or some other crucial information.

In a recent study,[4] we tested the effects of ads on in-store attention. Participants were exposed to a series of ads, but were told that this was only for calibration purposes before the real in-store test. One group saw a series of commercials prior to entering the store, and two groups saw exactly the same commercials plus an ad (15 or 30 seconds duration) for a particular paint brand. After this, all participants were sent into a retail store and given a series of different tasks that they could perform in their own order and pace.

At the same time, we recorded their eye fixations using mobile eye-tracking, and mobile EEG to track their cognitive and emotional responses. Crucially, one of the tasks participants were given was to purchase paint for their living room.

What we found was that compared to participants who were not exposed to the crucial ad (70% purchase), the groups that had seen the ad were much more likely to choose the product for which they had just seen the ad (90% and 100% for the 15- and 30-seconds ad, respectively).

When we asked people after they had left the product shelves and were about to buy the paint, all participant had explicit reasons for why they had chosen the product. Upon asking participants, only a few participants recalled they had seen an ad for the paint ad, and no participant believed that they had been affected the ad.

Nevertheless, when we assessed where participants had looked when they entered the store section relevant to the ad, there was a clear and significant difference between the groups. Compared to those who had not seen the paint commercials, the groups that had been exposed to the commercials demonstrated a much stronger and more distributed eye-fixation pattern. This clearly demonstrates that visual attention was significantly affected by the prior ad, even thought they were unaware of such effects, and claimed autonomy in their own decision.

Figure   Effects of ad exposure on in-store attention. Those who were not exposed to the crucial ad only briefly focused on the shelf for the relevant brand, while those who were exposed to the commercial were exploring the shelves much more. The heat map was generated from infrared markers (small black dots in the image), which allows the tracking of eye fixations independent from where people are standing relative to the shelf. Courtesy of Neurons Inc (www.neuronsinc.com)

This study clearly demonstrates that certain kinds of attention cannot be neatly divided into a dichotomous divide between bottom-up and top-down attention. Rather, our attention can be biased in a way that is still not overly controlled in a top-down manner. As we will see in the chapter on emotions, our brains are wired to respond to particular events, as well as to relevant memory, that can trigger changes in attention without top-down influence.

Visual complexity – what distorts visual communication

If there is one thing that can ruin or facilitate visual attention, it is the degree to which other information is competing for attention. It is much easier to spot a product, a face, or some crucial information when it is based on a simple background, than when it is situated in a noisy environment.

Think of finding a crucial word on a busy paper page, or a particular product on a busy shelf in the store. Sometimes it feels virtually impossible to find a particular product or information in a search, and who have not given up finding Wally / Waldo? When the load of information is too high, individual pieces of information are hard to spot and find.

As we will see in the chapter about memory – where we are talking about cognitive load – we will see that information complexity is highly related to processing difficulty and lower likelihood of sustained attention to individual pieces of information

THE BOTTLENECK BRAIN

Here is a task to illustrate a crucial point: please look at the following picture for 10 seconds and find the items that are red...

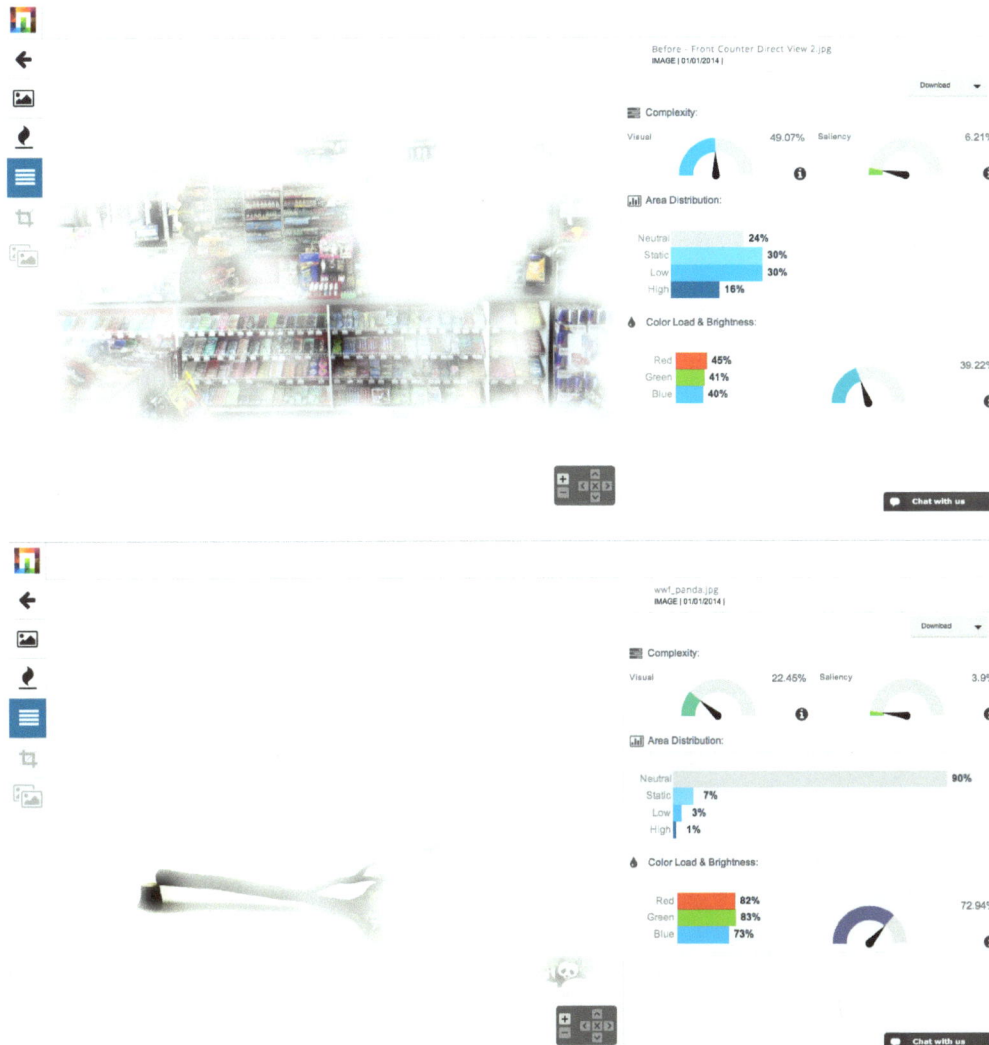

Figure High visual complexity is seen in retail environment (top), and when the score is high (NeuroVision complexity score = 49.07%) it suggests that there is a lot of competing information in the image. The likelihood that one particular piece of information will be seen – automatically or during a search – becomes lower when visual complexity goes up. In comparison, for information such as an ad (bottom) with lower complexity (NeuroVision complexity score = 22.45%) there is less competition from other kinds of information, and a higher likelihood that the crucial information is seen and attended, ensuring that the message comes across (e.g., combining the broken tree with the panda being hit, and the brand WWF). Courtesy of Neurons Inc (www.neuronsinc.com)

Now, without looking at the objects again, please name the red objects you found.

After this, please list any animals you saw. How about furniture...did you see any?

While this task in many ways also tap into your immediate memory – or, working memory – it is also a demonstration of the limits of your attention. At any one time, you will be unable to take in all available information. Your brain has a significant limitation how much information it can actively process at any one time.

The brain is *limited* in its ability to process information. This is often referred to as a bottleneck problem: imaging a bottle, where you are pouring out water. Due to the neck of the bottle, water cannot flow freely out of the bottle. In very much the same way,

our brains have biological limitations to how much information that can be processed at any one time, and therefore, adding information is not increasing any chances that more information can be processed.

As we will see in the chapter about memory – and in particular the section about working memory and cognitive load – there are severe limitations to the amount of information that we can hold on to at any one time. This puts constraints on any attempt to communicate with consumers, as we cannot assume that we have the luxury of pouring whatever information we have into their open and undivided minds. Rather, with consumers, companies have a severely limited time and an extremely competitive landscape to get their message across. When watching TV, being at

the movies, browsing the web and strolling down the street, we are exposed to pandemic amounts of information, and marketers cannot assume that consumers will pay particular attention to their material. Understanding this can indeed come as a shock to creatives and companies alike, but can provide a fertile soil for growing a new and better way to communicate with consumers.

Bottleneck examples

There are a few notable examples of the limits of our attention. One such example is **inattentional blindness**, which is the case in which we fail to notice crucial events even when they are within our line of sight, and that is typically happening when our attention is fully occupied with something else. Take for example the *invisible gorilla test*. Most of you have already seen this, but look at this video to see an alternative take on the same kind of illusion. Inattentional blindness is a prime example of how *focusing your attention on something is at the cost of attending other things*. We simply cannot attend it all at once. As William James noted more than a century ago, attention means "wit-

hdrawal from some things in order to deal effectively with others".

Another attention phenomenon is **change blindness**, in which we fail to notice otherwise large changes in an image or scene. In the following images, please identify 10 changes...

As you have discovered, you have to actively seek through the images to find the differences. They rarely if ever pop out. There are indeed ten differences between the images, and the differences are shown on one of the coming pages. Please note that *even though you know that there are differences* you have to actively seek for them. This is an effort that we as consumers rarely, if ever, do. Attention is an extremely valuable resource and consumers rarely devote much time to products, ads or communication if ever.

Blindness and the attentional blink

One interesting phenomenon is that after something has grabbed our attention, we tend to have a "blind period" in which nothing is really noticed. That is, if you see something on the screen that grabs your attention, and then stops, there is a period up to 500 millisecond

– half a second – in which whatever is presented will go unnoticed! This phenomenon is called the **attentional blink** (Shapiro 1994; Vogel 1998; Shapiro 1997), and it suggests that there is an attentional latency and recuperation period in which attention needs to "reset" in order to take in new information.

This phenomenon is highly interesting and relevant for communication purposes. Imagine that you have an emotional event during an ad and then present the product or brand. Chances are that the first 500 milliseconds of this product or brand exposure are wasted time: people are still resetting after the atten-

tion grabbing event! This therefore suggests that when designing commercials, movies, or any other kind of material wherein one can expect something to grab a strong hold of one's attention, then one should also be very careful with what one present after this event.

So, did you see the ten changes I made to the street image? If not, no worries – you are not alone! If you saw all of them, good job! In any case, the process of spotting such differences should demonstrate that you had to actively mobilise your mind to keep on looking and comparing the two images. If anything else distracted you – a mobile phone, an incoming email, or a sound

Figure Ten differences between the two images – were you blind to the changes?

outside – you lost track and had to start where you left off. Attention is a precious resource, and not something consumers or marketers can take for granted.

HOW TO ATTRACT ATTENTION

So what grabs attention? Unless you have convinced somebody that they need to actively search for your product or brand, you do not have the luxury of having customers engaging in top-down attention after your product. Even if they are actively engaged in searching for your product, you could consider making the task easier on them. But what can you possibly do? Here I list four specific examples:

- **Make it salient!**
  - Visual saliency is key: work on increasing brightness, contrast, density and other features that increases the visual saliency of an object
  - Auditory saliency can be achieved by having a high amplitude, but as we will see in the chapter on emotion, making a sound more unpredictable!
- **Less is more!**
  - Making something more salient also means making everything else *less* salient. In the visual domain, try to dim other visual stimuli, make them darker, more diffuse or anything that can help make the object at hand stand out. A single item on a plain background is an insurance that it will be seen. When it comes to grabbing attention, the old dictum of "less is more" really applies!
  - Lower the visual complexity of the image – the less competing information is available, the more likely it is that your crucial piece of information is seen.
- **Prime to search**
  - Present information that can be recognised and trigger emotional responses at a later stage – use

outdoor and entrance sign ages that capture attention and that can make an impact
  - Make powerful ads and commercials and other outlets that ensure recognition and can trigger responses upon facing the crucial product choice
- **Convince to search**
  - The best case is where consumers are actively pursuing information about your product, and actively trying to purchase it. Most products do not have this luxury but it is still something one should strive to achieve
  - Become a "love brand" – brands that people have a high affection for are more likely to be actively sought our by consumers. Think of the active engagement that Apple lovers show upon the release of the next iPhone, iPad or Mac, or how some women are looking forward to the next Women's Fashion Show from Dolce & Gabbana. If anything, these consumers show an active, mental preoccupation and top-down search for all kinds of information related to the brand they love.

CONSCIOUSNESS: STATE AND CONTENT

Most terms in this book have an easy understanding and a hard understanding. The easy part is that for a term, we know what it is about, but it soon becomes harder when we try to define the term more specifically.

This is particularly obvious in the case of consciousness and awareness. If you try to define the term as you are using it in your everyday language, it might seem quite straightforward. But try to come up with a really detailed account of what consciousness really is, and you are soon stuck. Leading philosophers, scholars and scientists have worked hard on the problem of consciousness and awareness for centuries. Still, we have no clear definition of the terms. But as we will see here, through the combination of psychology and neuroscience there has been tremendous progress in our understanding.

Consciousness as a state

What happened this morning? You woke up after a combination of dreamless and dreamful sleep. The morning started, and although you might have started off a bit groggy, you were present, there in the moment. But what happened before is actually the crux of our definition: you were gone! When you are in a dreamless sleep, you are not conscious. Your brain even operates at a very different level, such as the slow

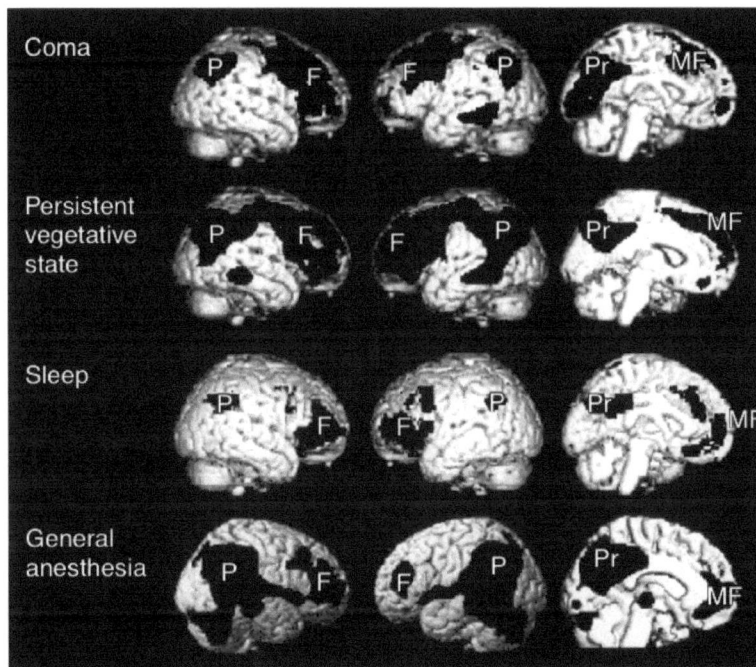

Figure   Four instances of unconsciousness and how the brain is deactivated (shown in black) and show lower metabolism. This suggests that across different conditions, unconsciousness has a common signature on brain activity. Abbreviations: F, prefrontal cortex; MF, mesiofrontal frontal; P, posterior parietal cortex; Pr, posterior cingulate/precuneus cortex. From (Baars 2003)

sleep spindles shown in an EEG seen when you are in deep sleep. When you woke up, that changed dramatically. Your brain altered its state altogether.

So one important definition of consciousness concerns which **mental state** consciousness is. Consciousness is the state in which we are awake, alert and present in the moment, as opposed to the states in which we are in (dreamless) sleep, unresponsive and "away". Think of whether you have been in general anaesthesia, or what happens if a person gets knocked out. They are "gone" and unresponsive to anything from the outer world. It also seems that no real information processing is going on.

A decade ago, I wrote a joint paper with Bernard Baars and Steven Laureys where we demonstrated what this meant in terms of the brain. When we compare different instances of unconscious states – such as coma, general anaesthesia, vegetative state and deep sleep, there is a persistent pattern of brain *deactivation* when we compare to the healthy and awake mind (Baars 2003). We see that large portions of the frontal and parietal cortices are disengaged and show lower activity than a healthy and awake person. Moreover, deep structures such as the thalamus has been found to be extremely important as the connecting *hub* for the global changes in brain state, and research into loss of consciousness in epilepsy suggests that loss of consciousness only occurs after the thalamus has "broadcasted" malignant

brain activation to these diverse brain areas (Blumenfeld 2005; Blumenfeld 2003). Together, this implies that consciousness is the result of a concerted and global brain state that is synchronised, at least in part, through the thalamus. As we will see, this yields an important indicator of the function of consciousness.

Consciousness as content

The second way to talk about consciousness is to look at what happens when we are awake and present, and then look at the difference between conscious and unconscious processing of information. For example, you are highly conscious (hopefully) about the content you are currently reading, but until this point you were most likely not aware of any other thoughts or sensations, such as sensing the seat you are sitting on, or thinking about what you are having for your next meal. Consciousness, when used in this way, pertains to your *experience* of something in the world, but as we will see, also a particular kind of cognitive processing of this information.

Think about having a toothache. There is something like having a toothache that is very different from eating an ice-cream, listening to a piece of music, or watching a piece of art. In all four situations, you are fully awake and therefore conscious in the *state* sense, but the content of your mind is extremely different.

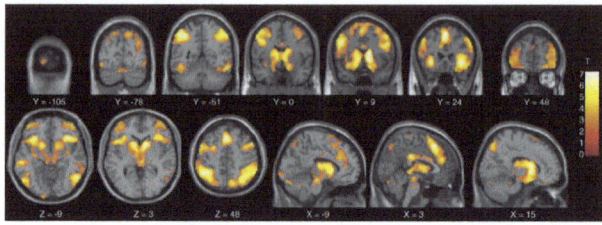

Figure Large scale activation of the brain when people are conscious of a stimulus, as compared to being unconscious of the stimulus. This suggests that consciousness is the result of a change in the "global brain state".

In this book we are not concerned about consumption during unconscious states. We need not to bother our minds with what happens when a person is sleeping or in a coma. Therefore, whenever we are talking about consciousness, we refer to the active content of a fully awake mind – the things that a person is experiencing.

Notably, our conscious experiences are closely linked to our ability to report this – we can ask ourselves or each other: what do you think about that movie? What were your thoughts when you picked up that product? How did that meal taste? In doing so, you will often get rich and detailed answers about a person's immediate experience, and possibly other kinds of information such as what the person thought about at the time.

Importantly, what we will see in this chapter, as well as other chapters, being unconscious about something is not the same as saying that what is processed unconsciously cannot have an effect on your thinking or behaviour. Quite the contrary, our unconscious mind constantly makes decisions for us, ranging from the control of the heart and breathing, all the way to social choices. Sometimes, we will even find that the conscious mind is hindering effective processing and action.

Interestingly, when we are comparing conscious and unconscious processing, we find that conscious processing requires a "global" engagement of the brain, and this activation pattern is highly similar to what we discussed earlier, in the differences between conscious and unconscious states. In a study by Christensen and colleagues (Christensen 2006), we used fMRI to compare the brain activation of conscious and unconscious detection and identification of simple visual stimuli. Here, we found that conscious processing, compared to unconscious processing of a stimulus, was associated with large-scale activation of the frontal cortex, parietal cortex, anterior cingulate cortex, basal ganglia, thalamus, anterior insula and many other regions.

Consciousness vs unconsciousness – gradual or distinct changes?

Is the difference between being conscious and unconscious of something based on some kind of dichotomous and distinct brain states, or is the change from unconscious to conscious processing more fluent? Leading scholars have suggested that there is a large-scale quantitative change in brain state when something is processed consciously as opposed to unconsciously (Dehaene 2001; Kouider 2007; Dehaene 2006). This model suggests a strict dichotomy between being conscious and unconscious about something. You are either seeing something you you don't.

While we certainly see gross changes in brain state, I have had a different take on this. There are many conditions in which we anecdotally know that "something was present" but where you are unable to determine what it was. It was there, but not really there as if it was completely conscious. This can go from having a vague thought or eerie feeling about something, to seeing something "out of the corner of your mind".

It soon turned out to me and my colleagues that we could be dealing with a measurement problem. If you ask people to report whether they are conscious or unconscious about something, you are surely going to get answers that fall neatly into categories of either seen or unseen. But when we opened up and asked people to verbalise more and construct their own categories, we soon found out that there were other kinds of experiences that did not neatly fall into the categories of conscious vs unconscious.

Indeed, we found that people often responded that they had a "vague feeling" that there had been something present, but that they could not identify it. When exploring this condition closer, we soon found out that when people reported having this "vague" sensation, they were much better than chance levels at guessing what they had seen (Overgaard 2006; Ramsøy 2004). What we had found, also called the "sensory fringe" by some scholars (Mangan 1993), had already been documented by scholars a century earlier. For example, William James (James 2008) famously coined the so-called *fringe* in the following way:

> "What must be admitted is that the definite images of traditional psychology form but the very smallest part of our minds as they actually live. The traditional psychology talks like one who should say a river consists of nothing but pailsful, spoonful, quartpotsful, barrelsful, and other moulded forms of water. Even were the pails and the pots all actually standing in the stream, still between them the free

water would continue to flow. It is just this free water of consciousness that psychologists resolutely overlook. Every definite image in the mind is steeped and dyed in the free water that flows round it. With it goes the sense of its relations, near and remote, the dying echo of whence it came to us, the dawning sense of whither it is to lead. The significance, the value, of the image is all in this halo or penumbra that surrounds and escorts it, -- or rather that is fused into one with it and has become bone of its bone and flesh of its flesh; leaving it, it is true, an image of the same thing it was before, but making it an image of that thing newly taken and freshly understood."

James has also been said to think of the fringe as an "active void". Think about the tip-of-the-tongue (TOT) phenomenon. You are *just* about to remember the name of something or someone, but it just does not appear. This is exactly what James meant by the "active void". You are not conscious about the actual name, but the process of finding it. The TOT phenomenon is thought to represent a *non-sensory fringe*, while the vague impression of having seen something is called a *sensory fringe*.

When we studied people's brain responses to seen and unseen stimuli, as noted earlier, we also included the possibility of reporting these sensory fringes. Here, we made an exciting observation. Compared to when people reported not seeing anything, sensory fringes were associated with a widespread network of activation. It was almost identical to the network when we are conscious of something, albeit at a much lower power. Still, this provided further clues to us that consciousness is not an either-or, but rather a continuous (or at least semi-continuous) phenomenon.

More recently, we have argued (Baars, Ramsøy & Franklin 2013) that conscious contents are distributed throughout the brain through what is called *cortical binding and propagation*, which depends on a fluctuating shift in "functional hubs" that distribute signals throughout a limited amount of brain regions. As seen in our prior fMRI findings on conscious processing, one such hub can be the thalamus, and we argue that

other regions with a broad connectivity, such as the hippocampus, can act as such a hub. This proposal suggests that consciousness is a highly dynamic and ever changing phenomenon in the human brain. What happens in a smaller region "A" of the brain can be "broadcasted" to a broader set of brain regions, leading to both the experience of something, as well as the cognitive availability of what happens in brain region "A" to affect our behaviour.

A note on terminology

The words conscious and unconscious have many descriptions and even cultural flavours. In Sweden, the term "medvetande" indicates a "shared knowledge", in German "Bewusstsein" as in Danish and Norwegian ("bevidsthed" / "bevissthet", respectively), is a vague reference to "being aware". If we then add terms such as subliminal, unconscious, non-conscious and so on, we may soon end up with confusion. Here, we are going to use the terms in the following way:

- **Conscious**: the condition of having clear and vivid experiences about particular events, stimuli or thoughts (while being awake and alert).
- **Fringe**: the condition of having vague experiences about events, stimuli or thoughts (while being awake and alert).
- **Unconscious**: the condition of not having experiences about particular events, stimuli or thoughts (while still being awake and alert). Notably, unconscious processes *can become conscious* at some time.
- **Non-conscious**: refers to processes that are completely outside the realm of our consciousness that still operate as brain states. Examples include the control of respiration and pulse, body temperature and digestion, which can only be consciously controlled and affected through indirect means.
- **Subliminal**: stimuli or processes that operate under the limen of consciousness, that can still have an effect on our behaviour and conscious thought.

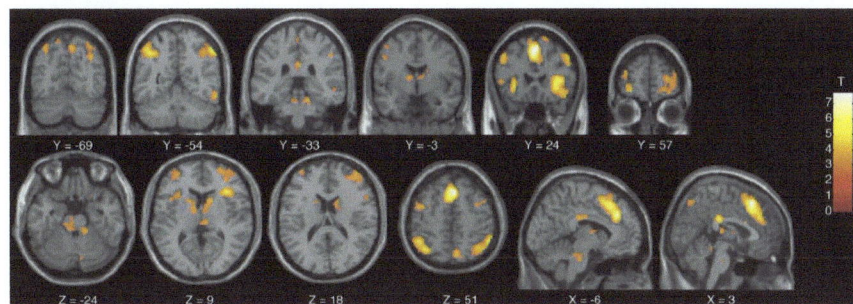

Figure Brain activation when comparing vague experiences and unconscious processing shows a widespread pattern of activation. Although the activation level is not the same as when people are fully conscious about a stimulus, it is notable that the same brain regions are engaged.

BS claims by Chopra, nothing new, but I'm compelled to reply

Deepak Chopra has moved into the domain of the brain... and it should come as no surprise that what he claims is not only erroneous, but also deeply misleading and ignoring the vast amounts of data for the view he so eagerly wants to discard.

Why do I even bother? Because I care! I care that people are not misled by what I believe to be baloney!

I jotted a decent reply to Chopra's recent text at Huff Post, but it was obviously too long for being a comment, so I'm posting it here, hoping that many of you will make it from my direct link at the HP site. Thanks for clicking.

Mr. Chopra,

The level of BS in this assertion is so high that I don't even know where to start! We now have a whole century (actually, much more, but let's leave it at that) of evidence providing a very close link between the mind and the brain. I am utterly puzzled at how one can even make such claims as you do, and feel compelled to do some debugging of your text:

- The starter dish fallacy: The brain does not "light up" – what you see is a statistical representation of the change in signal intensity that (for fMRI scans) represent changes in oxygenated blood, which is an indirect measure of brain activation. Dark regions are still active, but not particularly for the task we have chosen to focus on (or rather, the tasks that researchers have decided to compare). This is a non-trivial distinction, because the link suggested by Chopra to a radio tuning in is simply erroneous. See more below.

- The big leap of reason is the semantic trick of saying that neuroscientists (including myself) believe that the brain is in charge, and not you... I thought Chopra just agreed that neuroscientists believed that the brain IS you? Actually, most scientists I know believe that the brain and you are indeed the same! What happens in the brain is part of you as an organism, as a person, and often as a sentient being. The activation of hypothalamic nuclei can help control hunger, thermoregulation etc.; the response of the amygdala can help you become aware of specific events; the activation in the medial orbitofrontal cortex does indeed reflect quite closely how much you enjoy reading this paragraph, the taste of that chocolate you're having (lucky you) or the music you have playing in the background.

- The great news is: this takes NOTHING away from the wonderful richness of your conscious life! But we understand so much better now HOW it is that the mysterious wet matter of the brain can even produce such magic. And the best part is...no supernatural explanations are yet needed. No need to evoke additional dimensions, pseudoscientific explanations or altogether magical mental bypasses.

- "Brain activity isn't the same as thinking, feeling, or seeing". True enough as a general statement BUT brain activation in regions such as the amygdala, striatum, orbitofrontal cortex and insula ARE equated to emotional responses. Thinking...well that's too mongrel a concept to start with so our ability to "think twice", i.e. control our impulses, is well known to be closely related to activation in other brain regions, such as the anterior cingulate cortex and other parts of the prefrontal cortex. And hey, if you suffer a lesion to any of those regions, you DO lose your ability to respond emotionally, control your actions etc. SO instead of infusing wannabe scientific explanations, why not start with the obvious? The brain is the organ of thought and emotions. Just because we have not understand all of it in minute detail, this explanation exceeds and outperforms any other alternative explanation by zillions of miles!

- "No one has remotely shown how molecules acquire the qualities of the mind"... I don't even know what that statement means. Obviously, nobody have ever proven the mind to exist outside the brain. Our best guess is...the mind and the brain is the same thing! I've seen far too many neurology patients (and psychiatric for that matter) to believe that the brain is not the culprit.

- "It is impossible to construct a theory of the mind based on material objects that somehow became conscious." I love the "impossible" statement here. See, this argument goes straight in the face of your own claim: if you believe that there is an immaterial mind and a material brain, and there is no interaction between them, you're in BIG trouble. Why do we talk about the brain at all, then? Why does a lesion to the brain in any way lead to a change in the mind? How can I physically simulate your brain using electrodes and Transcranial Magnetic Stimulation and make your hand move, make you partially blind for a split second, or alter your social decision? Your claims must be backed up with facts. Claims are not facts!

- Radio analogy: this is just a pretty darn good example that your choice of analogy is wrong, and not that there is anything wrong with how we view the brain!

So it is NOT a massive struggle to neuroscientists to "see those flaws". Indeed, I do see the flaws, but as I've put out above, the flaws are on you, Mr. Chopra. Please consult a neuroscientist next time, I'd be happy to discuss this at any time!

Sincerely, and with my mindful brain intact

Thomas Zoëga Ramsøy, PhD & neuropsychologist

(Note: I did in fact receive a dinner invitation from mr Chopra)

Cognitive differences between consciousness and unconsciousness

One highly relevant issue concerns what consciousness is good for. What, if anything, can the conscious mind do that the unconscious mind cannot? Here, leading thinkers such as Bernard Baars have demonstrated the cognitive capacity of consciousness. He has summarised the differences as follows, by comparing consciousness to controlled processes and unconsciousness to automatic processes:

From this description it is clear that consciousness is *for* something. Conscious processes are slow and effortful, but seem highly flexible and great for solving general and domain-unspecific problems. Think about solving a novel problem, or coming up with a great new creative idea! You cannot do this on autopilot.

By contrast, think about having to concentrate hard on taking each step, or every move when riding a bike. Unconscious processes are great for automatisation: they are fast, rule based processes that need little effort. On the other side, these processes are so rule based that they are also inflexible. When you ride your bike downtown and you want to change the route, you have to actively and consciously change the route, or your unconscious process will just keep you going in the same direction you were originally going. Who have not tried to end up in the living room with your toothbrush and asking yourself "what was I about to do now?" Here, your auto pilot has brought you to a situation and you have lost conscious control over your behaviour (hopefully just for a moment).

Consciousness is extremely important for problem solving, for creativity, for changing one's behaviour. Conversely, unconscious and automatic processes are extremely good at relieving our conscious minds from tedious and repetitive things. We can drive for long distances while thinking about other things; we can walk down the road having a deep and philosophical conversation; we can chew a sandwich without having

to think about every movement of the jaw. It seems almost annoyingly simple to do these things, but that is also exactly the point! The unconscious autopilots in your brain make your life much easier.

Automaticity and consumer choice

It may therefore come as a surprise that we think that we can just ask people why they are choosing one product over another, and why they are behaving in the way they do. Self reports are always going to be imperfect approximations to the true drivers of our behaviours. As we will see, even when we think we are in complete control, our unconscious brains have made decisions for us.

Think of the way in which we saw in the prior chapter that your sensory system selects and represents the world for you. Your experience of the world is an incomplete image of what the world really is, but you experience it as something coherent and a "true" representation of the world. How does this translate to consumer behaviour?

Automaticity – how we behave automatically – is an important feature of consumer behaviour. It has long been known that certain deep brain structures, such as the basal ganglia, are responsible for automatic behaviours and habits (Graybiel 2008; Lang 2010; Yin 2006). Habits and other automatic behaviours are the true opposition to conscious processing: they are processes that operate completely as a rule-based computer script that, when executed, runs to the end. They are extremely good at solving a specific problem (e.g., walking) once the behaviour has been learned, but is at the same time highly inflexible.

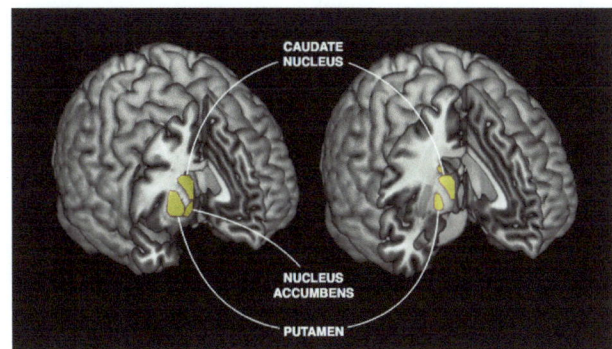

Figure Structures of habit. The deep, subcortical structures of the basal ganglia are an essential part of the brain's habit system. Here, the basal ganglia structures (putamen, caudate nucleus and nucleus accumbens) are essential parts of a dynamic interplay in both the acquisition, execution and maintenance of habitual behaviours. As such one can say that the basal ganglia is the "auto pilot" of the brain's behavioural system. In this figure, we show two different cutouts of the right side of the brain. In the brain shown to the left the slice is a bit more anterior than the right side.

• AUTOMATIC PROCESSES	• CONTROLLED PROCESSES
• riding a bike	• learning to ride a bike
• driving a car	• driving a car in a snowstorm
• being with a close friend	• learning to know a new person
• Features:	• Features:
• fast, effective	• slow, ineffective
• problem specific	• problem general
• little effort	• effortful
• low flexibility	• high flexibility

Interestingly, in a review we have found that brand loyalty – the repeated purchase of products from the same manufacturer repeatedly rather than from other suppliers5 – is related to stronger engagement of the basal ganglia (Plassmann 2012). Although further studies are needed, this may suggests that brand loyalty may be tightly connected to habitual behaviour, rather than overt, conscious deliberation at all times.

What are automatic behaviours in relation to consumer choice? Here, we can propose the following division of behaviours:

- **Instinctual behaviour** – certain aspects of human choice are driven by human instincts, the prime example being human caregiving of infants, which drive caregiving behaviours and therefore purchases (Cryer 1997). Another instinctual trait is how women's natural ovarian cycle affects male preference and clothing choices (Jones; Farage 2008; Jones 2008; Havlicek 2005).
- **Habits and automatic behaviours** – repeated purchase and use of same products, ranging from choice of milk brand in the store (product purchase), to riding a bike (product use).
- **Evaluation and labelling** – the brain responds emotionally early on to events and creates emotional "tags" or "labels" that guide further attention and behaviour (Bechara 2000; Northoff 2006; Bechara 1999; Knutson 2007).

Together, this suggests that the counterpart to conscious processing – unconscious processing – has a deep impact on our everyday lives, and may even be driving a large part of our everyday behaviours.

So what, then, is consciousness for? This is a question with deep roots in philosophy that has yet to be resolved (Seth 2009), and while some scholars have suggested that the element of *experience* should be dissociated from any *function* that consciousness has (Block 1996; Block 2005), other scholars have argued that the experience and function elements of consciousness are two sides of the same phenomenon (Cohen 2011). Here, although not believing that the debate is resolved, we will take the stance that our experience of the world ("phenomenal consciousness") is just another element of the functional role that consciousness has in our brains and on our behaviours ("access consciousness").

Advantages of conscious processing

Consciousness research during the past several decades has demonstrated several significant cognitive differences between conscious and unconscious processing. One of the main findings emerging from this research is that unconscious processes exhibit a limited and rigid information processing capacity, make use of specialised processors, and exert little effect on overall thinking and action (Baars 1993; Shanahan 2005; Baars 2013; Baars 1997; Baars 2002; Baars 2003).

By contrast, conscious processes are associated with a large and flexible information processing capacity, make use of dynamic processors, and exert a much larger effect on thinking and behaviour (Baars 2002). In other words, whether a stimulus is consciously seen or not has a tremendous influence on the effects it can have on thinking and behaviour. Although there is robust evidence showing that subliminal perception – stimuli processed by the sensory system but not consciously experienced – can influence thoughts, feelings and actions (Friese 2008; Fitzsimons 2008; Chartrand 2008; Fitzsimons 2002) – consciously perceived information enables a much stronger effect on flexible and creative behaviours (Shanahan 2005).

As a consequence, stimuli that reach consciousness are more likely to affect behaviours relevant to consumer decisions, including memory, preference and decision-making. In the informationally crowded environment of today's product and media landscapes, only a fraction of the stimuli being processed by the perceptual system gains full access to consciousness and the benefits of such processing. Therefore, understanding the factors and mechanisms responsible for whether or not a given stimulus gets access to consciousness is of great value for academic consumer researchers as well as for marketers.

Preference, attention and selection to consciousness

Imagine that you are walking in a store and you are looking for a new pair of shoes. What brands and products are you most likely to see first? This was a question that my colleague Martin Skov and I asked, studied and reported (Ramsøy 2014). Prior studies have demonstrated that negative stimuli can make their way to consciousness easier than neutral events. This makes good evolutionary sense, as threatening and other dangerous events need to be dealt with as effectively as possible.

However, what about positive stimuli – are they not more likely to be seen? From an evolutionary perspective, it would indeed make good sense that a piece of meat was easier to detect than a pile of dirt. Similarly, could it be possible that associated preferences could be harder or easier to spot? In the study we

Figure Brand preference affects conscious detection. In a recent study we demonstrated that detection of brand names that were presented very briefly and with pre- and post-masking of the words (left), was highly influenced by people's preference for the brands. Overall, both liked and disliked brands were more likely to be seen (a), while for different durations this effect showed an interesting effect (b): at very brief durations brands that were disliked were more likely to be seen, while at middle and long durations, liked brands were more likely to be seen.

tested this by showing brand names very briefly, and found that brand preference indeed had an impact on whether the brand was seen or not. The more you disliked or liked a brand, the more likely it was that you detected the brand, regardless of how well you knew the brand.

This study suggests that our conscious experience of things in our surroundings as consumers can be highly affected by an evaluation process preceding even our experience of the world. Why did you see just that Nike shoe in the store, or spot that McDonald's on the street? Chances are that your brain has made a rapid value computation that has affected your visual attention and increased the likelihood that you will see the product, and even rising the chances that you will inspect it more and eventually buy it. In this way, brands imbue value to our experiences long before we are aware of having such experiences.

Optimal effects of advertising – conscious or unconscious?

There is an ongoing interest in the role of consciousness, communication effects and consumer choice. Indeed, one may ask the question about whether communication effects, such as advertising, has it's strongest effect when we are *not* aware of it, and only partially pay attention to it. This point has been raised by thinkers and practitioners, such as Robert Heath, in his great book "Seducing the subconscious" (Heath 2012), in which he argues that the optimal effect of advertising is not when we are fully conscious of the ad, due to a set of "natural defences" that come up when we are conscious about an ad, as opposed to advertising that we are now fully aware of, and thus where our defences are

lower. As we are seeing throughout this book, this suggestion is highly likely to be true, and it suggests that consumer behaviour is not at all a serial, thoughtful, conscious process in which we can ask people about their preferences and expected purchases.

COSTLY SENTIENCE

The body has an energy budget. It cannot spend enormous amounts of energy without taking in at least the same amount of energy. In general, there is a tremendous variation in how much a body spends in energy per day, but an approximation is about 2.900 kcal for a man, and 2.100 kcal for a woman.

The brain takes up only around 2% of the body's overall weight but is known to consume around 20% of the body's energy. Thus, the brain is in all thinkable ways an energy cost to the body. The energy budget of the brain is negative: the more we spend our brain in thinking hard about something, the more energy we use! The energy budget of the brain has been documented and demonstrated by leading scholars such as Marcus Raichle (Raichle 2007; Raichle 2002; Raichle 2006; He 2009; Raichle 2010), who has argued that the brain may also have a kind of unconscious "default mode", leading to a widespread study of the brain during rest (Esposito 2006; Raichle 2007; Chen 2008; Raichle 2001; Gusnard 2001; Greicius 2003; Harrison 2008).

Figure Consciousness is costly to the brain and body. Compared to unconscious processing, becoming conscious of something is associated with a global activation of the brain. From (Christensen 2006)

Less conscious, less costly

Thus, the cost of being conscious may be highly related to why unconscious functions are so important. Consciousness seems to produce demands on our energy expenditure, and if we were to be conscious about every single step in our processing of information (leaving the physical impossibility of this aside) then we would soon use more energy than the body could use.

Unconscious and automatic behaviours use much less brain activation, and by this token, we could suggest that auto-pilot behaviours are less resource demanding and there fore energy efficient than conscious

$$I^{(C,U)} = \left(\frac{\eta^C}{E^C} \right) - \left(\frac{\eta^U}{E^U} \right)$$

Figure A simplified model of cognition and energy budget. The choice of processing information (I) consciously (C) or unconsciously (U) depends on the relative energy expenditure (E) of the activity and the efficiency (η) that this activity has had. If the final number is positive, conscious processing will be most optimal; if the number is negative, unconscious processing is most optimal.

processing. However, this is *only* true if the automatic behaviour can solve the problem at hand effectively.

A tentative model can be set up to capture this phenomenon:

In this model, we want to address the efficacy with which conscious and unconscious processing occurs for a given task. Here, the key concepts are: whether the process is conscious (C) or unconscious (U); the efficiency that the process has in solving a given task (η), meaning whether the task is sufficiently and satisfactorily solved; and the energy expenditure for solving this task (E).

The more efficient the conscious processing is (left side of equation, where η is high and E is low, compared to unconscious processing (right side of equation, which would mean a relatively lower η and E is higher) the more likely it is that conscious processing needs to be employed. Conversely, if a given task can be solved unconsciously, even if η is the same between conscious and unconscious processing, then due to differences in the energy spent, unconscious processing will be a more cost-optimal solution.

Let us take a learning of how to juggle, which has an active learning stage, and a second phase where learning has occurred. During learning, there is no automatic scheme, and unconscious processes are less efficient at solving the task. Conscious processing, by comparison is costly but also has a relatively good performance. In time, however, unconscious proces-

ses become better and even superior at the task, and due to their lower energy expenditure, become more efficient at solving the task. This can be calculated with the following numbers:

1. For the **early learning stage**, unconscious efficacy is low and conscious efficacy is relatively higher. The energy expenditure for conscious and unconscious processing are different due to different levels of neural activation levels. This leads to a positive score, indicating that conscious processing is optimal for solving the task at hand:

$$I^{(C,U)} = \left(\frac{10}{5} \right) - \left(\frac{1}{1} \right) = 1$$

2. For the **acquired skill stage**, unconscious efficacy has become higher due to the generation of automatic motor programs from the first stage. Conscious processing efficacy, as well as the energy expenditures for conscious and unconscious processing, remain the same. However, the generation of unconscious skills leads to a switch in which unconscious processing becomes the optimal solution, as indicated by the negative score:

$$I^{(C,U)} = \left(\frac{10}{5} \right) - \left(\frac{3}{1} \right) = -1$$

This model is not prescriptive, in the sense that this is what the brain has to do. Rather, it is a model of how the brain processes information consciously vs unconsciously, and the ramifications this can have on processing success and behaviour. As we know, auto pilot behaviours are not always appropriate. In such cases the model could include a loop function, in which learning induced changes in how the brain uses the energy to solve a task, but also at what times conscious processing should be selected over unconscious processing.

Forgetting mind wandering...?

A recent post at Neuroskeptic discusses whether neuroimaging studies may provide a misleading picture of the brain. The issue is made relevant due to recent studies that demonstrate that for simple tasks, the brains were more or less globally active:

> "Both studies found that pretty much the whole brain "lit up" when people are doing simple tasks. In one case it was seeing videos of people's faces, in the other it was deciding whether stimuli on the screen were letters or numbers."

The big surprise – should we take their word for granted – is that the whole brain is active whenever people do these simple tasks, and that it most likely only can be found when looking at a lot of people (most studies use around 20 people in fMRI studies).

There are several problems with this "big problem", and just to name a few:

- **Task unrelated images and thoughts (TUITs) and Mind wandering:** since the 60s and 70s, psychology has studied what happens whenever people are relaxing, or doing very repetitive tasks. These studies uncovered that these states were highly active, not "passive" in any sense. This fact seems to have been forgotten in so many studies on the brain's "default mode" and "resting state", which surprisingly has uncovered increased activation in a number of widespread brain regions for "less" active tasks. Thus, having your subjects doing a highly repetitive – and even very boring – task is related to mind wandering. That such an active state would produce large-scale activation throughout the brain should come as no surprise.
- **The conscious brain:** being conscious about something seems to be related to large-scale "global" activation in the brain, including the parietal, prefrontal, temporal cortices along with structures such as the thalamus. Should we be surprised that such regions are largely activated when 1.000+ subjects are scanned while conscious?
- **Individual differences:** yes, even large individual differences between subjects may – when you are testing 1.000+ people – provide the false impression of a general large-scale activation of the brain "in all people"

So I don't buy it: I think we can trust the fMRI data we have thus far. There are many challenges in using these measures, and many studies fall prey to a lot of the validity, reliability and sanity checks one can (and should) apply. But the purported problem by Neuroskeptic is, IMO, misfiring.

SUBLIMINAL PERCEPTION

It's an urban legend, but yet it bears truth to it! Subliminal messages can affect thought, emotions and choice. As noted earlier, in 1957 James Vicary reported that he had successfully affected 45,699 movie goers to buy more popcorn. He claimed that he had inserted brief flashes of the text "Eat popcorn" and "Drink Coca-Cola" into the regular movie, and observed a tremendous (57.5%) increase in the sales of popcorn after the movie. The Coca-Cola sales effect was less impressive (18.1%) but still statistically significant. The story soon got press and fame, and Vicary was figured in numerous talks, seminars, and even political debates. The unconscious persuaders soon became a theme, both every advertiser's wet dream and every consumer's great fear.

Only a few years later, Vicary admitted that the story was made up. He never actually did the popcorn "experiment", but was inspired by contemporary talks about the whole topic, and even led to the urban legend of hidden cinema ads. This was also later referred to in movies such as "Fight club", in which the main character inserted snips of pornographic images into children's cartoons which led the children to cry during the movies.

Subliminal perception itself is still a robust phenomenon. In a very early study by Sidis (1898), subjects were shown small cardboard cards, each containing a single printed letter or digit. The distance between the person and the cards was such that the they often complained that all they could see on each card was a dim, blurred spot or nothing at all. Based on this, Sidis assumed that the participants were unaware of perceiving either digits or letters. However, when he used a second measure, forced-choice guessing, he discovered that his subjects were able to guess the category of the card (digit or letter). Furthermore, he discovered that the subjects were better than chance at guessing the precise identity of the card. Thus, Sidis uncovered a dissociation between two measures of perception. The subjective, verbal measure from the subjects suggested that they did not 'see' the critical stimuli, while the behavioural measures from forced-choice guessing suggested that the subjects indeed had perceived the stimuli. Hence, the findings also provoked a theoretical discussion about the relationship between perception and consciousness.

Subliminal perception has also been reported in many special cases. For example, Merikle and Daneman (2000) gave patients undergoing general anaesthesia earphones and a tape recording of repetitions of a series of words. After the surgery, the patients were presented word stems such as 'gui' or 'pro' and asked to complete these stems to produce a common English word. While these word stems have many possible completions, patients more often used stems of words presented during general anaesthesia (e.g., 'guide' and 'proud'). Merikle and Daneman concluded that "memory for specific stimuli presented during anaesthesia shows that information is at times perceived without any awareness of perceiving during general anaesthesia" (p. 498).

Subliminal effects are real

After Vicary, the phenomenon now known as *subliminal advertising* became increasingly used by advertisers and popular science alike in arguing that the human mind can be affected without awareness. But since Vicary's admitting the fraud, the pendulum swung the other way, where people believed that subliminal perception did not exist at all! However, as we will see, this is not true, either.

Subliminal perception is a term that covers where a stimulus (word, image etc) is presented below the threshold for conscious detection, but still has an effect on conscious thoughts, feelings or actions. A few notable examples can be mentioned here:

- Going to the idea of popcorn in movies, a study has recently shown that eating popcorn (or chewing gum) during movie watching can *reduce* the effect of advertising (Topolinski 2014). Although not being strictly a subliminal stimulus, this study suggests that a less attended stimulus or behaviour can have an impact on communication effects.
- In a study by Mathias Pessiglione and colleagues, it was demonstrated that people could learn to associate an abstract symbol with an outcome (win or lose) and act accordingly, without any conscious awareness of such a relationship (Pessiglione 2008). This suggests that decision making and risky choices can be affected by subliminal cues without our awareness.
- Being subliminally exposed to stimuli makes us rate these stimuli as more likeable, a phenomenon known as the *mere exposure effect* (Monahan 2000).

These studies demonstrate that subliminal effects are in fact real. Information that is processed unconsciously can nevertheless impact our thoughts and choices. But from this, we still know little about the effects that brands may have on our preference when they are only perceived unconsciously. There has a need for marrying findings from subliminal perception with the studies of brand effects on preference and choice.

Unconscious framing of choice

To alleviate this, my PhD student Dalia Bagdziunaite and I did a small but impactful study that was presented at the annual convention for the 2014 Association for Psychological Science in San Francisco (download PDF from here). In the study, 30 women were subliminally exposed to brands prior to evaluating fashion clothing. What we found was that women's own brand preferences impacted on their ratings of the clothing – despite that they had not seen the brands consciously.

By using eye-tracking running to assess arousal (pupil dilation) during the test, we found that brand preference was related to changes in arousal already during subliminal brand exposure. Pupil dilation showed a non-linear response pattern, in which pupil dilation was strong for both extremes of brand liking. In other words, when subliminally exposed to brands they either loved or despised, women were aroused. This effect was carried over and even exaggerated during the rating phase, where they evaluated the clothing. This study demonstrates that brand equity – a term we will return to – can trigger emotional responses even for completely unconscious exposure, and then carry on to affect the preference and rating of products.

Subliminal problem solving

Have you ever pondered about a problem but come short of a solution, only to have the solution "pop into your mind" when you were doing something completely different? Or how about remembering a person's name, only when it was too late? Research on problem solving has long recognised the fact that unconscious processes can be "wise", to the extent that they can process information in a meaningful way that can yield answers to problems that you would otherwise think that only the conscious mind could deal with (Dorfman 1996; Simon 1977; Smith 1991).

Problem solving, while not being subliminal perception, is a testament to how processes and thoughts

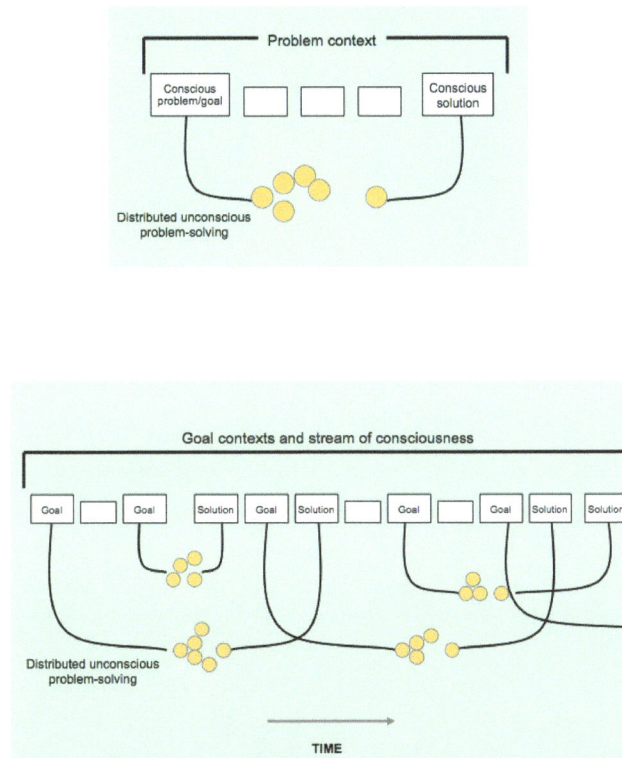

Figure Subliminal operation of problem solving. TOP: In a single case, a problem can be consciously defined and thought about until a certain point, but may be disturbed by other more pressing problems, after which the ongoing problem solving can be left to a more distributed unconscious processing. Nevertheless, this process can be productive and eventually lead to a "popup" of a solution for the problem at hand. BOTTOM: In a more real-life setting, problem solving occurs at many multiple and parallel levels. One problem might be solved directly and immediately, while other problems reside unresolved and may pop out as solutions after subliminal processing. From Baars & Ramsøy (in progress) "Is there a resting state in the brain?"

that the conscious mind is not privy to, can nevertheless go on, relatively undisturbed, often until a solution comes up, and often at the most expected moment. Niels Bohr was famously known to walk or work in the garden when a tough problem occurred to him, only to find the problem solved when he reached the desk again; Albert Einstein took breaks from his fanciful streams of thought to play the violin or take long walks; Charles Darwin took walks in the garden to distract this thinking about evolution, even ranking his problems as one-stone, two-stone and three-stone problems as he put a stone on a pile as he reached the same point in the garden.

As part of an attempt to better understand how unconscious processes can affect problem solving, one of my in-progress papers illustrates just how exactly such processes can occur. Problems need to be consciously defined, but can then take on a completely unconscious life while the person is occupied with other, more pressing matters. The ongoing and persevering unconscious processing eventually leads to solutions that, once reached, will pop into your mind. So often, we know of or hear about the great ideas coming right at the brink

of falling asleep, or in the shower, or being in some other strange and completely unrelated situation.

SENSE OF AGENCY

Do you feel in control? Sure, you probably think – sure I'm in control. Why are you asking? It turns out that our feeling of being in control – also called the *sense of agency* – is not a guarantee for us being in control! Rather, while the multiple brain regions implicated in our sense of agency are highly engaged (Farrer 2008a; David 2008; Kircher 2003; Moore 2009; Farrer 2003; Rao 2008), the actual underlying drivers of our choice may lie at completely different places.

Our feeling of being in control can be an illusion. For example, in a recent study my colleagues and I[6] reported the findings from a study of the experience of control in financial transactions. Here, participants were asked to first make a series of choices of buying and selling a stock while the stock price was fluctuating. Unbeknownst to the participants, some trials were rigged so that upon stock purchase the price would go either up or down, leading to a net win or loss to the

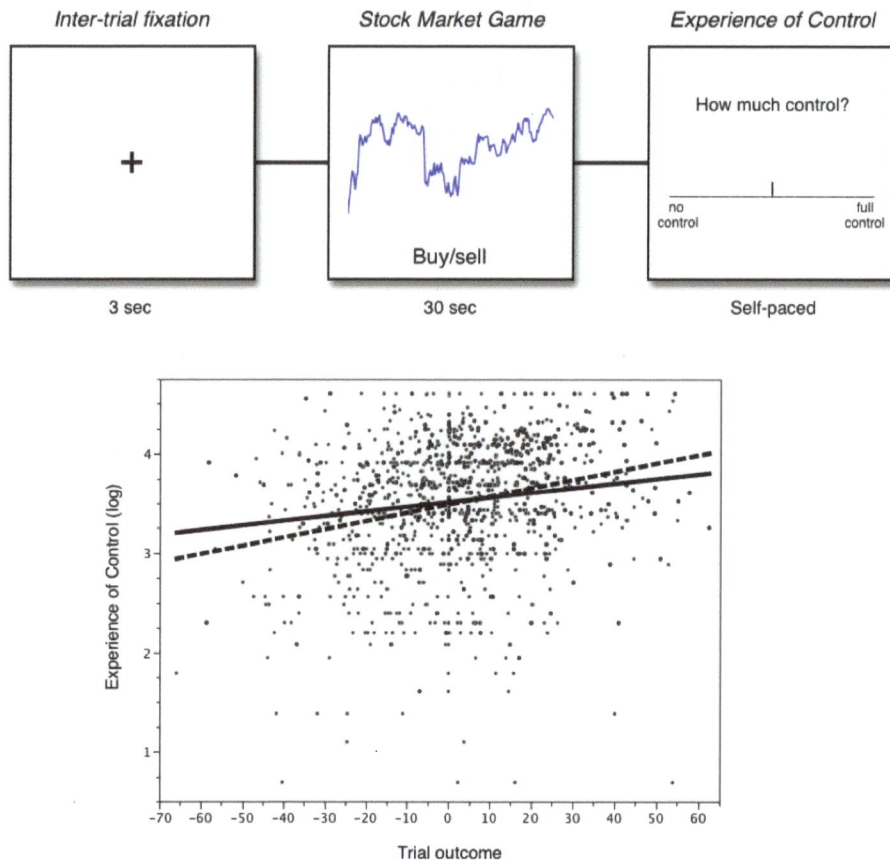

Figure The experiment on choice and experience of control. TOP: participants first saw a fixation cross, after which they were asked to make multiple choices to buy and sell stock while the stock price fluctuated. After each trial they were asked to rate, among other things, their experience of control over the outcome. BOTTOM: For non-manipulated trials (solid black line) we found a positive relationship between outcome and self-reported experience of control, while for manipulated trials (dotted black line), this function was even stronger, shown as a steeper linear function.

participant. After each trial, participants reported on a number of different measures, including how much they felt like being in control over the end result.

What we found was very exciting: first, when taking the manipulation into account, we found that compared to neutral outcome, stock market wins were associated with a higher self-reported control, while losses were associated with lower self-reported control. This finding confirms prior results, but the exciting finding was when we looked at the manipulation. Here, we found that in cases where the game was rigged so that participants would win, they reported even larger experiences of control. Conversely, when the game was rigged so that their choice led to a definite loss, they reported even lower scores of self-reported control. Interestingly, none of the participants ever detected the manipulation of the game.

This finding suggests several things about our experience of control, or agency, when making choices:

- Our sense of agency depends on the outcome of our choices
- When we are in fact in control, this effect of outcome is even stronger, even though we do not notice the added control

The illusion of control is a profound phenomenon in consumer behaviour. Just take the example of Pessiglione and colleagues (Pessiglione 2008) mentioned earlier in this chapter. Here, it was found that people's risky choice depended on subliminal cues. This means that participants only experienced their choices as "random guesses", while in fact their subliminal mind had learned to associated certain unconscious cues with outcomes. Here, the experience of control was none, while true control was high. Indeed, the experience of control and the mind's actual control is dissociated, and we cannot take a person's word for feeling in control or not as a token of them actually being in control.

CHAPTER **6**

Emotions & Feelings

How do you feel right now? You probably have some kind of experience about the state you are in, something that puts a flavour to your experiences. It can be positive, like joy; or negative, like sadness; or even neutral, like the feeling of dullness. But how do you get these emotions? Emotions and feelings may to many seem like two words that describe the same. But here, we will demonstrate that they are in fact two very different phenomena, where one depends on the other. Here, let us start off by making the first definitions of the two terms:

- **EMOTIONS** – an organism's expression of an inner/bodily state; a bodily response to an event with a mechanical, stimulus-response basis. It is typically occurring before or without consciousness.
- **FEELINGS** – an organism's (person) experience of being in a certain emotional state. It is always associated with consciousness / experience. It is "introspective", in the sense that it is something we can look at and explore, as in "Today I woke up feeling happy".

From this definition, we can say that the distinction between emotions and feelings cannot be more profound. One is conscious and related to our experience of the world, the other is more or less completely unconscious. How can we know this?

Take the example of watching a horror movie. You are seeing the movie at home, feet crumbling up under your blanket, you feel anxious, and you jump when you are surprised in the movie. At the same time, your body shows a whole range of symptoms of stress: your pupil dilates, your heart rate and respiration goes up, your digestion system stops momentarily, your palms start sweating, and much more. In fact, even before you feel anxious and you jump in your seat, your brain responses has produced these bodily changes.

The example shows you what the distinction between emotions and feelings are: your feeling of being frightful is accompanied (and preceded) by the bodily

Figure Watching horror movies are associated with significant changes in your brain and body. Anything from the social and behavioural changes such as postural changes and facial expressions, to physiological changes such as increased respiration and pulse are clear signals of increased arousal and fear responses. Image courtesy of blackstock / FreeDigitalPhotos.net

responses. At the time you experience fear, your brain has already determined that something is important and dangerous, leading to a cascade of events in your brain and body.

Stupid emotions?

Why do you get fearful when watching a horror movie? In reality, it does not really make any sense: you are sitting in your safe home, watching at flickering lights from a box. What's to fear about that?

Let's face it: today, emotions can look pretty stupid. We get frightened when watching a movie, we can fear small and insignificant spiders or mice, and get angry when our sports team does not win. At the positive end of the spectrum, emotions can also drive unwanted behaviours, such as playing too much Candy Crush or Flappy Bird, eating too much or even lead to substance abuse. How does this all happen?

We can say that our brains are not perfectly wired for a modern life. We still live with brains that have evolved to a life in the woods or at the savannah; approaching calorie rich food whenever it was available, or fleeing from a predator. In these situations, our emotions have been nothing like stupid: they have been powerful survival engines that have helped us (and our ancestors) survive!

But today, many places around the world have an abundance of food, and there are rarely any dangerous animals around (except that pit bull terrier down the road). At the same time, we see an increasing pace with which we are exposed to new inventions that our brains have not been equipped to deal with. The abstract sum of money, now just a number on a bank account, is one such example, but the many sources of digital interfaces – from smartphones to the web – are so new we have to rely on our existing neurobiological architecture to deal with them.

So, our emotions are not "stupid" – they are the only responses we have available, that have originally evolved to deal with highly specific real-life situations. Nevertheless, these emotions are what drive our cognitive and behavioural responses. As a central theme throughout this book, emotions drive consumer behaviour – for good or bad – and in under search for understanding consumer choice, we need to understand what emotions and feelings are, and why they are there in the first place.

The term "emotions" stems from the latin word "emovere", which is closely linked to the term "to move". This is very fitting: emotions are indeed very closely linked to action and movement. The stronger an emotion is, the more likely it is that an organism will be acting on this emotion, whether it means approaching an delicious meal or fleeing from a predator. Emotions are closely linked to *meaningful survival behaviours*. It is no surprise, therefore, that emotions are not uniquely human – they are shared throughout

Figure   Eating candy and calorie rich food – or even just watching others do it – is associated with increased engagement of the brain's reward networks. Image courtesy of blackstock / FreeDigitalPhotos.net

much of the animal kingdom. Rats display freezing behaviours upon stress and fear; doves can learn to associate a symbol with a reward; sea slugs can learn to fear an otherwise meaningless stimulus; macaque monkeys show increased emotional response to distress calls from their conspecifics.

Being hedonic about it

At its core the brain's evaluation system – regardless of whether it is conscious or unconscious – is driven by a simple rule:

Maximise reward, minimise pain

However you twist and turn it, our valuation systems are driven by this plain insight. For an organism to survive, it must avoid the dangers in life and approach and approach items and situations that increase the likelihood that it will survive. This goes from the avoidance of unexpected noises to the approach towards specific social situations.

The brain – just as the body proper – is a "hedonic machine". Organisms are always driven to maximise reward and minimise pain. But this does not necessarily mean that we are always driven by immediate results. Rather, we should think about rewards and pains as something that can occur either immediately or delayed. Let's look at this in a figure related only to healthy choices:

	INSTANT	DELAYED
REWARD	Eat chocolate now	Keeping healthy
PAIN	Physical exercise	Poor health

From this table, it is clear that rewards and pains can have at least two dimensions. You may be very attracted to eating a chocolate (an instant reward) but over time, you also have see the reward of staying healthy and not gaining weight (a delayed reward). Conversely, you may dread the hard training (an instant pain) but can also see that if you do not do this your overall fitness and health may be worsened (delayed pain).

Interestingly, emotions and feelings are connected in complex ways to this division. While on the one hand, emotions are tightly connected to the early reward and pain operations associated with instant judgments, feelings seem to be connected to both the instant feelings of reward and dread, as well as the thinking about delayed outcomes.

The incomplete model of hedonic choice

As we will see in the chapter about wanting and liking, motivation does indeed have two distinct systems in the brain that operate in parallel, and sometimes in conflict.

The incomplete model of hedonic choice

In our description of a consumer neuroscience model of choice and branding effects, we have referred to the model by Plassmann, Ramsøy & Milosavljevic (2012). However, upon closer inspection of this model, we can see that it does not provide a clear answer to the topic of emotions and feelings. Here, let us focus on the aspect of predicted value and experienced value, as highlighted in the model:

If you talk to a person at a restaurant who has just ordered her favourite meal, she will probably say that she is looking forward to her meal. She is explicitly stating her *predicted value* of the meal. Upon receiving the meal, she may express dismay about the meal, and thus stating her *experienced value* of the meal. These are both examples that the model is all about.

But what if at least one of the components have an unconscious side? What if predicted value indeed could also be unconscious? Here, we can recall the study by Pessiglione and colleagues (Pessiglione 2008), who found that despite their participants stating that they only guessed – a conscious predicted value – they nevertheless learned to act upon subliminal cues to increase their betting. This suggests that predictive value also has an unconscious side. Your brain can actually predict what will happen, even though you are not consciously aware of this prediction. Of course, our experienced value is closely related to our conscious

brain processes, but one may still say that these experiences are caused by processes we are not privy to (as we saw in previous chapters).

In any case, this suggests that our consumer neuroscience model should be modified. We should not only talk about predicted value in one sense, but in two senses – an unconscious and a conscious aspect. While this does not change our model fundamentally, it provides a crucial nuance that we will dig deeper into in later chapters.

THE THERMOMETER MODEL OF EMOTIONS – FEELINGS

Can we have emotions without feelings? Yes
Can we have feelings without emotions? No

It's that simple, really. But for us to expand our understanding of this relationship a bit, let us use the analogy of a thermometer. Think of this as the "Emotional Thermometer". Increases and decreases are related to the relative engagement and disengagement of the emotion system. The higher the score on the Emotional Thermometer, the stronger the emotions:

The actual point at which emotions also trigger feelings is unknown, and as we saw in the discussion about changes from unconscious to conscious processing, it is likely that the change is not an either-or, but rather a more smooth transition. Have you ever had an eerie feeling that was something was not quite right? That could serve as an example of a "fringe feeling", where emotions are working up the Emotional Thermometer, but not quite crossing the point at which you have a fully fledged conscious feeling.

The Emotional Thermometer model is good for illustrating the relationship between emotions and

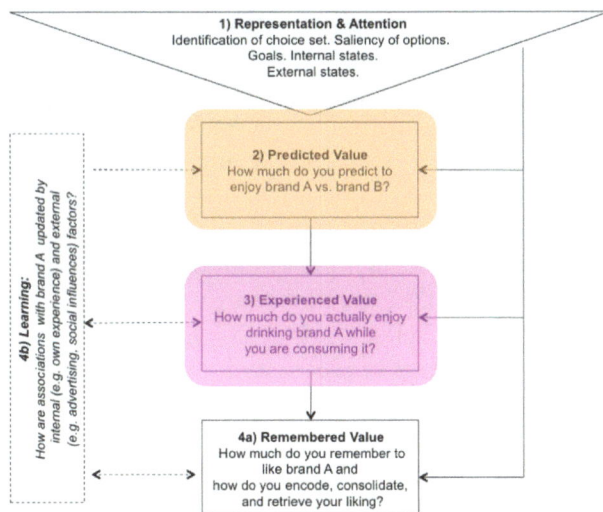

Figure Highlighting predicted and experienced value in the consumer neuroscience model. How much is conscious and how much is unconscious? Adapted from (Plassmann 2012)

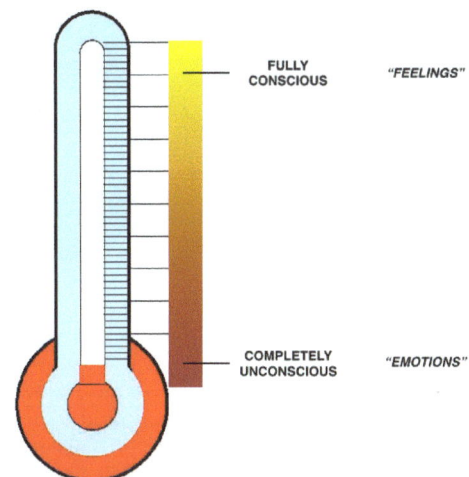

Figure The Emotional Thermometer. At some level of activation, emotional engagement becomes so strong that it is also accompanied by conscious experiences, i.e., feelings.

feelings. Put simply, you cannot have feelings without emotions, but you *can* have emotions without feelings. This means that your brain (and body) can indeed show a change in response pattern to a particular event, without you noticing it consciously. Moreover, these changes can affect your perception and action, as we saw in how brand preference altered their likelihood of being seen (Ramsøy 2014) and in how subliminal cues can affect risky choice without awareness of such effects (Pessiglione 2008).

Emotions, action, feelings!

Rene Descartes, the famous Renaissance philosopher, described how the body would respond to a burn, and how you experienced the burn. He correctly identified the main mechanism with which the burn could trigger a response and a feeling – namely via the brain and spinal cord – although his invoking of the human soul through the pituitary gland is not supported by today's neuroscience.

Nevertheless, think about the way we respond to being burned or stung by a bee. If it happens to our hand, we immediately withdraw the hand from the pain, and at the same time experience the pain. We do not first experience the pain, then think "I should probably remove my hand", and then remove our hand... This small thought example suggests that at the time we have the painful sensation of being burned, our brains have already made the decision to move our limbs.

Similarly, in our example of watching a horror movie, our experience of fear is embedded in a host of bodily responses of sweating, trembling, increased pulse and respiration. Again, this demonstrates that our emotions *precede* our conscious feelings about an event.

In the above model, we can see how emotions now seem to simultaneously lead to action and feelings.

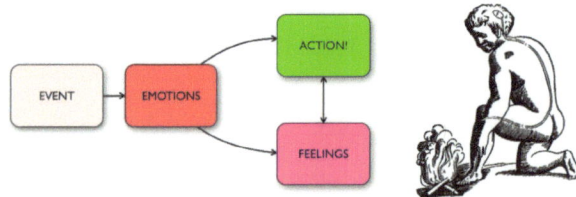

Figure A tentative model of emotions and feelings. LEFT: An event leads to an immediate emotional response that "labels" the response and initiates action. Conscious feelings typically occur at the time that emotions have already lead to actions, and although they can modulate actions (e.g., inhibit ongoing behavioural tendencies), they are not the core cause of these actions. RIGHT: The original drawing from Rene Descartes' thinking about how the brain received signals from the external peripheral senses, leading to both the withdrawal from the heat and the sensation of being burned.

When you watch a horror movie, you are predisposed to take actions to reduce the stress, such as hiding (from the TV...). As we saw with the Emotional Thermometer model, emotions do not necessarily lead to feelings, but can still drive our actions.

DIMENSIONS OF EMOTIONS

There are two well established dimensions in emotions and feelings, pertaining to the strength and direction of emotional responses. Here, we will call them *arousal* and *valence*, which is what a major part of the scientific literature is also doing (Olofsson 2008; Kousta 2009; Nasrallah 2009; Gorn 2001; Wright 2003; Dolcos 2004; MickleySteinmetz 2010; WilsonMendenhall 2013; Kron 2013; Bradley 2008; Hamann 2012). Here, the terms will mean the following:

- **Arousal** – Bodily responses of general excitement to "relevant" cues, ranging from low to high.
- **Valence** – Emotional responses indicating the relative hedonic evaluation, ranging from positive – neutral – negative.

Crucially, all emotions must be ranked according to both dimensions, and knowing only one dimension is always insufficient to understand people's emotional responses. For example, knowing that someone has a high arousal is often not indicative of whether that emotion is positive or negative. Similarly, although you may be able to demonstrate a positive valence, the strength of this response is unknown without the arousal measure.

Arousal – from low to high

We have all felt the bodily responses to anxiety – from the stress prior to an exam to how we feel speaking in front of a crowd. The bodily responses are easy to recognise: increased pulse and respiration, dryness of mouth, sweating in palms, and mental blocking.

Arousal is a key function in our evolutionary past: it has allowed the brain to engage bodily functions that can improve our ability to respond to the situation at hand. Dilation of the pupils is related to increased intake of light and better visual acuity; increase in heart rate and respiration is related to better responses to threats such as "flight or fight" responses; sweat in the palms is related to improved grip, and so on. Emotional responses are there to serve a purpose, increasing an organism's overall survival chances. Interestingly, although consumer behaviour rarely if ever relies on "fight or flight" responses, we nevertheless rely on the same basic mechanisms when making choices. We do not have a "shopper brain centre", or an "insurance decision region". To solve the tasks of dealing with money, choosing products and services, consumer choices rely on an existing set of brain regions that were originally evolved for more primitive but real-life situations.

Arousal measures have a drawback – they are bivalent. This basically means that arousal can be high for both positive and negative events. This leads to an *inference problem*: we cannot use arousal as the only index of emotions! For example, if you measured someone's sweat response, and you concluded that the person showed an increased arousal, you could **not** at the same time conclude that the arousal response was positive or negative. A strong arousal response to an ad can mean that the person enjoys the ad, but it can also mean that the person becomes very sad or annoyed by the ad. From the arousal measure, you are unable to tell which from which.

This suggests that arousal can tell us something about the *amplitude* and strength of an emotional response, but not the direction of the emotional response. The arousal response can in many ways be seen as a relevancy response: the stronger the change in arousal, the more relevant it is to the organism.

The aspect of arousal is nicely covered in the study by Andrea Groeppel-Klein entitled "Arousal and consumer in-store behavior" (GroeppelKlein 2005), where she demonstrated how arousal at the point to sale (POS) could inform our understanding of arousal and consumer choice. By running three consecutive studies while participants' arousal responses were recorded with GSR, she made the following observations:

- An experimental store with a richer store environment, including multiple products and sales materials, lead to a higher overall arousal compared to a less stimulating control store
- Increased in-store arousal was associated with subsequent reports of "joy"
- When comparing those that buy products with non-buyers, buyers demonstrate a higher POS arousal

Taken together, the study by Groeppel-Klein demonstrates that in-store arousal is highly related to actual choice. The study furthermore clearly demonstrates that measures of arousal – even in challenging conditions such as a mobile setting – can provide novel insights into how consumers make choices and how particular emotional responses (both to the store environment and to particular POS situations) can drive their choices. Similarly, although not addressed by Groeppel-Klein, a high arousal can be related to ne-

gative emotions such as frustration. Here, no studies to date have focused on in-store arousal for negative emotions.

Valence – from negative to positive

We have a second emotional dimension that is about the direction of the response. Here, emotional valence is a measure of whether a stimulus is evaluated as positive or negative. This covers the spectrum from strong negative fear responses to extreme positive experiences of euphoria.

As with arousal, valence is one-dimensional and can only point to the direction – from positive to negative – of an emotion. What it does not tell us is the amplitude of this emotion. However, as we will see, arousal and valence are not completely orthogonal, but do display a systematic relationship.

As an exercise, try to mention as many emotions you can, and range them on a line from maximally negative to maximally positive. How many emotions do you get, and what are the extremes? Any "dull emotions" too? At the same time, try to notice what actually happens and how you feel when thinking about these items – in particular, how does your body respond? Take a highly aversive event, such as thinking about breaking your leg – it *feels* aversive, and your bodily responses can range from tension and facial expression, to increased sweating and pulse.

Do we have a valid valence measure? Besides self reports, it has proven harder to find a crystal clear brain index of valence, but one region seems to be highly relevant: the medial orbitofrontal cortex (Sescousse 2010). That is, the brain region just above the middle (medial) of your eyes (orbits). Here, several studies have demonstrated that this region is involved in appetitive and aversive goals (Plassmann 2010), the enjoyment of wine (Plassmann 2008a), the pleasure of art (Kirk 2009), as well as the enjoyment of unbranded soft drinks (McClure 2004). As we will see, there is a distinction between feelings and motivation, and even some indices of hedonic pleasure may be closer associated with motivation rather than associated feelings.

Motivation – approaching or avoiding?

While decades of research has focused on finding the brain correlates of subjective pleasure, it was discovered that some metrics rather track what the organism will do. In particular, there is a phenomenon called *frontal asymmetry* that seems highly related to feelings. For more than two decades, researchers have consistently found that higher engagement of the left relative to the right frontal brain is related to positive feelings. Therefore, for a long time, it has been thought that the left frontal hemisphere was responsible for positive feelings (Davidson 2004; Schaffer 1983; Davidson 1992; Davidson 1988; Sutton 2000; Coan 2003).

However, recent evidence suggests that the asymmetric engagement of the frontal lobe is more related to what has been coined *approach-avoidance* behaviour. The key finding was that aggression, a negative emotion but with "approach" properties, was associated with a stronger left than right activation (HarmonJones 2010). If indeed the left-right asymmetry index was related to emotional valence, then aggression should show a stronger right than left activity. If the index was a measure of approach vs avoidance motivation, then it would show stronger left than right engagement. Since the latter was found, the frontal asymmetry index has been coined an index of approach vs avoidance, or a *motivation index*.

The Arousal–Motivation Matrix

As noted, arousal is bivalent – you cannot just read out an arousal measure and then be able to tell whether the emotional response is positive or negative. On the other side, valence and motivation have the good side

of being able to tell the positive–negative direction of the emotion, but not the strength of the response.

When we put these scales together, we see a very interesting phenomenon. It produces a fine model for seeing the relationship between arousal and motivation. For the extreme motivation scores, arousal is high, but for relatively neutral motivation scores, we see a lower arousal. This is illustrated as what we call the Arousal-Motivation Matrix:

In most instances of consumer neuroscience research, we rarely test extremes: we rarely test erotic stimuli, and even more rarely we employ stimuli that are disgusting or horrible. Thus, while the principal function is a neat U-shaped curve (right side of figure above), in most cases we are studying the neutral-to-positive aspect of consumer responses (left side of figure above). Nevertheless, we can see that the most intense motivational responses are related to the highest arousal scores, compared to neutral motivation that produce the lowest arousal scores.

As we will see, the Arousal-Motivation Matrix can provide one of the most important metrics in neuromarketing and consumer neuroscience research. If anything, having and using a brain metric that is related to actual choice may turn out to be a magic bullet both for predicting and understanding consumer choice.

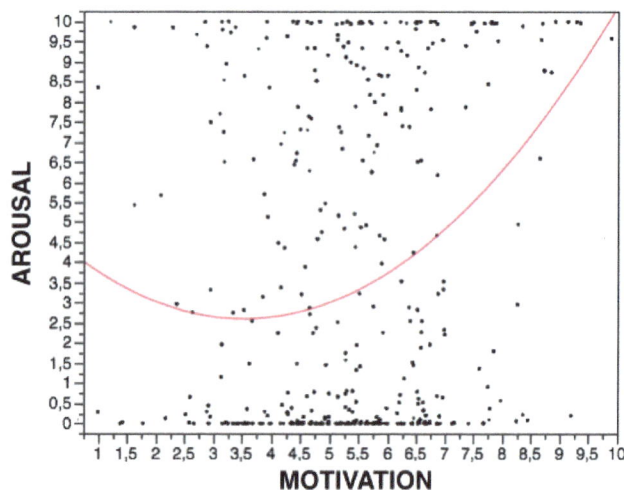

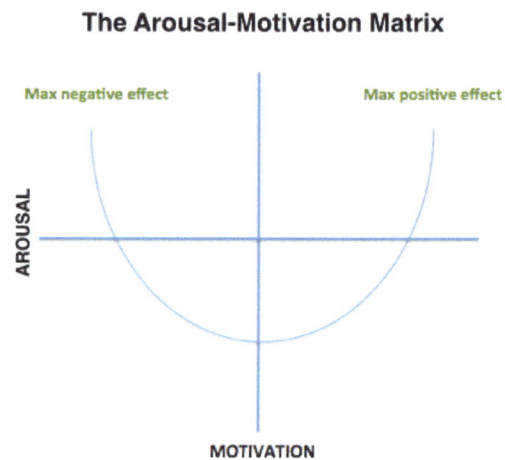

Figure The Arousal-Motivation matrix. In many studies, the typical relationship between emotions and motivation is nonlinear (left), showing higher emotional arousal for strongly positive and negative events (red line), but often with a skew towards the positive emotions. Most often, studies of consumer behaviour are not testing the effects of highly aversive events, which would produce a more balanced model (right). The data used here are relating EEG measures of emotional arousal (y-axis) and motivation (x-axis).

Fear and pleasure in the amygdala

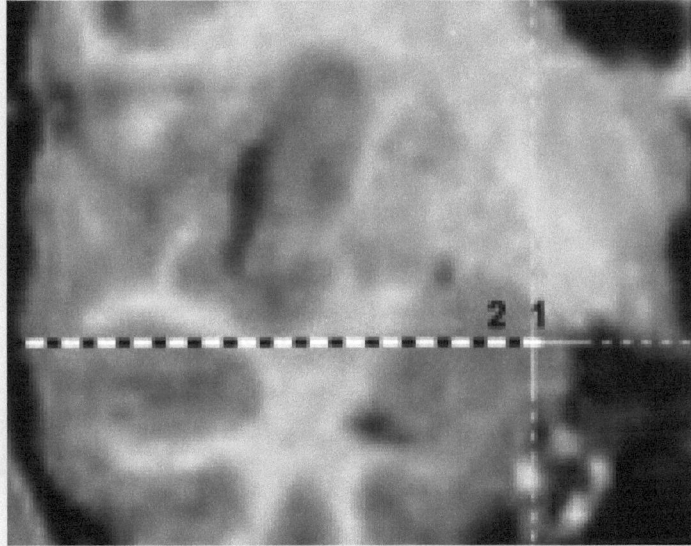

Here is a heads up for a recent study demonstrating – again – that the amygdala is not merely a "fear centre" in the brain. I have previously blogged about the amygdala, first not being a single structure, and that it is not only involved in fear.

In 2007, a team of French researchers demonstrated that direct stimulation of the amygdala did evoke emotional responses, but that there was a difference between which hemisphere was stimulated. Right amygdala stimulations induced aversive responses, in particular fear and sadness. In contrast, left hemisphere stimulation induced either positive (happiness) or negative emotions (fear, anxiety, sadness). As the abstract reads:

> "Very few studies in humans have quantified the effect obtained after direct electrical stimulation of the amygdala, in terms of both emotional and physiological responses. We tested patients with drug-resistant partial epilepsies who were explored with intracerebral electrodes in the setting of presurgical evaluation. We assessed the effects of direct electric stimulations in either the right or the left amygdala on verbally self-reported emotions (Izard scale) and on psychophysiological markers of emotions by recording skin conductance responses (SCRs) and by measuring the electromyographic responses of the corrugator supercilii (EMGc). According to responses on Izard scales, electrical stimulations of the right amygdala induced negative emotions, especially fear and sadness. In contrast, stimulations of the left amygdala were able to induce either pleasant (happiness) or unpleasant (fear, anxiety, sadness) emotions. Unpleasant states induced by electrical stimulations were accompanied by an increase in EMGc activity. In addition, when emotional changes were reported after electrical stimulation, SCR amplitude for the positively valenced emotions was larger than for the negative ones. These findings provide direct in vivo evidence that the human amygdala is involved in emotional experiences and strengthen the hypothesis of a functional asymmetry of the amygdala for valence and arousal processing."

Interestingly, there is more to say about this study. First, it may be that there is a systematic bias introduced by the way the researchers did the study. By using high-frequency (50 Hz) stimulation in 1 second, they might have induced one characteristic response of the amygdala. This structure is often seen as having quick "on-off" responses. Thus, one second pulse trains is actually a long duration for the amygdala. So a pulse of 20 milliseconds could be hypothesised to produce different responses. Also, the researchers found that GSR responses were actually larger for positive emotions, when they were reported. As the amygdala has often been implicated in unconscious emotional responses (mostly aversive responses) one may speculate that the left-hemisphere amygdala involvement in positive emotions may be related to conscious emotions.

As always, new findings leads to numerous novel questions, ideas and hypotheses. Which is why science is so much fun. But it is important to note the change we see today the role of the amygdala in emotional responses. We are moving away from the LeDouxian paradigmatic focus on fear (and some aversion)as the sole emotion of the brain, and more towards a balanced view towards a similar focus on positive emotions and (hopefully) more complex human emotions. Through this development, we can see that novel findings are breaking down the old ideas of neo-phrenology, breaking single structures into smaller parts, and into parts of a larger network of convergence and divergence structures. Keep your eyes open, more is on the way.

EMOTIONAL BRAIN REGIONS

A number of brain regions are implicated in emotions and feelings. In the most traditional sense, we have regions implicated in emotional valence. Regions such as the **amygdala** have long been known to be involved in negative emotions such as fear and stress; the **insula** has been shown to be involved in disgust. Conversely, the medial **orbitofrontal cortex** has been shown to be involved in enjoyment, and the **nucleus accumbens (NAcc)** has been shown to be involved in desire and approach behaviour.

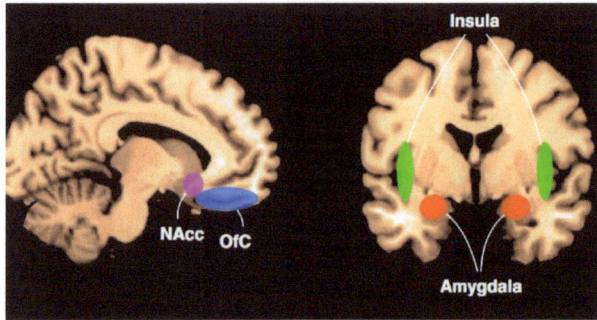

Figure   Brain regions of emotions and feelings. LEFT: medial view of the brain Abbreviations: NAcc = Nucleus Accumbens; OfC = orbitofrontal Cortex

So that's it – case solved: we have brain centres for enjoyment, and brain centres for negative emotions. Right? If this is true, then we can just find reliable indices of those brain regions and *voilà*, we can measure what people feel and will act upon. Unfortunately it is not that simple, and several studies have provided a more nuanced picture of these brain regions. For example, many regions show a bivalent response function, while other regions have shown a differentiated signal within smaller clusters of the same region. This is summarised in the table below:

As the table shows, we have moved from a relatively simple view of correlating one mental function to one brain structure; to a view that encompasses multiple functions. This is interesting for our understanding of the human brain and emotions, but it does not help if we are interested in finding brain regions responsible for only one emotional valence, or motivation.

As we will see in the chapter about wanting a liking, there are brain correlates of actual choice. These are highly related to certain responses of certain brain functions. As we will see, they will allow prediction of consumer choice, even in the absence of conscious feelings of making such choices.

REGION	TRADITIONAL VIEW	NEW FINDINGS	References
Amygdala	Fear, stress, aversion	Negative AND positive emotions	Murray, Elisabeth A. "The Amygdala, Reward and Emotion." Trends Cogn Sci 11, no. 11 (2007): doi: 10.1016/j.tics. 2007.08.013.
Nucleus Accumbens	Reward	Negative AND positive emotions	Levita, Liat, Todd A Hare, Henning U Voss, Gary Glover, Douglas J Ballon, and B J Casey. "The Bivalent Side of the Nucleus Accumbens." Neuroimage 44, no. 3 (2009): doi:10.1016/ j.neuroimage. 2008.09.039.
Orbitofrontal cortex	Hedonic feeling of reward	Differentiated responses: - immediate vs delayed reward - concrete and abstract rewards - positive or negative outcomes	Kringelbach, Morten L. "The Human Orbitofrontal Cortex: Linking Reward to Hedonic Experience." Nature Reviews Neuroscience 6, no. 9 (2005): 691-702.
Insula	Disgust, negative emotions	Positive and negative emotions Is highly engaged in most emotions	Craig, A D Bud. "How Do You Feel--now? The Anterior Insula and Human Awareness." Nat Rev Neurosci 10, no. 1 (2009): doi: 10.1038/nrn2555.

WHAT EMOTIONS "DO"

As noted, emotions are tightly connected to motivation and action. But this link is not necessarily direct, and emotional responses are related to a host of different behaviours. Below, we list four main effects:

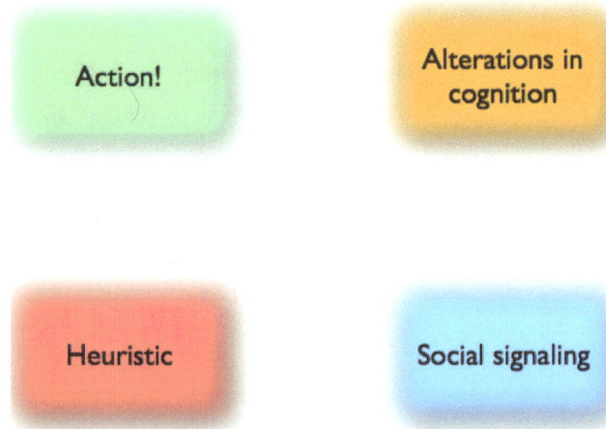

| Action! | Alterations in cognition |
| Heuristic | Social signaling |

Figure What emotions "do" can be seen from at least four different perspectives: leading to action, alterations in cognitive processes, behavioural heuristics and shortcuts, and social signalling.

Emotions can therefore be seen to have at least four different functions:

- They lead to action
- They can change the way we apperceive and process things
- They function as heuristics and behavioural shortcuts
- They provide a signal to our conspecifics
  We will go through each of those in turn here.

Emotions leading to action

If there is anything that emotions "do", it is to lead to choice and action. Emotions can make us run away, fight back, approach something positive, or allow us to relax. As the example with watching a horror movie, our emotional brain systems are constantly evaluating our surroundings and make us ready to act upon these, often long before our conscious awareness of such.

In a series of books and science papers, Antonio Damasio, Antoine Bechara and colleagues demonstrated how emotions are there to "label" events and provide input for our decision making system (Bechara 2000a; Bechara 2000b; Damasio 2005). Indeed, they provided a clear image that the ventromedial prefrontal cortex is highly implicated in connecting emotional responses to behaviour – patients with lesions to this region can show emotional responses, but are unable to make use of those emotions in guiding their beha-

viour (Bechara 2000; Northoff 2006). Conversely, they demonstrated that emotional responses were guiding choice long before people consciously felt like making their choice (Bechara 1997).

In a study by Mathias Pessiglione (Pessiglione 2007), it was found that the size of a subliminally presented monetary reward was associated with the strength at which people responded during a choice task. That is, the higher the amount of money they could win (which was only shown subliminally) the stronger the hand grip response was. This suggests that emotional responses are directly related to our behavioural responses.

Alterations in cognition

Emotions are directly related to changes in many brain regions. As such, emotions are not responses that only lead to changes in action – they also change the way in which we may perceive something. For example, in the previously noted study about the detection of brands, we found that individual preference for brands was related to increased chances of stimulus detection.

In a notable study, Sabatinelli and colleagues (Sabatinelli 2005) studied how arousing stimuli could engage the primary sensory cortices. Here, the researchers found that the amygdala and infertemporal cortex were closely co-activated, suggesting that the amygdala can modulate the engagement of earlier stage processes. Similar findings have been reported by other researchers looking at even earlier visual processes, including the primary visual cortex (Serences 2008).

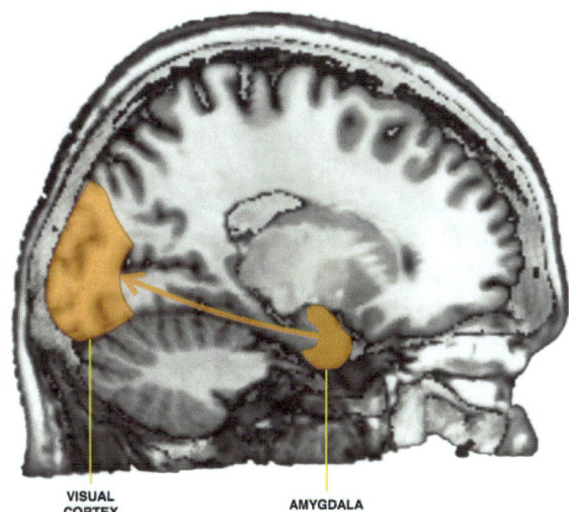

VISUAL CORTEX   AMYGDALA

Figure The amygdala has strong projections back along the ventral stream all the way to the primary visual cortex. This allows the amygdala to modulate the processes along the ventral stream. The brain is seen from the medial side, nose tip is to the right.

In a very similar vein, my own lab recently tested whether unpredictable sounds could lead to changes in cognition and ultimately preference. Prior studies had shown that unpredictable (but otherwise meaningless and simple) sounds could engage the amygdala, and lead to avoidance behaviour (Herry 2007). In the same way, we demonstrated that unpredictable sounds could lead to changes in preference, and that this was driven by changes in pupil dilation – an indication that visual processing was enhanced by the emotional trigger (Ramsøy 2012).

Several studies have demonstrated that our emotional state can affect how we perceive situations and stimuli, ranging from risk assessment to evaluation of novel events (Peeters 1990; Wright 1992; Isen 1983). Notably, a phenomenon called the **mere exposure effect** is the phenomenon where a stimulus that is repeated leads to a more positive evaluation, compared to comparable stimuli that are not seen multiple times (Zajonc 2001; Zajonc 1968; Falkenbach 2013). While many explanations have been offered, one likely scenario is that novel items – that is known to engage the amygdala (Burns 1996; Wilson 1993; Blackford 2010) – is associated with a relative negative emotional response, and that multiple presentation of the same item will be related to a waning of emotional negativity – the object turned out not to be dangerous after all.

Can biological differences and fluctuations affect our emotional responses and related cognitive and behavioural responses? In a review article, Martin Skov and I demonstrated how individual differences in the genetic makeup that we have can affect the way in which our emotional system responds and drives choices (Ramsøy 2010). For example, we reviewed the finding from Boettiger and colleagues (Boettiger 2007) that compared subjects according to their genotype, and reported a positive correlation between the choice of immediate reward and the magnitude of BOLD (blood oxygen level dependent) fMRI (functional magnetic resonance imaging) signal during decision making in the posterior parietal cortex (PPC), dorsal PFC, and rostral parahippocampal cortex. Conversely, the tendency to choose larger, delayed rewards was correlated with activation in lateral OFC. By dividing their subjects into which genetic type they were on the so-called COMT gene, the researchers were able to look at differences in brain activation and behaviour. It was found that the genotype at the val158met polymorphism of the COMT gene predicted both impulsive behaviour and activity levels in the dorsal PFC and PPC during decision making. In particular, this difference was driven by the homozygote val/val genotype compared to homozygote met/met and heterozygote val/met genotypes. That is, val/val subjects more often displayed increased levels of activity in dorsal PFC and PPC, and more often chose immediate rewards. In this sense, individual differences in choosing immediate or delayed rewards were shown to be shaped by each individual's genetic make up.

In a similar vein, in a recent study in my lab we found that women in different stages in their ovarian cycle showed substantial attentional, emotional and cognitive differences in the responses to erotic content in advertising.[7] Women in the high fertility follicular stage were paying more attention to and responded with stronger arousal responses to erotic content than women in the low fertility luteal phase. The former group also demonstrated higher preference and memory for these ads.

Together, these studies clearly demonstrates that emotional responses can drive cognitive processes, perception and eventually choice. *Emotions are there for a reason*, and they are part and parcel of our everyday choices.

Heuristics

Emotions are tightly connected to behavioural shortcuts, and they can help us make the right decisions even when not knowing so. The groundwork of this insight was provided by psychologists as long as a century ago, but most notably today we find the contributions from psychologists like Daniel Kahneman and Amos Tversky on the modern view of economic behaviour (Kahneman 2003; Kahneman 1979; Tversky 1992; Kahneman 2003a; Tversky 1986). In particular, they ran a host of studies that demonstrated that human decision making was profoundly affected by contexts and cues, and that most if not all decisions were deviating from what would otherwise be seen as "optimal" and rational choices.

What neoclassical economics had thought was that human decision making was rational, conscious and occurred after the agent was fully informed and had the time, resources and means to evaluate these options in turn and find the optimal "rational" choice.

The studies in behavioural economics – combining psychology with economics behaviour – clearly demonstrated that our minds make shortcuts. What seems to happen is that emotions operate as a guiding principle for such shortcuts. If the immediate emotional responses evaluate something as threatening, our choices will be affected accordingly, even in the complete absence of conscious processing. This was clearly demonstrated in the aforementioned study by Pessiglione and colleagues (Pessiglione 2008), in which subliminal cues affected choice behaviour by modulating the response of the nucleus accumbens.

In this sense, emotions are behavioural guides: they assist us in making rapid evaluations and choices, and they even affect the way we see our options and how we perceive the world. Understanding the roots of emotional responses is a powerful means to understand what drives consumption behaviours – ranging from their ability to predict our willingness to pay for products, to why some people develop behavioural addictions such as "shopaholism" and pathological gambling.

Social signalling

A final yet highly adaptive function of emotion is that they work as a signal mechanism to others of our kin. Think about how we use facial expressions to communicate our innermost feelings – either as the controlled greeting smile in a conversation to the uncontrolled bursting into tears. Facial expressions are so universal that many scholars have suggested that they have a universal language. Facial expressions are powerful devices for communicating an internal state. It is relatively easy for most of us to recognise whether a person is happy or sad, in pain or enjoying herself. Similarly, we use facial expressions – consciously or unconsciously driven – to communicate our innermost states.

Beyond facial expressions, we have a host of other emotional expressions, such as:

- **Language** – communicated as text or speech, what we say is ultimately a statement of emotions and feelings, such as "I feel sad"
- **Intonation** – we use our voices to underline or even contradict war we say. Even shouts and cries that have no language contents per se, can function as meaningful emotional expressions
- **Gestures** – the way we use our hands and body can be used to appear threatening, angry, happy or sad
- **Blushing** – related to facial expressions, but can work as a powerful social cue

Interestingly, emotional expressions like these always seem to have some unconscious basis. They are simply not initiated and executed as conscious plans. Rather, most emotional expressions are the result of completely unconscious processes and evaluations. As we will see, this poses challenges when we want to assess emotions – we simply cannot just ask people.

A final interesting finding about emotions is that they seem to "mirror8" other people's responses. That is, when we see and understand other people's emotional responses – such as seeing a smiling, fearful or sad face – this also leads to an increased engagement of brain structures that are involved when we display the same expressions. This has also been shown in other primates, such as a study by Kuraoka and Nakamura (Kuraoka 2007), where the researchers showed that individual neurons in the amygdala showed increased activation to visual and auditory representations of emotional expressions from other monkeys. These were the same regions of the brain that the monkey uses when it expresses emotions itself. Thus, we can say that the amygdala seems to be an emotion system evolved to both expressing and understanding emotions.

Why gory marketing will never work

Do the gory warning pictures put on cigarette packages work? Do disgusting images work as intended?

Health warnings are indeed a hot issue these days, and one particular theme on the use of gory images of health related diseases and disorders that stem from smoking. One can understand the basic intention of such warnings: scare smokers (or wannabes) to avoid smoking. For example, tips for quitting safely can be found here.

In a recent blog post, Roger Dooley gives his take on the reason he thinks that these warnings will not work:

> "I don't think these images will actually increase the desirability of smoking, and perhaps a few non-addicts will be dissuaded from starting. As repulsive as we find the images, though, we shouldn't expect them to have much impact on long-time smokers."

I'd love to have Dooley go a bit more in depth about his take on this. I do believe that he is indeed on the right track, and I believe that I can explain why it is so.

And let me be right up front and remind you that the explanation offered by Martin Lindström is really not tenable. Rather, it is a prime example of pseudoscientific babble stemming from a misuse and misunderstanding of neuroimaging results.

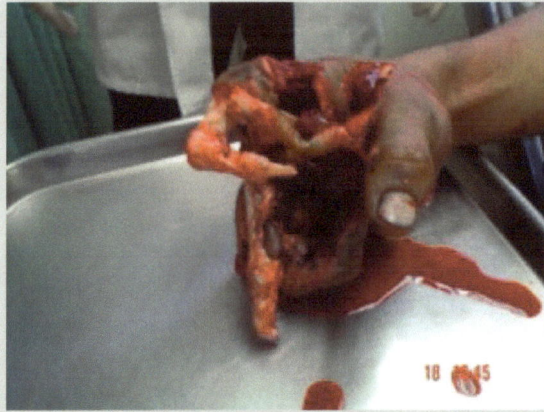

Let's first look at the basics: The main assumption in these warning ads is that when seeing these images, smokers (and non-smokers alike) will be motivated to avoid smoking. In some way, smoking will be associated with bad health, and motivate us to stop smoking, or never start smoking. Another possible assumption is that this relies on a very overt and rational process. We have to connect the bad outcome to smoking, and consciously make the decision to avoid smoking. Alternatively, one may wish to prime smoking to bad health (and disgust).

It strikes me that several years ago, a study by one of my colleagues, Maurice Ptito, reported that these warning labels did not work at all. In fact, when studying smokers' behaviour, they realised that smokers did not pay much attention to the warnings at all. Interestingly, instead of using the packages, smokers had their own packages where they put the cigarettes. The original package with the warnings was thrown away.

Why? Here's the intersting story… and let me start by asking you to look at this image…please try as much as you can:

Want to look away? Did you even go beyond this image to read on? Scrolled past the image? How did you feel when you looked at it?

Sorry to put you through it. But I felt it crucial for you to understand my point – to feel it yourself.

Here's what happens: disgusting images have a primary motivating force on us: it makes us want to look away. Indeed, several studies have now demonstrated a role of the insula in disgust (see review). Disgusting images recruit this region involved in processing and producing visceral responses. It is also involved in our ability to understand facial expressions of disgust in others.

Indeed, in the study by Ptito et al, it was found that the disgusting warnings recruited the insula. The role of the insula has also been shown to be related to motivating behaviours. One example is the study by Knutson et al. who demonstrated that increasing product prices were related to increased engagement of the insula, and a lower likelihood in purchasing that particular product.

In other words, disgusting images of all kinds makes us want to look away. We are primarily motivated to avoid that situation. We look away, walk away, close our eyes. Indeed, this primary motivation is the reason that the warnings do not work. First, they lead to reduced attention to the information. Second, the primary and direct motivation to avoid the information/picture never becomes connected to the cigarette packages themselves. Warnings of this kind could only work if they were so disgusting that people would not even touch them…

So what could work instead? *Positive and direct motivation!* Some opportunities would be to say that quitting smoking makes things better for you. This could include better skin, body odour and oral hygiene; improved health and so on. The more one can get at direct motivating factors, the better. The image on the right side may be in the right direction, but not quite there…

MEASURING EMOTIONS AND FEELINGS

Since emotions and feelings are now deemed to distinct phenomena, we must also expect that they cannot be measured with the same device. Most notably, we often can hear people stating their feelings about something, such as "this food is delicious!" or "I was really scared during that movie". However, when we measure emotions, we can get a much more nuanced picture of exactly when something occurred.

Think about the last movie you saw: you probably have some gut feeling about what you feel about the movie: it was good etc, and you probably have some episodes you recall vividly. However, reporting your conscious feelings about something is simply not sufficient if we want to understand your responses to each and every segment of a movie. This is hard for four reasons:

1. **The sentience fallacy** – Emotions are by their nature unconscious; we cannot use our conscious

minds to directly observe what goes on in our unconscious minds.

2. **The memory fallacy** – Our memory is not infallible. It is virtually impossible for you to recall in vivid details the two hours' duration of the movie, and how your responded.

3. **The observation fallacy** – It is well known that observing your own emotions changes the actual nature and strength of these emotions.

4. **The social fallacy** – Our willingness to report all feelings we have – especially personally relevant and socially inappropriate ones – is often low, leaving an incomplete picture of our responses.

Taken together, we need other measures to be able to assess the second by second changes that emotional responses show. Below, we show some measures that can be used to assess emotions and feelings, as well as their behavioural effects:

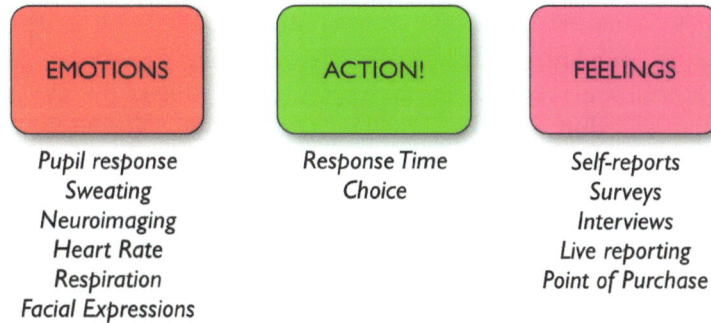

EMOTIONS	ACTION!	FEELINGS
Pupil response	Response Time	Self-reports
Sweating	Choice	Surveys
Neuroimaging		Interviews
Heart Rate		Live reporting
Respiration		Point of Purchase
Facial Expressions		

Figure Measures of emotions and feelings. The diversity of tools allow for the construction of a complete toolbox for the academic and commercial study of consumption behaviours and communication effects.

Measures of emotion

We have a host of behavioural responses that we can use to measure emotional responses. Purely automatic responses are pupil dilation responses, respiration and pulse, and sweating, and beyond this we can see that facial expressions can work as an external image of inner emotions. Many researchers even claim that we can display rapid and almost undetectable emotional responses, called microexpressions (Pfister 2011; Matsumoto 2011; Ekman 2009), although this has yet to be fully validated by the literature.

The following table has a more comprehensive – although not exhaustive or complete – overview over different methods that can be used to assess emotional responses:

MEASURE	DESCRIPTION	COST	PROS	CONS
Pupil dilation	Pupils are dilated due to arousal, cognitive load and luminance. If one controls for the two latter effects, then one can reasonably argue that one has a measure of arousal. The analysis is done by looking at the relative difference in pupil size from onset of a stimulus.	Medium	Pupil dilation is very accessible from most high-end eye-trackers. When performed correctly, pupil dilation can be a reliable index of arousal. Pupil dilation is, compared to GSR, a rapid response.	Pupil dilation responses are caused by 1) brightness; 2) cognitive load; and 3) arousal. Unless one has the sufficient expertise in controlling for the two first factors, pupil dilation is not a reliable measure. Does not assess valence. Pupil dilation cannot be measured in mobile settings with any known devices.

MEASURE	DESCRIPTION	COST	PROS	CONS
Posture	This is a completely novel measure that comes "for free" when doing many eye-tracking experiments. Changes in posture is known to signal engagement and disengagement, and recent studies suggest that this may operate as a low-cost valence measure.	Low to Medium	The method comes "for free" through eye-tracking studies, but can also be measured in other ways, such as through video observations.	The method is not yet fully validated. Recent studies have found a link between changes in posture and motivation, but replication is needed.
Galvanic Skin Response	Sweating is a natural arousal response. By employing two electrodes with some distance, typically on the hand, one can measure the electrical conductivity between the two electrodes. Higher arousal is associated with increased sweating and higher conductivity. The analysis looks at the relative change from stimulus onset.	Low	GSR is easy to apply and can be used in many situations, even mobile settings.	The GSR response is sluggish and subject to distortions as well as "natural drift". Unless one knows how to control for this, GSR is not a valid measure.
Respiration	Changes in respiration can be done by measuring with a respiration belt or other precise measurement of respiration. Increased arousal is associated with an increase in respiration, but also other hallmarks, such as a brief pause in respiration and shorter breaths.	Low	A relatively easy to use measure that indexes changes in arousal. Can be used in mobile settings.	The response is sluggish and has a relatively low sensitivity, statistical power and temporal resolution.
Pulse	Changes in pulse is associated with changes in either physical exercise or arousal. If one holds physical exercise constant, heartbeat variation can be a reliable index of arousal.	Low	Relatively easy to use measure of arousal. It can be used in mobile settings.	The response is sluggish and has a relatively low sensitivity, statistical power and temporal resolution.
EEG	The most reliable and well validated EEG measure of emotional response is the change in raw EEG signals as the person goes from a drowsy (high alpha) to an alert state (high beta and gamma). Besides this, the novelty response found in event-related potentials (ERPs). Other measures such as the frontal asymmetry also has a couple of decades of scientific evidence, and is indicative of approach-avoidance behaviour.	Medium to High	EEG has a superior temporal resolution. Can be used in stationary and mobile settings. Hardware cost is coming down, making EEG measures more accessible.	Requires expert knowledge about data processing and analyses. Noisy data can corrupt insights.

MEASURE	DESCRIPTION	COST	PROS	CONS
fMRI	Studies that have looked at the neural bases of arousal have consistently found that the amygdala, insula and ventral striatum (NAcc) are closely related to arousal. These regions all show the bivalent response profile of associated with arousal. Valence responses are demonstrated in the medial to lateral orbitofrontal cortex.	High	High spatial accuracy in determining arousal response, and when modelled correctly can provide reliable index of arousal and valence.	Expensive method per subject. Requires sophisticated research team to maintain and run scanner protocols, and analytics team to run analyses. Requires sophisticated understanding of the brain and mathematical modelling. The temporal resolution is relatively low (about 2 seconds), and there are substantial limitations to experimental design (noisy and artificial environment, need for repeated measures)
Facial expressions	A relatively new method that analyses the changes in facial mimicry as an index of emotional valence. The method is based on the assumption that facial expressions are universal and that they are reliable indicators of emotional states. The method often used is recording of faces with web cams, although other methods such as electromyography is used.	Low to Medium	Relatively low-cost solution for assessing emotional valence. Suggested to work in mobile settings and be scalable as an online solution.	The basic assumptions of facial coding are currently being criticised and debated. Not known whether facial expressions are that reliable. Often problematic recordings in online settings due to requirements on hardware. Suggestions of "micro-expressions" not yet validated The value of facial coding as a predictive or even diagnostic tool is yet not validated by independent academic research.
Blush response	A relatively new method that analyses changes in blood flow through the skin of the face. While this also allows the measurement of pulse, it also provides an index of increased blood flows to the face region. This is highly associated with blush responses and increases in arousal.	Unknown	Holds the potential for being a low-cost solution for assessing blood flow and thereby both arousal and social emotions (blush responses).	The method is still under development, and neither the validity or price point is yet known.

The table shows that we have a host of measures of emotions. One of the notable things from this table is that despite that we now seem to have a toolbox to assess different aspects of emotions, a main requirement is still that we need to have a sophisticated understanding of both the complexity of emotions and feelings, and not the least a good understanding of how we can test and analyse emotions with each tool. It is simply not straightforward to run an eye-tracking study and make claims about emotional responses, nor is it straightforward to set up a valid test and make valid generalisations about findings from a study. All these methods make demands on how we set up an experiment, our understanding of their uses and limitations, and our knowledge of statistics. We can summarise this as a list of demands that any researcher or lab – academic or commercial – must be in possession of in order to run studies with these methods at hand:

- **Experimental design** – It may seem simple, but setting up an experiment that addresses the key questions and has a high validity is a complex task.
- **Brain insights** – One's knowledge about the brain cannot be stressed enough – it is impossible to do proper consumer neuroscience research with only a crude understanding of the brain or neurophysiology.
- **Statistical skills** – Most if not all consumer neuroscience methods provide an abundance of data, even for a single person. Knowing how to handle, preprocess, analyse and interpret the results from data analysis is a crucial skill. There are no magic bullet that can provides a clear-cut answer about emotional responses, that do not also demand a certain degree of statistical analysis.
- **Analytical skills** – A perfectly executed and analysed study might still be interpreted horrendously. This may be particularly prominent in commercial studies, in which statistical power is high, study execution may be good, but the need for actionable insight can drive vendors and clients to simplify, misrepresent and misinterpret the results. Having the knowledge about what a result means, and having the guts to "kill the client's darling" is a challenge, but nonetheless an important skill.

In addition to this, other features may apply for academic and commercial research, such as one's ability to understand clients' key questions. Often, clients

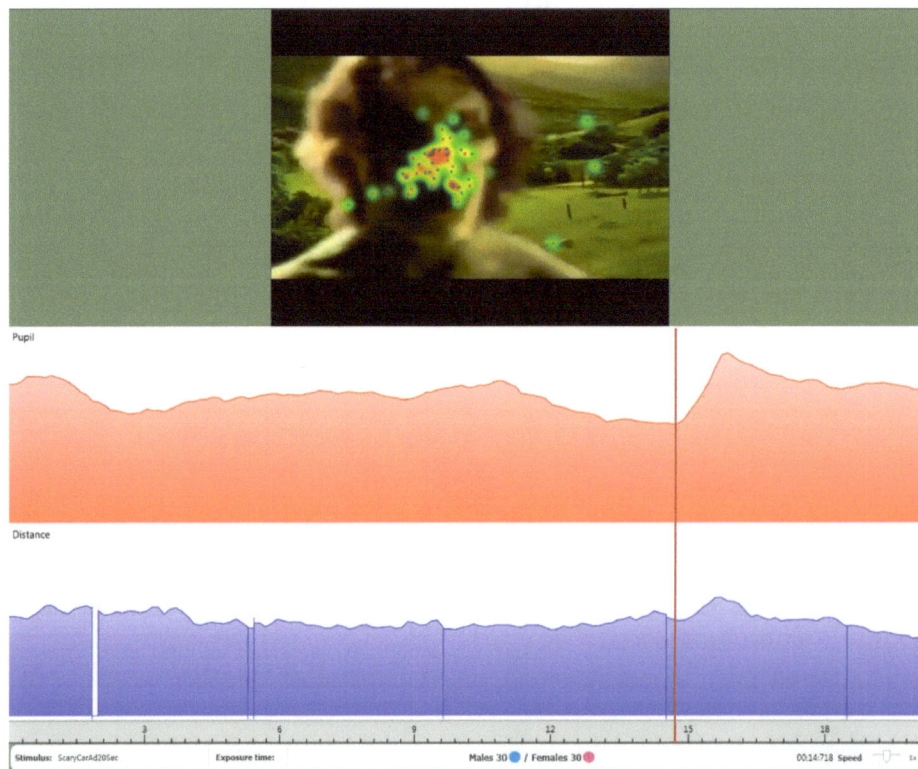

Figure Emotional responses to a scary event. In a move watching a car driving down the road, a highly unexpected events occur: a zombie-like person jumps in front of the screen and screams. Bottom: When assessing pupil dilation (red) and distance to the screen (blue) it is obvious that both scores demonstrate a strong emotional response. Increased pupil size is, among other things, an index of arousal, and increased distance is related to leaning backwards and "avoiding" a stimulus or event. The careful observer will note that the measures are slightly delayed relative to the real event. Courtesy of iMotions (www.imotionsglobal.com)

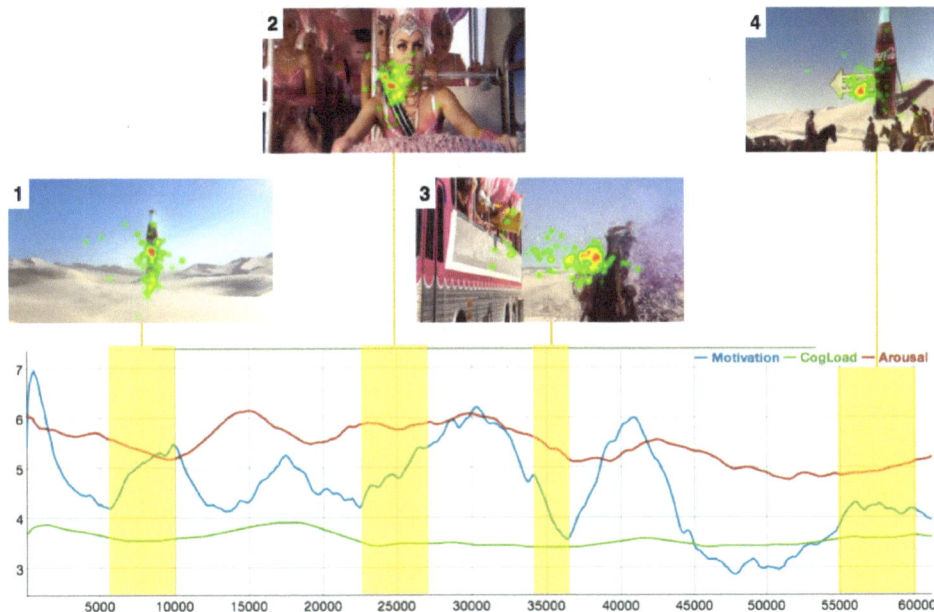

Figure Motivational response to a commercial, using the Neuromarketing Omnibus. In testing a Coca-Cola commercial we saw that product placement lead to a significant increase in the motivational response (1). Also, note the strong attentional focus on the product (heat map), suggesting that the product received much visual attention. The display of colourful and beautiful women was also associated with an increase in motivational response (2), while a situation in which a character is shot at and dodges is associated with a negative deflection in motivation, suggesting avoidance response (3). At the end, where the main message comes across and another product is placed in plain sight, we see a moderate positive deflection (4). Courtesy of Neurons Inc (www.neuronsinc.com)

do not have any sophisticated insight into statistics, the brain or experimental design, and their research questions may be poorly defined and operationalised. The task of commercial researchers is therefore first and foremost to help the client define their questions in a form that can be tested, and then help the interpretation of results in relation to this. Academic researchers may be more free to define these issues, but are more often challenged by finding the questions that are both relevant and that help shed light on important theoretical questions.

Let us take an example from a recent ad test in the US. In this Neuromarketing Omnibus test, 100 people from a nation-wide sample were tested with EEG and eye-tracking while they saw different ads, tasks and documentaries. Here, we will focus only on the motivation response, where one sees that the fluctuations can be used to "diagnose" the effects of separate ad components, to questions such as "do participants demonstrate an assumed positive response to seeing a product?" and "are viewers responding to aversive components in the ad?"

Notably, the use of these kinds of methods allow researchers to go beyond self-reports, and most methods provide a high temporal resolution. This makes it possible to assess the responses to specific elements and group differences, including:

- **Manipulation effects** – comparison of manipulation of specific properties of an ad, such as the use of a particular scene or not. This requires that one either test ads on two comparable groups (between-subject design), or test both versions on all participants (within-subject design).
- **Inherent group differences** – comparison of naturally occurring differences between groups, such as men vs women, or young vs old.
- **Performance group differences** – comparison of groups that are different on task performance, such as comparison of those that like an ad vs those that dislike the ad; or those who remember vs forget the ad.

Measures of choice

As we have noted many times, emotions are closely connected to action. By assessing particular aspects of our choices, we can make assumptions about their emotional properties. For example, the more we like something, the faster we will approach it and grab it, and the more we will work to obtain it. In many ways, behavioural responses are closely correlated to our emotional responses. As we saw in the study by Pessiglione and colleagues (Pessiglione 2007), our unconscious emotional responses can affect how much effort and power we put into our choice, even though we consciously do not experience any such relationship.

We will dig deebodyper down into response times and other behavioural indices in the chapter about wanting and liking. As we will see, response time and other similar indices can be used to understand both emotional responses and emotional conflict.

Measures of feelings

The most obvious way to measure feelings is to just ask people what they think about something. Did they like the taste of the product? How do they feel about the brand Nike?

A number of studies have demonstrated two particular features that are closely linked to self-reported liking. First, prefrontal asymmetry has been closely linked to what people report to like, as well as what they subsequently choose (Davidson 2004; Hewig 2006; Schaffer 1983; Berkman 2010; Ohme 2010a; Ravaja 2012; Sutton 2000; Pizzagalli 2005; Coan 2003). Second, changes in the medial orbitofrontal cortex has consistently been implicated in "liking" responses, and in self-reported hedonic experiences (Kühn 2012; Smith 2009; McClure 2004; Plassmann 2008a). This suggests that to some extent, we can explore these responses as probable indices of preference and liking.

Indeed, there are instances where the assessment of feelings cannot naïvely use self-reports, such as when you test movie preference. You may recall the last time you actually watched a movie, and how you feel about it. You may recall certain episodes from the movie, but overall you probably have a poor recollection or impression of what actually has driven your liking or disliking of the movie. With measures that have a higher temporal resolution this is now possible. Think about using EEG, or fMRI, to assess valence responses such as frontal asymmetry or orbitofrontal responses while people are watching the movie. This allows you to both track the emotional valence responses over time, and the relate these responses to subsequent self-reports or choices. This allows us to better understand how different emotional responses during movie or ad watching can drive subsequent feelings and behaviours.

What if you do not have neuroimaging at your disposal or at your expertise? Here, eye-tracking can provide some clues. In many eye-trackers it is possible to track changes in both pupil dilation and postural changes (distance between the eyes and the screen). In a recent study, we found that the combination of measuring incremental changes in pupil dilation and posture allowed us to tentatively assess arousal (pupil dilation) and valence (posture)[9]. Here, we found that when people lean forward, stronger pupil response was indicative of a positive valence response, more positive ratings and a higher chance of subsequent purchase. Conversely, when people leaned backwards, stronger

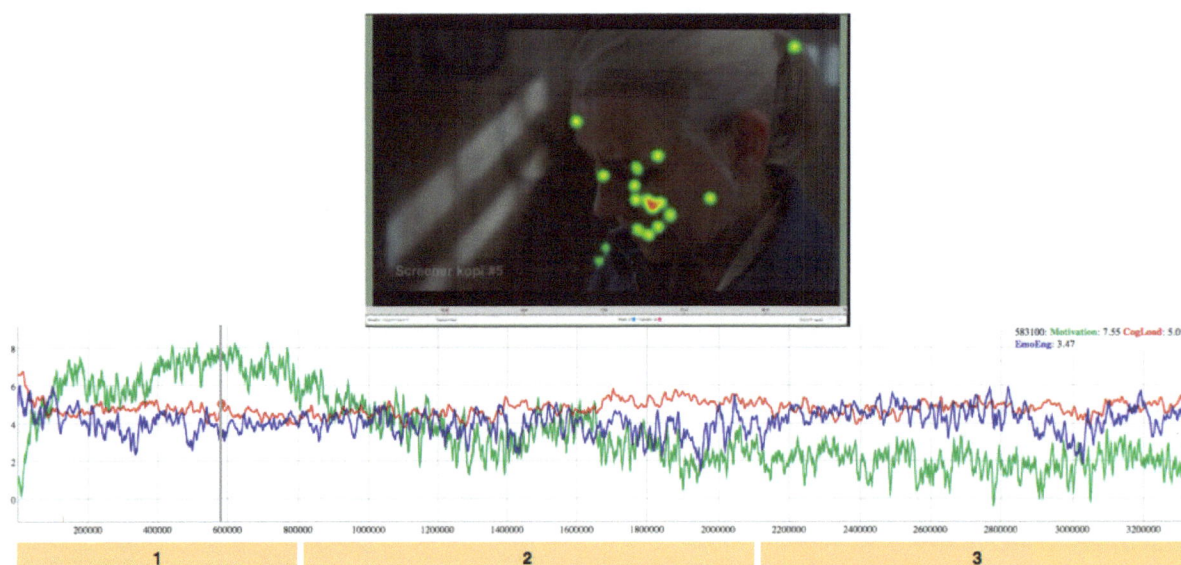

Figure Motivational phases in a movie of approximately 60 minutes. In pre-testing a TV drama using EEG and eye-tracking we found that there were three gross phases corresponding to the emotional content of the drama. First, an increase in motivation score (green line) was related to anticipation and self-reported enjoyment of the episode (1), which was followed by a plummeting of the motivation response in relationship with aversive events of severe illness, death and grief (2). This was then followed by an increase in arousal (blue line) in the third phase, which corresponded to self-reported tension, which was thematically related to interpersonal conflict, motivation score was low and arousal was high (3).

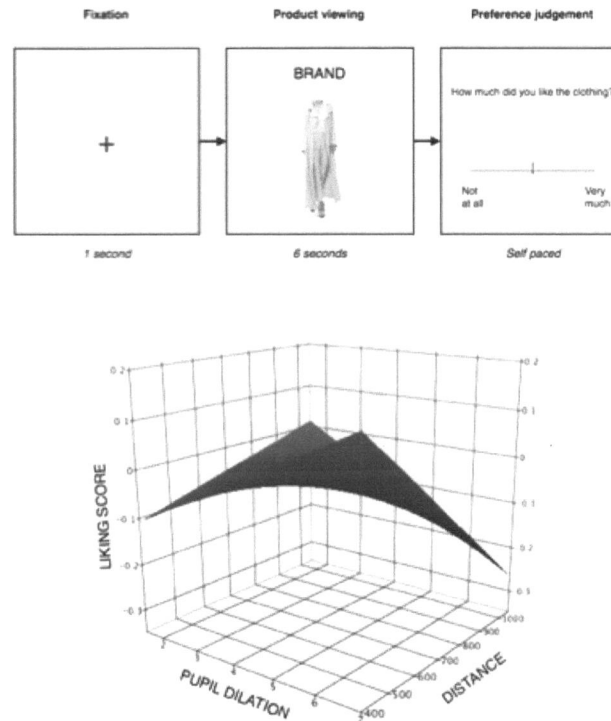

Figure   Relationship between pupil dilation and posture. TOP: Participants were exposed to different fashion items and were asked to rate each item. When combining the assessment of pupil dilation with posture as responses to fashion clothing, we found that we could better predict subsequent evaluations and consumer choice. BOTTOM: Model of the relationship between posture, pupil dilation and raring. When subjects leaned forward ("Distance" score is low) we saw a positive relationship between pupil dilation and subsequent image rating. Conversely, when people leaned backwards ("Distance" score is high) this relationship flipped, and increased pupil size was related to lower image rating.

pupil response was related to negative valence, and ultimately lower ratings and less purchasing behaviour. While these findings are still the first of their kind, they hold the promise that eye-tracking can get to grips with both emotional arousal and valence.

THE EMOTION MATRIX

Taken together from this chapter, we can provide a measure of emotional responses on three dimensions. The first dimension concerns the **valence** of the response – is the response positive or negative; are we showing approach or avoidance behaviour or tendencies of such? As we have seen, valence ranges from the most extreme pleasures to pains, with neutral and dull responses in between. Please note that arousal is not shown, but that the insights from the Emotion-Motivation matrix applies here, in that arousal is expected to go from low (neutral valence) to high (increasingly positive or negative valences).

The second dimension concerns the degree of **sentience** associated with a response – is the emotional response completely unconscious, or is it fully conscious? This employs the Emotional Thermometer model, and from our discussions about consciousness vs unconsciousness, please note that we do not expect a distinct transition from unconscious (emotions) to conscious (feelings), but rather some degree of smooth transition.

Finally, from the consumer neuroscience model (Plassmann 2012) we could see that our responses could be ranked according to their **stage**. That is, whether they were at the prediction stage, to the specific outcome of an event, or whether it was more associated with learning from events (and possibly the confirmation or violation of expectations) The Emotion Matrix allows us to better describe the emotional responses we are studying. However, this also provides a complex matrix – a so-called 2 by 2 by 3 factorial matrix – with 12 different combinations. The most notable difference is between conscious and unconscious processes. Whenever something is **conscious**, we can make explicit statements, including positive and negative statements. Conversely, a similar listing of when something is unconscious, we cannot rely on self reports, but need to assess responses by using neuroimaging, neurophysiology or specific behavioural measures:

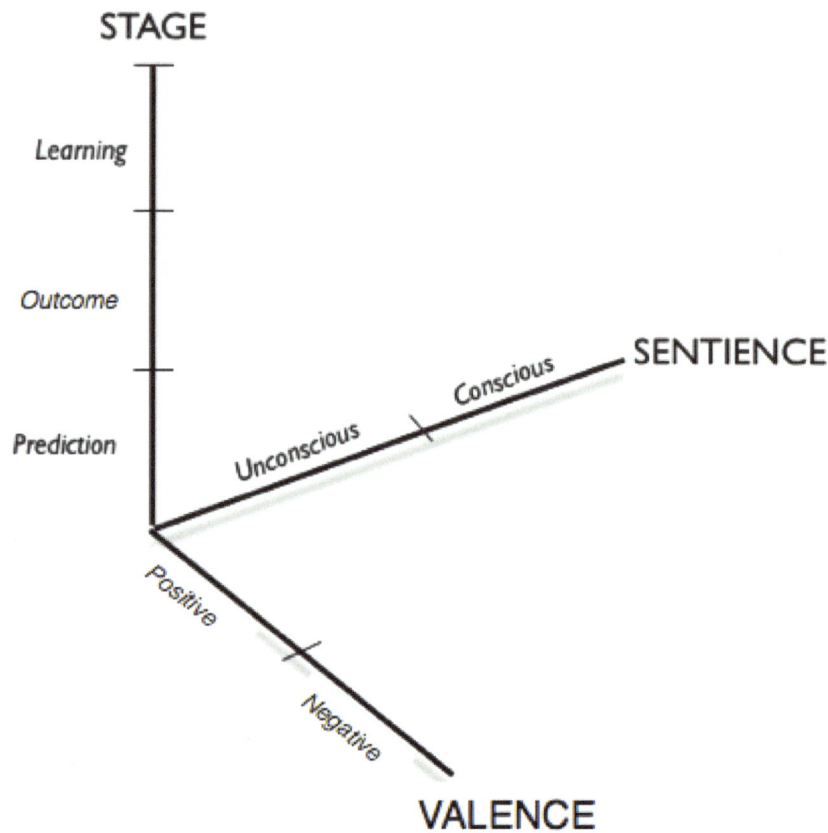

Figure The Emotion Matrix. By drawing on the consumer neuroscience model (Plassmann 2012), it is possible to disentangle different emotional and responses. First, we must determine the valence of the response from positive to negative. Second, we can categorise responses as to whether they are conscious (emotions+liking) or unconscious (emotions only). Finally, we can look at the stage of the response at hand – is it a prediction and expectancy, is it a response to a specific event, or is it the subsequent learning from an event?

The Emotion Matrix can also be used to discuss the aims of communication and the *actual effects* of communication. In a commercial example, what is the intended effects of a particular commercial, and do customers actually respond in this way?

One prominent example is in the case of the now so abundant use of obscene images put on cigarette packages to motivate smokers to quit smo-

king. The reasoning behind this can be summarised as follows:

1. Watching horrible images of smokers' health is associated with negative emotions
2. Smokers who see these images will be motivated to avoid the effects of smoking, as illustrated by the images
3. Smokers will be consciously motivated to quit smoking

STATE & VALENCE	PREDICTION		OUTCOME		LEARNING	
	positive	negative	positive	negative	positive	negative
Conscious (self reports)	I'm looking forward to this meal"	"I dread the consequences of this choice"	"You look beautiful!"	"This movie makes me sad"	"Last time I was here the staff was excellent"	"My last phone from brand X was so bad, I'll not buy another one from them"
Unconscious	Increased engagement of the ventral striatum and amygdala, postural changes (leaning forward)	Increased engagement of the ventral striatum and amygdala, postural changes (leaning backwards)	Increased arousal and valence scores, such as GSR and prefrontal asymmetry	Increased arousal and decreased valence scores	Engagement of the ventral striatum in operational learning, behavioural changes showing learned rewards (approach behaviours)	Engagement of the ventral striatum in operational learning, behavioural changes showing learned pains (avoidance behaviours)

This all sounds quite reasonable, but consider the following case: in a now decade-old study my colleague Maurice Ptito discovered that smokers in fact did not reduce their willingness to smoke after seeing these images. It was only when he made careful observations about their actual behaviours that he understood what happened: upon purchasing cigarettes, smokers took out their own cigarette case and moved the cigarettes over to this new case, disposing of the original one with the horrific images. That is, smokers were not motivated to quit smoking – they were motivated to avoid the stimulus in the first place! It was simply much easier to avoid exposure to the images than to change one's smoking habit entirely.

Based on this, there was a profound difference between the intended effects of the cigarette package information and the actual effects. Instead of the motivation assumptions seen above, consider the actual result:

1. Watching horrible images of smokers' health is associated with negative emotions

2. Smokers who see these images will be motivated to avoid being exposed to the images

3. Smokers will be motivated to develop automatic strategies to avoid image exposure

The Emotion Matrix provides the means to visualise and discuss the intended and actual effects of communication, and an as such be a powerful pedagogical tool to understand the relationship between different types, stages and levels of awareness of emotional responses.

EFFECTS OF EMOTION MODULATION

How can we control our emotional responses? In many situations of everyday life, we need to modulate and inhibit our emotions. It can range from the inhibition of eating unhealthy foods to avoiding to act socially inappropriate. If we had not emotional control, we would act solely on impulses. While this may happen in many cases when we are influenced by psychoactive substances such as alcohol, or suffer from particular

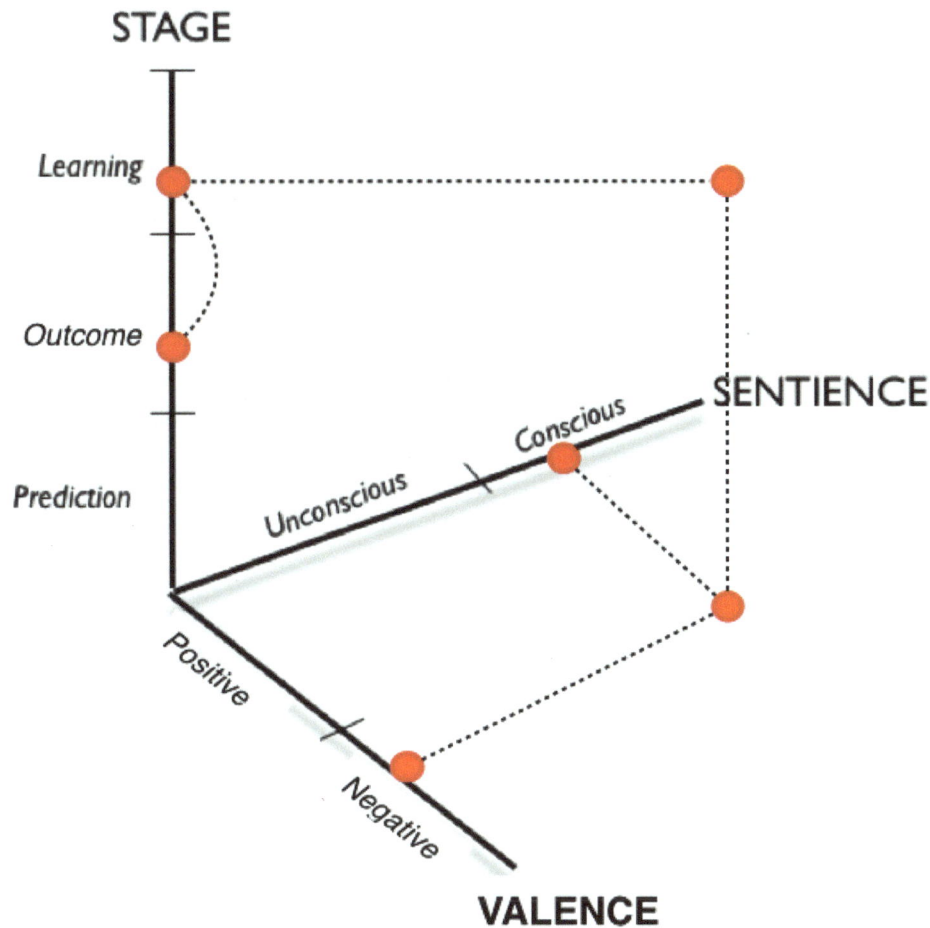

Figure Usage of the Emotion Matrix can be used as appoint of departure for discussing the intended effects of communication, both at the planning/idea stage but also when comparing intended and measured effects.

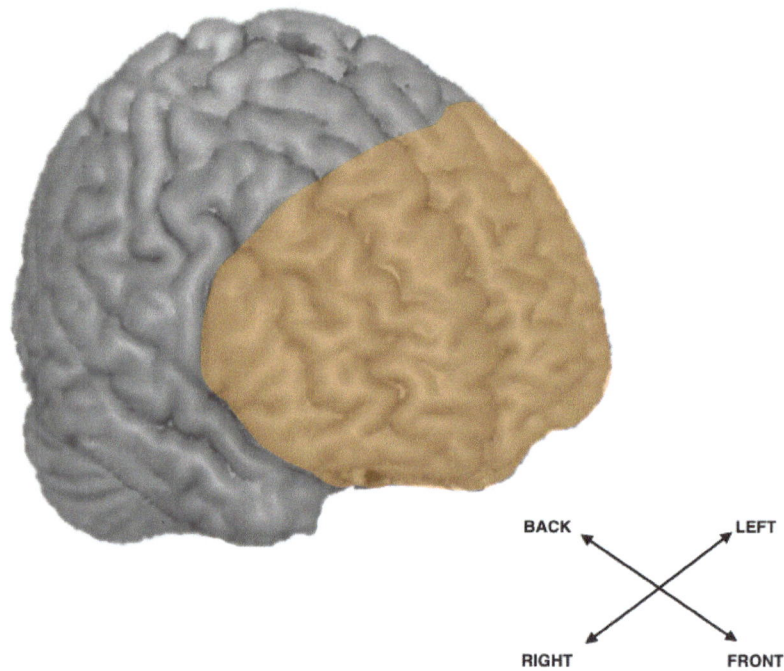

Figure   The frontal cortex has been implicated in the modulation and inhibition of emotional responses.

injuries or dysfunctions in the frontal lobes, healthy behaviour entails a mixture of impulses and control .

How does the brain actually allow such modulation and inhibition of emotional responses? Several studies have looked at this and implicated a role for several frontal regions of the brain. For example, in an fMRI study by Banks and colleagues (Banks 2007), participants were trained to either passively perceive stimuli with different emotional contents, or actively reappraise the images. That is, in the so-called *Maintain task*, they were asked to "attend to, be aware of and experience naturally (without trying to change or alter) the emotional state elicited by the pictures; they were told to maintain the evoked affect for the entire task block." Conversely, in the *Reappraise task*, they were asked to "voluntarily decrease the intensity of their negative affect by using [a] cognitive strategy of reappraisal; they were told to reinterpret the content of the picture so that it no longer elicited a negative response."

When participants were reappraising the emotional contents, compared to the Maintain condition, the researchers found a stronger coupling between the amygdala and the dorsolateral, dorsomedial, anterior cingulate, and the orbitofrontal cortices. Furthermore, the researchers found that the connection strength in a subset of this network – between the amygdala and the orbitofrontal and dorsolateral prefrontal cortex – predicted successful emotion regulation. Together,

this study suggests that emotional appraisal involves a modulatory role of the frontal cortex on emotional responses that, if successful, can alter our emotional and behavioural responses to emotional events.

This finding is relevant in a broader context of so-called *executive functions*. Here, we will see that the role of impulse control can be a crucial building block in normal consumer behaviour and that when it fails, particular consumption disorders can manifest.

Similarly, in a study by Goldin and colleagues (Goldin 2008), participants were instructed to either reappraise or suppress emotional responses. That is, the researchers compared the effects of reappraisal, as employed in the study by Banks et al. (2007) with the effects of suppression, in which participants were instructed to "keep their face still while viewing films so that someone watching their face would not be able to detect what was being experienced subjectively." Here, they found differential effects on emotional modulation as well as two different ways in which the brain performed the modulation response:

- **Reappraisal** was associated with a fast (0 - 4.5 sec) response in the prefrontal cortex, which was then associated with decreased amygdala and insular responses, and ultimately decreased negative emotion experience.
- **Suppression** was associated with a slower (10.5-15 sec) prefrontal response, an interesting *increase* in

amygdala and insula activation, but still decreased negative emotion behaviour and experience.

This finding suggests that the strategy we employ to modulate our emotional responses – whether we try to suppress our emotional responses or reappraise their contents – can have a substantial effect on brain operations and our subjective feelings related to events.

Utilising micro-effects on emotions to assess brand equity

Emotional responses are rapid, and as a consequence, with the right measuring tools we can assess the emotional fluctuations that different events bring on. We even know that these responses are extremely fast, and can affect subsequent behaviours, such as response time. In a study presented at the 2010 NeuroPsychoEconomics conference[10], Martin Skov and I tested whether subliminally presented brand names could affect evaluations of emotional words. The task people had, was merely to report whether a single word on a screen (e.g., "HAPPY") was positive or negative (obviously, it is a positive word). Prior to this, they were exposed to brief visual blinks on the screen, and were instructed that these were just there to "clean" their visual system before each new word. Importantly, we often presented brand names for 30 milliseconds during those brief glimpses, making sure that they were not consciously detected.

To assess brand liking, all participants rated each brand they had been exposed to (after the full test) for their level of liking of the brand and level of knowledge of the brand. We looked at the different levels of liking and used brand knowledge as a constant variable (a co-variate of no interest). For this study, we expected that brands that people had a positive association to would lead to a micro version of a "positive mood" that could affect the subsequent conscious judgment of words. That is, if the brand was liked, it would be associated with a positive emotional response, and subsequently it would be easier for people to make positive judgments and harder to make negative judgments of a word. The latter would impose a readjustment of the answer from a positive to a negative, which we expected would be associated with a slightly longer decision time.

This was exactly what we found! When people were subliminally exposed to a brand name they liked, they made faster responses to positive words than negative words. They also made fewer response errors. Conversely, when the brand shown was disliked, people made faster responses to negative words and slower responses to positive words.

This finding suggests that our emotional connectedness to brands are so profound that they can be invoked unconsciously and affect subsequent judgments and choice behaviours. Beyond being used as a mere experimental study, one can easily propose that a si-

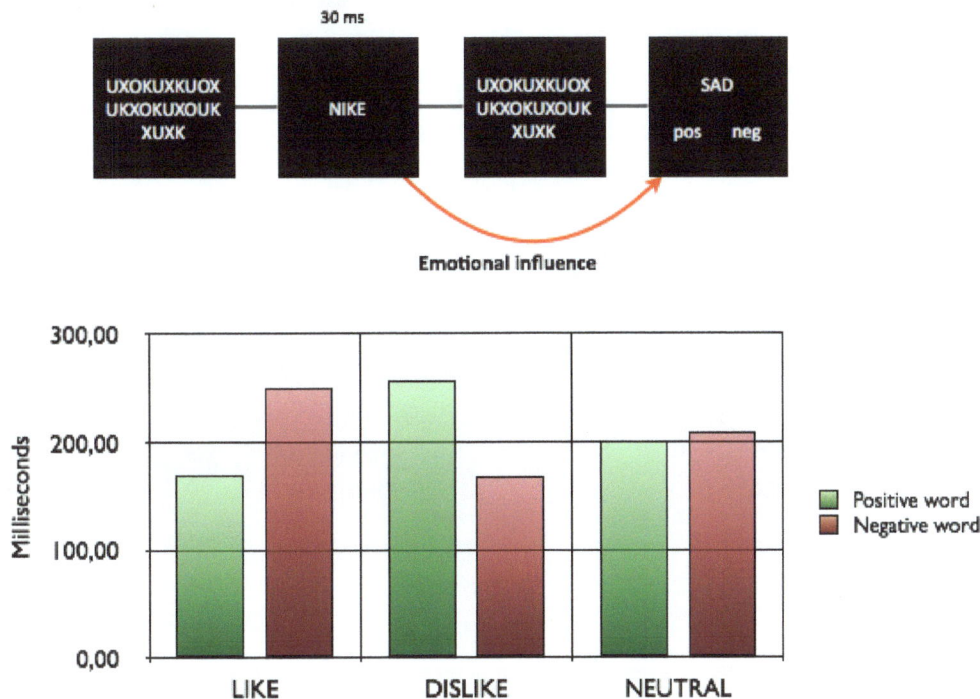

Figure Effects of subliminally presented brands on behavioural response times to word categorisation. In this study, using purely behavioural methods, we demonstrated that a person's brand preference – even when only invoked unconsciously – could have a significant effect on response time. TOP: the paradigm, wherein we used a forward- and backward-masking paradigm to mask the brand name. BOTTOM: The effects of brand preference on response time for word categorisation, measured as differences in milliseconds of response time.

milar method is use this as an index of emotional "gut reactions" to brands. While many market research companies are using large scale Brand Trackers to assess people's conscious feelings and thoughts about brands, this proposes an alternative avenue of making an "Unconscious Brand Tracker" that allows companies to not only assess people's or groups' direct emotional responses and affection towards them, but also how specific periods or communicative efforts can affect unconscious emotional responses. In this way, one could imagine asking "does this ad lead to a more positive emotional response to our brand?".

Moods and choice

While emotional responses are immediate, another yet untreated topic relates to "moods". Moods can be described as a "temporary state of mind", and thus reflects an emotional type of state that is longer lasting than the immediate responses that emotions represent.

How do moods affect the way consumers experience products, brands and situations? Despite that thinkers have suggested that the field of consumer neuroscience and neuromarketing should focus on moods , little research has yet been done.

In a recent study in my lab[11], we explored the effects that moods could have on decision making and experiences of the outcome of our choices. Here, we investigated the correlation between affective states, risk taking and the illusion of control, trying to understand how emotions and mood affect risk taking behavior and decision-making. In a trading game, participants could win or lose money, and reported their level of control and both prior and subsequent subjective mood.

We found that moods had a dynamic relationship to the experience and illusion of control. For example, moods such as aggression, anxiety, surgency and egoism showed a negative impact on the experience of control, whereas concentration, social affection and skepticism showed a positive effect on experienced control. Thus, the results support our expectation that moods would be positively correlated with the experience of control and show positive correlation between emotions and risk taking. Taken together, these findings suggest that an increase in arousal leads to individuals becoming increasingly risk seeking for positive emotions and less risk seeking for negative emotions. High levels of arousal were primarily found to lead to higher risk taking and vice versa for both positive and negative moods. Risk taking was found to correlate negatively with the experience of control contradicting the extant literature.

These findings can be extended by an influential paper by Di Muro and Murray (Di Muro 2012) who demonstrated that moods indeed affect consumer choice. By running three consecutive studies, they demonstrate that when consumers were in a positive mood, they were motivated to choose products that were *congruent* with their state of mind. Conversely, when they were in a negative mood, they tended to choose products that were *incongruent* with their state of mind. This suggests that as consumers we are motivated to maintain a positive mood, and that consumption behaviours can be a vehicle for maintaining or boosting our moods.

Emotions, moods and work performance

In a related study[12], we recently also tested whether exposure to emotionally laden words could impact work performance, by running two related studies. In Study 1, which was a lab based test, we demonstrated that being primed with positive or negative words affected the moods of participants to become more positive or negative, respectively. In Study 2, we provided positive or negative words in a morning communiqué to employees at a financial institution, and measured the work performance over three consecutive days. Here, we found that those who received positive words in the morning had an increased work performance over the course of all three days.

Taken together, while confirming other related studies (Monahan 2000), this study shows that emotional responses on the short term can have longer-term effects on subjective moods as well as behaviour. While few studies have looked at this effect, our results are nevertheless in line with prior studies that have shown effects of moods on decision making and behaviour (Dunn 2009; Berridge 2009; Gorn 2001; Meloy 2000; DiMuro 2012; Mayer 1992; Harlé 2012; Wright 1992; Raghunathan 1999).

Two ways of reducing fear

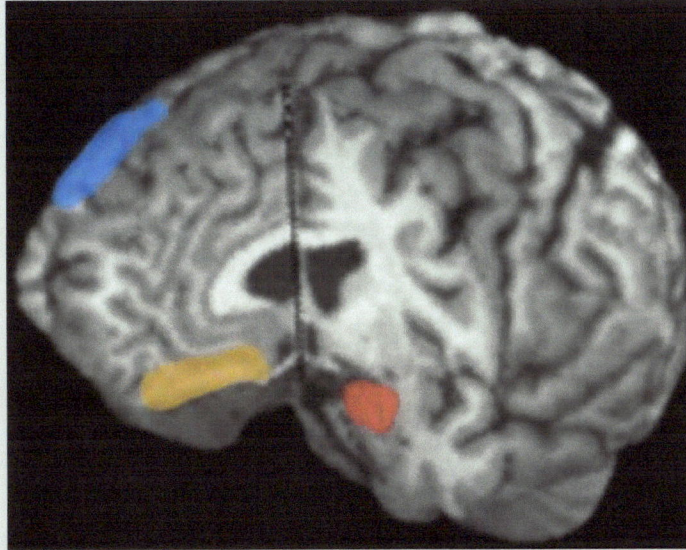

Emotional reactions may come in many forms and have different causes. But one of the main responses is the fear response, which has been shown to involved the amygdala. Different nuclei of the amygdala may contribute differentially to the fear response process.

One vital feature of emotion and amygdala is that emotional responses can be reduced, and eventually diminish. This is one of the basic mechanisms at play when we habituate to (or even extinguish) fearful stimuli. But is is also possible to reduce fear responses through more controlled processes, what has been termed cognitive emotion regulation. Such basic cognitive mechanisms underlie the psychological treatment of, e.g., phobias. In other words, there are two ways of reducing fear responses of the amygdala: 1) through habituation/extinction and 2) through cognitive ("rational"?) processing.

However, the exact neurobiological nature of these processes have been unknown. In a recent paper in Neuron, authored by Mauricio Delgado, and including prominent emotion researchers such as Joseph LeDoux, Elisabeth Phelps, looks at precisely this relationship. Using an emotion regulation strategy, the researchers compared the brain mechanisms (using fMRI) for conditioned fear regulation and for classic extinction.

From the methods section, one can read:

> "Each trial began with the presentation of a word cue, presented for 2 s, which instructed the participant on the type of trial. It was followed by either a blue or yellow square that served as a conditioned stimulus (CS) and was presented for 4 s. A mild shock to the wrist served as the unconditioned stimulus (US) and was administered during the last 200 ms for six of the CS trials. During one experimental session, a specific colored square (e.g., blue) was paired with the US, thus serving as the CS+, while the other square (e.g., yellow) served as the CS−. This contingency was counterbalanced across participants. The trial concluded with a 12 s intertrial interval.
>
> When instructed to "attend," participants were asked to view the stimulus and attend to their natural feelings regarding which CS was presented. In these Attend trials, for example, participants might focus on the fact that they may receive a shock (if the cue was followed by a CS+) or would never receive a shock (if the cue was followed by a CS−). When instructed to "reappraise," participants were asked to view the CS and try to imagine something in nature that was calming, prompted by the color of the CS. During these Regulation trials, for example, participants could think of an image of the ocean or a blue sky when viewing the blue square, or they could think of the sunshine or a field of daffodils when viewing the yellow square."

During both cases of fear reduction, the amygdala (red in top image) activation level went from high to low, for both What the researchers found was that during extinction learning, the ventromedial prefrontal cortex (orange in top) showed a higher activity, and this was thought to cause the observed reduction in amygdala activation. In contrast, cognitive emotion control lead to a higher activation in the dorsolateral PfC (blue in top image).

So this is a very nice demonstration of two different mechanisms of emotion regulation. However, it stills seems open to me whether the two are overlapping or very different mechanisms. One way of assuming the relationship is that the dorsolateral PfC works through the ventromedial PfC on regulating the amygdala. However, it may also be possible that the dorsolateral PfC bypasses the ventromedial PfC altogether. By comparing the activation patterns of all three structures, the findings suggested that the dorsolateral PfC works on the amygdala through the ventromedial PfC. Or, as put by the authors:

"Our results support a model in which the lateral PFC regions engaged by the online manipulation of information characteristic of cognitive emotion regulation strategies influences amygdala function through connections to vmPFC regions that are also thought to inhibit the amygdala during extinction. These results are consistent with the suggestion that vmPFC may play a general regulatory role in diminishing fear across a range of paradigms."

The implications of these findings may be plenty, but a few immediately comes to mind: first, the identification of the dorsolateral PfC in controlling emotions may, in general, be used as a marker for emotional regulation in different psychological states. Lie detection may be one issue, and studies of implicit racism seem to suggest the same. Another interesting consequence is in the modelling of the phylogeny and ontogeny of emotion regulation in primates. The present results may suggest that the dorsolateral PfC role in emotion regulation has occurred later in primate evolution, and that it works through a more "ancient" ventromedial PfC basic regulation of the amygdala. It may even be possible that developmental studies can show that the later maturation of the dorsolateral PfC also corresponds to the development of emotional control. Finally, this idea may also serve as a good model for studying brain injury and the consequences of emotion regulation.

CHAPTER 7

Learning & Memory

During my clinical years as a neuropsychologist, I saw several patients with brain injuries or disorders that left their memory capacities severely affected. Take one particular woman in her late 40's, who had consumed too much alcohol over the course of many years. This is well known to lead to a deficiency in the extraction and consumption of particular B-vitamins, including thiamine. This deficiency, which can also occur in eating disorders and in certain heritable digestion disorders, is severe for certain cells of the brain. In particular, the structure called *hippocampus* is affected.

This was the case for my patient. As an effect of her drinking, she had destroyed much of the functions of her hippocampi[13], a phenomenon sometimes called the Wernicke-Korsakoff syndrome. Her symptoms were obvious: when I met her and brought her to my office, she greeted me and could say her name. I then went out for a cup of coffee for her and myself, a mere five footsteps away and lasting less than a minute. When I came back to my office, she stood up and greeted me as if she had never met me before.

What her symptoms tell us is that her ability to hold on to information for more than a few seconds was severely affected. That is, she could understand where she ways and who she was, and she had preserved memories that were either stored a long time ago, or for information that was present right in front of her. But she was *unable to store new information, and lost information that required maintenance for more than a few seconds.*

My patient had lost the ability to store long-term (and intermediate) memories. She lived in 20-second bubble, where she could not hold on to information beyond this time limit. And as her brain scan showed affection of the hippocampi, we could have reason to assume that this particular brain structure is responsible for the demise of her memory.

Indeed, the hippocampus has repeatedly been shown to be an important structure for certain kinds of memories (Kumaran 2005; Ramsøy 2009; Battaglia 2011). Today, we know it as the crucial structure for **declarative memories**, that is, memories that we can declare that we have, such as "I remember what I did yesterday", and "I know the name of the current US president" and "I know that 'etymology' is the study of the origin of words and the way in which their meanings have changed throughout history".

However, as we will see, not all memories are created equal, and memories can range from the inborn and instinctual to the fully sentient and wilfully acquired.

City maps, taxi drivers and brains

How important is the hippocampus for spatial navigation? In a famous study by Eleanor Maguire and her colleagues (Maguire 2000), London taxi drivers were

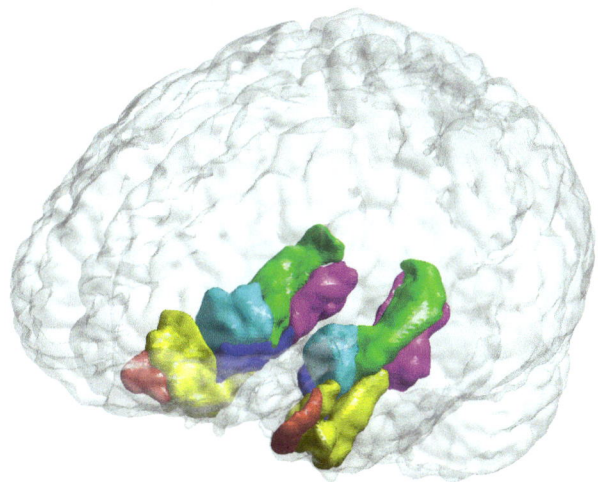

Figure The medial temporal lobe (MTL) region is a densely packed collection of structures in each hemisphere that together mediate the perception, storing and recollection of events. The structures are the temporopolar cortex (red), perirhinal cortex (yellow), entorhinal cortex (blue), amygdala (cyan), hippocampus (green), and parahippocampal cortex (purple).

studied with structural MRI scanning and neuropsychological tests. One thing that taxi drivers should obviously be good at is to navigate around the city, and consequently one could assume that there could be structural differences in regions responsible for spatial navigation compared to non-taxi drivers.

Indeed, this is what the researchers found: taxi drivers showed an enlargement of the posterior (back) part of their hippocampus, compared to control participants. Similarly, they also performed better on neuropsychological tasks that involved spatial problems. Interestingly, the researchers also found that the improved skills and enlargement of the posterior hippocampus, was at the cost of a relatively smaller anterior hippocampus, and a lower performance on tasks that involved object (but not spatial) memory.

In a follow-up study, Maguire, Woollett and Spiers (Maguire 2006) compared taxi drivers to bus drivers and found that compared to bus drivers, taxi drivers had larger posterior hippocampi and smaller anterior hippocampi. Moreover, the researchers found that years of experience as a driver correlated with the size of the structure, but only in taxi drivers. That is, the longer they had been taxi drivers, the larger

the posterior hippocampi and the smaller the anterior hippocampi.

WHAT IS MEMORY?

To have a working definition of memory, consider this from Endel Tulving, one of the fathers of modern cognitive neuroscience:

- **Learning** is the acquisition and containment of information
- **Memory** is a living organism's ability to contain and make use of information

Memory is notion of processing and retainment of information, broadly defined, in a way that allows subsequent use. However, this definition is pretty broad and all-inclusive, so we need to work out some subcategories from which we can talk about memories. Learning a new motor skill such as how to ride a bike is not the same as the vivid recollection of a movie. For brands and products, there is a large difference between having positive associations to a particular brand, and knowing how to operate a particular smartphone. Memories also have different durations – spanning from milliseconds to years. This allows us to make the following distinctions:

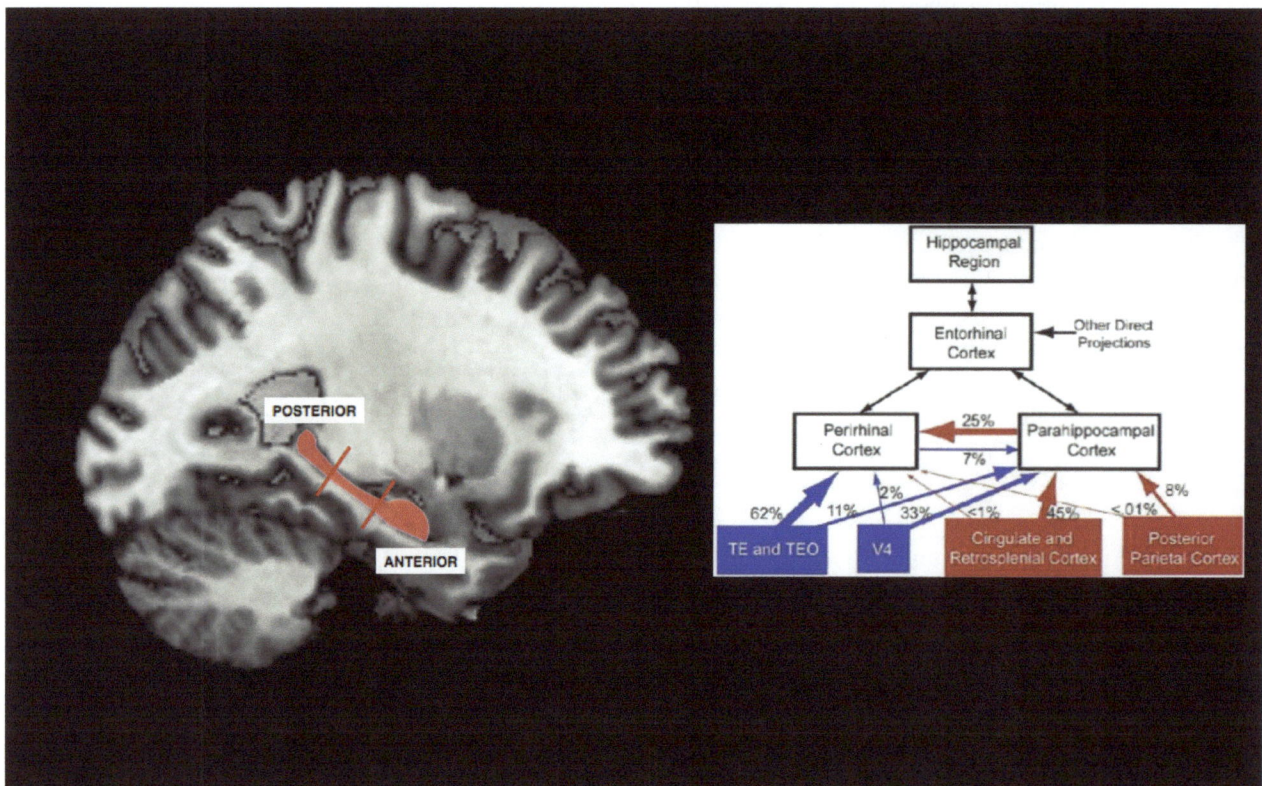

Figure In the London Taxi driver study, the comparison group had larger anterior hippocampi, while the taxi drivers had larger posterior hippocampi, as shown on the structural brain scan (left). The anterior and posterior hippocampus receives very different input (right).

TYPE	DURATION	KIND	EXAMPLES
Sensory memory	Milliseconds to seconds	Iconic, Echoic	Spots in your vision after a flash Hearing what a person says after some time
Working memory	Seconds	All contents (?)	Remembering a written sequence of word Keeping an oral sequence of numbers in mind
Intermediate memory	Seconds to minutes (hours?	All contents (?)	Remembering your line of thought a few minutes ago
Long-term memory	Hours, days, years	All contents (?)	Having a flash of recall from your past

DIFFERENT KINDS OF LONG-TERM MEMORIES

Cognitive neuroscience textbooks often use a neat distinction between particular kinds of memory, and organise them in nice flowcharts. Here, we will call these models collectively as the *Squire-Zola* models of memory, after the works by Larry Squire and Stuart Zola. Suffice to say that recent studies have challenged both whether memory work in this way, and whether the brain regions are indeed correctly implicated in these functions. Alternative models have been proposed, but here we will focus on the Squire-Zola model and then point on the problems with this model.

The Squire-Zola model of memory

The memory model was originally based on rodent and imate research, and in combination with studies of brain lesioned patients. Based on these findings, researchers suggested that there is an overall distinction between conscious and unconscious memories, or declarative and non **declarative memories**. In this scheme, declarative memories are those which we can explicitly state that we know something. Declarative memories were further subdivided into **episodic** and **semantic** memories: memories for events and facts, respectively.

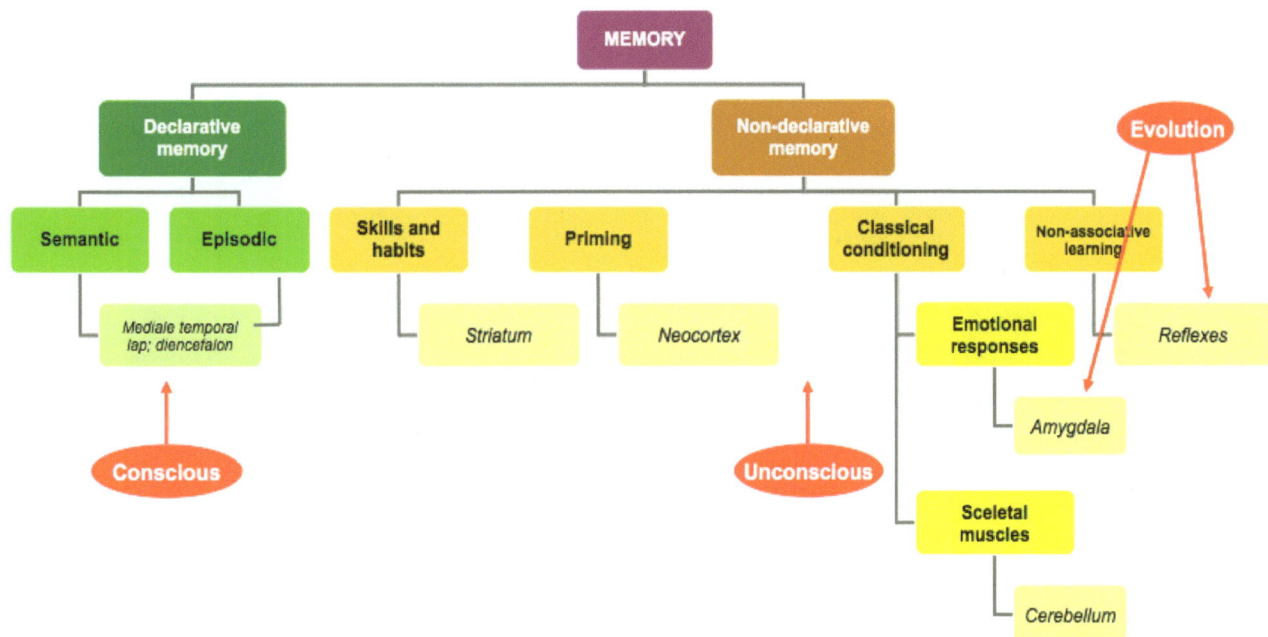

Figure The Squire-Zola model of memory and the brain.

The declarative memories were thought to rely on the medial temporal lobe region, from a very simple and powerful observation: people with no medial temporal lobe had lost their ability to form new semantic and episodic memories! My own Wernicke-Korsakoff patient clearly demonstrated this: she was perfectly able to learn new skills, but she would not remember that she had learned the skill!

Conversely, **non-declarative memories** are memories that we cannot state explicitly that we know, but where we can still demonstrably show that some kind of information has been retained. This contained a list of many different memory functions, including:

- **Skills and habits** – Acquired abilities and motor movements, ranging from the ability to juggle to the routine behaviours we build in our everyday lives.
  - This function is thought to rely on the striatum and basal ganglia.
- **Priming effects** – This phenomenon is well established in behavioural economics and psychology, and is crucial to branding effects to happen. If products are accompanied by a brand, the person's preference for this brand will imbue value to the product. For example, people tend to give higher rating to fashion, art, wine and coffee products if they believe that the product is from a high-value brand. Similarly, single words can alter the way in which people cooperate in social dilemmas and situations.
  - This function was generally assumed to rely on the neocortex (the 'higher' brain functions), but much research now suggests that the hippocampus and possibly the dorsolateral prefrontal cortex play important roles.

- **Classical conditioning**[14] – Discovered a century ago, our ability to learn associate an otherwise neutral item with a positive or negative outcome has been suggested as a powerful learning mechanism for marketing – brands imbue value to products that would otherwise be relatively indistinguishable from each other. Take, for example, coffee. People are bad at distinguishing between different brands of coffee. Nevertheless, they will be easily influenced to feel that their experience of a coffee from a preferred brand is much better than coffee from a non-preferred brand, even though we intentionally fooled them and gave them the same coffee on both occasions.
  - This function has often been linked to emotional regions such as the amygdala, and we know that we can now extend this to the ventral striatum, and possibly extend it to the hippocampus and medial temporal lobe region.
- **Non-associative learning** – This covers behaviours that are the result of learning over evolutionary time. Behaviours that have increased our likelihood to survive have been carried on to the next generation. Today, we rarely think about having these behaviours, but we know that several human behaviours are indeed reflexes. For example, the suckling behaviour of infants is well known: you can stroke a finger on an infant's chin and it will turn towards the cause of the stroke and start sucking. Similarly, completely newborn infants demonstrate a similar trend of looking up towards the caregiver to obtain eye contact, and even be able to mimic facial expressions, all in the name of increasing the likelihood that it can achieve a dyadic relationship with its caregiver.

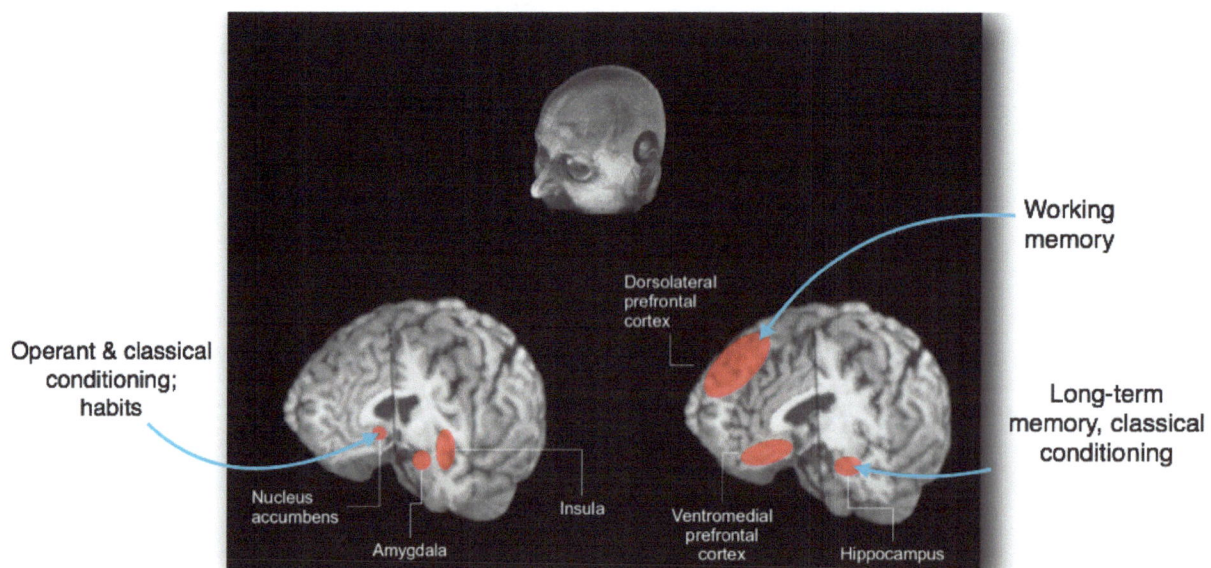

Figure   Brain regions associated with different types of memory.

- These functions are thought to rely on anything from spinal cord reflexes (as when you pull away your hand from a hot stove) to basal ganglia functions and maybe even cortical functions.

We will treat many of these behaviours more in the chapter about decision making and executive functions. Below you can find an overview of the most prominent brain regions related to memory

Problems with the Squire-Zola model

The Squire-Zola model has been very influential for decades, but has also attracted much criticism, especially during the past years. A few prominent critical issues include:

- The medial temporal lobe is not only about conscious memories – studies have implicated regions such as the hippocampus in implicit memories (Chun 1999; Wang 2010). This goes against the assumption that declarative memories are tightly and uniquely driven by this region.
- The model has a simplistic neo-phrenological view of the relationship between brain and mind: many structures have multiple functions, and many functions rely on multiple brain systems – this breaks with a core assumption of the model.
- Learning mechanisms such as classical conditioning have been noted as non-declarative in the model, although research has pointed to a significant role of "declarative" regions such as the medial temporal lobe (Schmajuk 1992; Berger 1976).

Based on these and other problems, alternative models have been proposed. For example, Katharina Henke (Henke 2010) has suggested that memory systems of the brain should not be distinguished based on whether they are conscious or not, but rather in terms of their processing modes:

"Consciousness seems to be a poor criterion for differentiating between declarative (or explicit) and nondeclarative (or implicit) types of memory. A new model is therefore required in which memory systems are distinguished based on the processing operations involved rather than by consciousness."

Although the matter is in no way solved, it is likely that a more fruitful model of memory and the brain should focus on the actual memory processes that occur, rather than making clear distinction between whether the process is conscious or unconscious.

The omnipotency of binding

Is *binding* the single most important concept in neuroscience? I think it is, even without making the concept too general or vague. On the contrary, binding seems to be a general concept to understand the workings of the brain. No more need for modules of perception, cognition, memory and action. Binding is the solution.

More specifically, what is binding? Or, to reframe the question 100%: what happens when the brain works? To many, the brain *binds information* together at all levels throughout the brain. If you perceive an object, that particular object is a mixture between colour, form, position, movement etc., that is bound together. Because of you look at the early sensory processes in the brain, we know that the features of an object are treated by separate processes in the brain. Accordingly, they can be lesioned separately, leading to e.g. acquired colour blindness but with intact movement perception.

Later in the processing stream we find that lesions to e.g. the fusiform gyrus leads to agnosia, where patients lack the ability to recognise objects, despite being able to recognise their individual features. Here, one could say, we have an example of a binding mechanism gone awry. People with this problem cannot bind the relevant information together, because the biological wetware subserving it is damaged.

What, then, about memory? Memory is, as you (should) know, a mongrel concept. It covers a lot of different processes and types of memories. There is, for example, a general distinction between declarative and non-declarative memory. In declarative memory, the prime example being episodic memory, discrete events are coupled together into a scene and with a temporal frame (i.e. a storyline). As we remember such events, they are situated in both a spatial and temporal setting. This kind of memory is thus yet another example of how binding is a prerequisite for proper neurocognitive functioning.

Indeed, it seems that binding is a core function, or even lingua franca, of brain function. I have previously reviewed the release of the *Handbook of binding and memory*, from Oxford University Press. This book indeed taps into this very discussion, presenting nice overviews from different parts of the cognitive apparatus where the concept of binding is essential and influential. The mere treatment of binding as a single topic is probably one of the most important this year. It's a must-buy, must-read, and must-have-on-the-shelf. Go get it!

WORKING MEMORY

Keeping information active for a few seconds can be extremely important. Just imagine not having this ability and then try to do any mathematical calculation. Try to remember a person's name, or any crucial information, without having the ability to hold on to it for the first crucial seconds! It would simply not be possible. Not surprisingly, any ability to hold on to information for a few seconds will also be involved in ensuring long-term memory: if you cannot hold on to information for just a few seconds, there is no chance that the same information can go on to longer-term memory.

One function that is crucial to this relative short-term memory function – which we now call **working memory** – is the **dorsolateral prefrontal cortex** (or dlPFC for short). This region lies on the top and slightly to each side of the front of your skull. When you try to keep some information active during a few seconds, this is the structure that gets engaged. It is involved in *maintaining information* for a few seconds. Lesion to this region is well known to eradicate a person's ability to hold on to short-term information.

Interestingly, our working memory also has a bottleneck problem. You simply cannot remember an unlimited amount of information, even though you try your best. Try reading the following sequence of numbers, then wait a few seconds, and then write it down on a piece of paper, or name our the sequence loud. Then repeat with the next sequence, then the next and so on.

    4 7 3 1
    6 9 2 8 5
    2 9 4 3 9 0
    2 8 6 5 1 3 9
    1 5 3 6 7 4 2 9
    8 5 9 0 3 1 4 2 6

OK, so far so good: how well did you do? Did you get all these numbers in the right sequence? Typically, we have what we have a *forward digit span* of **7±2** items. This means that most people are able to remember seven items on average, and the normal variance is between five and nine items. If you remember less than this, chances are that you are stressed by other competing stimuli, but one cannot rule out brain disorder or injury. If you perform much higher, chances are that you are either employing a so-called *chunking strategy*, which allows you to conjoin items. But it might also mean that you just have a superior working memory span.

Now try to take the same sequence of numbers, and do as you just did, but with one crucial difference: as you have read a sequence, wait a few seconds, and then write down or say the sequence in the reverse order. As you will discover, this is a much more difficult task, and chances are that your performance will be much lower than the first time. We call this the *backward digit* span test, which is also often used to assess working memory capacity.

Working memory is a crucial component in consumer choice, such as when we want to study how much information they can process at any one time. Both in commercial and academic research, this factor is important but largely ignored. With a better understanding of the limits of our mental capacity, we can better understand how to communicate with consumers. Likewise, traditional academic models are, at best, ignorant about the effects of working memory on consumer choice.

Frontal theta and cognitive load

One measurement of cognitive load is found when looking at the EEG. If one uses the frontal (midline) electrodes and extract the theta band frequency (following a so-called Fourier transform of the raw EEG signal) it is possible to assess **cognitive load** reliably. Recent studies have demonstrated that the frontal theta is closely linked to cognitive load – that is, the active maintenance and recall of working memory representations (Jensen 2002; Onton 2005).

Recently my student Samir Karzazi and I studied cognitive load in relationship with reading of financial reports with different levels of information. By testing trained accountants and economists, we made three key observations:

- Cognitive load was higher for reports with high as opposed to low levels of information
- Participants who had prior experience with reading reports showed a (slightly) lower cognitive load response when reading the reports
- The relationship between cognitive load and evaluation of the report depended on the outcome stated in the report (good vs bad news)

This is illustrated below:

THE NEURAL BASES OF MEMORIES

What are memories made of? This question seems easy at the superficial level, but once we start looking at the core components of memory, things become extremely complex. Here, we will only superficially cover some of the more complex mechanisms, but refer to sections where more can be learned about it. We will

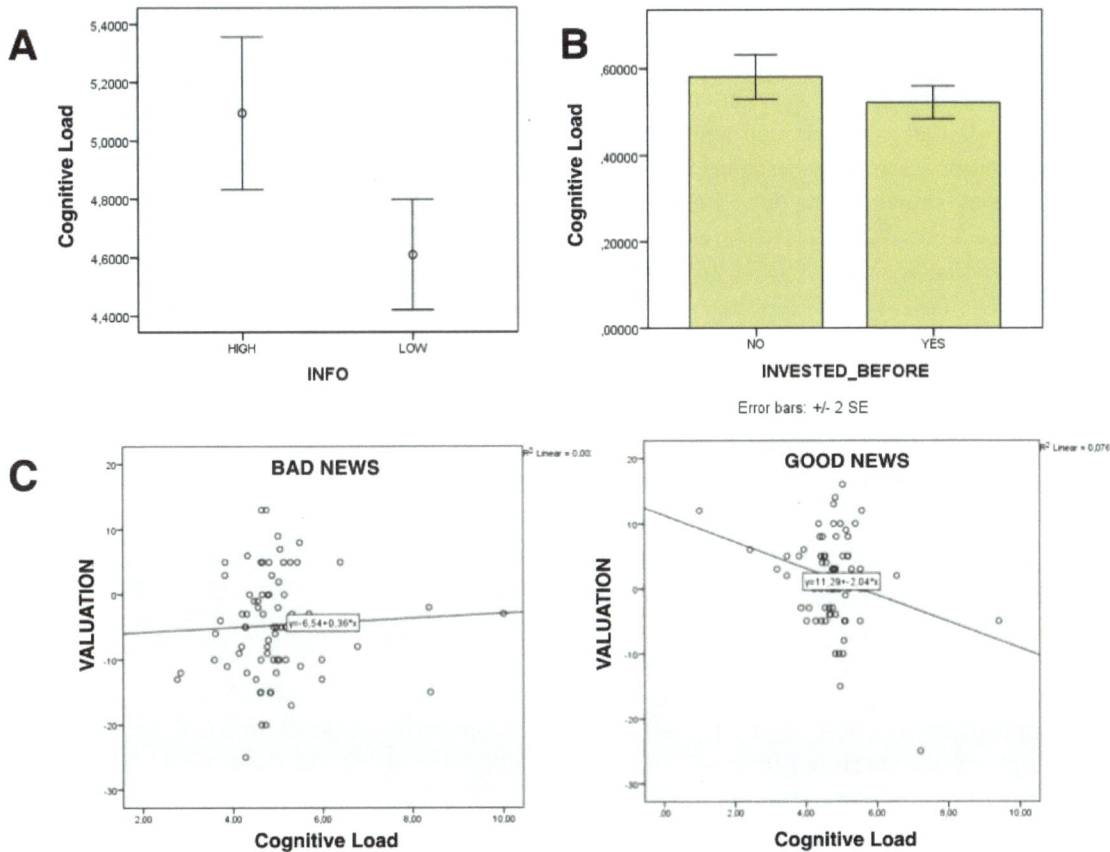

Figure The effects of information on cognitive load and preference. Financial reports containing more information are associated with higher cognitive load (A). If a person has prior experience in investing (and therefore tentatively more experience in reading financial reports) cognitive load is lower (B). The relationship between cognitive load and evaluation of financial reports depend on whether the financial report holds good or bad news: for bad news, cognitive load shows a slightly positive relationship to valuation, while for good news, higher cognitive load decreases valuation substantially (C).

then focus more on the brain regions associated with different kinds of learning.

Cellular learning

If we look at how cells learn, the neurons of the brain is indeed a unique species. This is not to say that learning does not occur in other parts of the body: the immune system is an extraordinary learning device for the body to learn how to protect itself. But neurons act in a very different manner. If we look into a single cell, there are mechanisms within the cell that provide the building block for learning.

Take one example, on how cells "bind" together. Famously, in 1949 the neuroscientist Donald Hebb stated that two cells that "fire together, wire together". In other words, if you have two neurons that are stimulated in such a way that they tend to fire together, they will evolve a connection in which they can mutually influence each other. Based on this observa-

tion, in 1966 Terje Lømo discovered and labeled the mechanism called **Long-Term Potentiation**, or LTP for short.

The mechanism underlying LTP stems from at least two different responses. As noted in the chapter about the brain, brain cells communicate via, e.g., axons-to-dendrite communication. Axons and dendrites that are consistently co-activated tend to show two responses:

- **More expression:** the axon transmits more neurotransmitters to the dendrites (or, rather, the synaptic cleft)
- **Higher sensitivity:** the receiving cell's dendrites can develop more receptors, which increases the sensitivity of the dendrite

There are also so-called retrograde messages, where the receiving cell has been shown to send messages back to the sending cell, and boost neurotransmitter production and transmission[15].

Taken together, these mechanisms allow cells to "bind together", just as Hebb had noted decades ago. LTP is a powerful mechanism that makes associations – learning – stick for hours, days, weeks and years. This also suggests that all memories are amenable to change, and no memory is immune to change.

This is exemplary shown by the opposite learning effect: Long-Term Depression (LTD) Just as we saw LTP is a mechanism that boosts the efficacy in communication between two cells, so LTD has the inverse effect: it decreases the effect that one cell can have on another. Why, you might ask, does it make any sense for two cells to have less communication? The answer is relatively simple: it improves communication and information processing overall. Here is what this kind of learning allows the brain to do:

- **Increased sensitivity** – The role of LTD is to allow a process to use less energy and time for a stimulus to trigger an effect. An over learned task, such as juggling, will require less effort and time, and be more coordinated after learning has occurred.
- **Increased specificity** – One role of LTD is for a process to not only be more sensitive, but also to inhibit other competing processes to be activated. For example, it is important that while we are riding a bike, our legs do not start running. So there needs to be an activation of the "bicycling program" but also importantly a de-activation of the "running program". This is stretching the analogy a bit, but the main insight is that LTD is an equally important mechanism that allows the brain to do things more efficiently.

Memory regions of the brain

It is reasonable to say that the whole brain is a learning organ: it is a retainer of information that is used

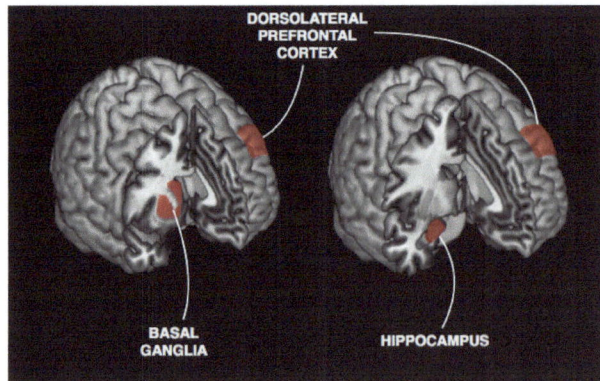

Figure Three learning and memory regions of the brain. The basal ganglia (consisting of the putamen, caudate nucleus and nucleus accumbens), the hippocampus, and the dorsolateral prefrontal cortex. Please note that the cutout in the left brain is more anterior (front) than the cutout in the right brain.

to guide behaviour efficiently, and to adapt according to input and effects of one's behaviour. Nevertheless, we tend to associate some brain regions with memory more than others. Here, we will focus on three regions: the dorsolateral prefrontal cortex, the hippocampus and the basal ganglia.

For working memory, we know that the **dorsolateral prefrontal cortex** (dlPFC) is crucial, possibly with a strong interconnection with the parietal cortex, and, depending on the content, regions of the visual cortex, auditory cortex, or inferior temporal lobe. Studies have consistently demonstrated that regions of the dlPFC are engaged when we have people perform a working memory task, or similar tasks that tax our short-term memory abilities. Similarly, when we have people working with increasingly *complex* information, we see an increase in dlFPC activation. This function, which we previously referred to as **cognitive load**, is a crucial component if we want to evaluate consumers' mental efforts in trying to understand particular contents. For example, is the plot in a movie too complex? Having an index of working memory or cognitive load will allow researchers to better pinpoint exactly where people are overloaded and eventually drop off in their information processing.

As we have seen in this chapter, the **hippocampus** and the surrounding medial temporal lobe region has been implicated in declarative memory function. Indeed, much evidence supports that this region is strongly involved in our ability to remember episodes from our past, as well as our ability to know the meaning of words and store facts and knowledge about the world. It is assumed that the hippocampus does this *by binding* regions of the brain together during learning, and that a memory is the *reactivation* of these memories. To point further to this role, think about how a single stimulus, such as an odour, can trigger a rich experience. This was nicely demonstrated by Barbeau and colleagues (Barbeau 2005), who performed deep brain stimulation to a patient's perirhinal cortex, adjacent to the hippocampus, and noted that the patient reported an episodic memory that started with a single item (a boot) that expanded to the full episodic memory replay of a neighbour wearing boots while driving a motorcycle.

Another notable memory region is the **basal ganglia** – a collection of structures encompassing the putamen, caudate nucleus and nucleus accumbens. This region has long been implicated in motor learning, such as learning how to ride a bike, and other motor behaviours. Due to its central positioning and strong connections to the motor pathways of the brain, the basal ganglia have been highly implicated in automated and habit behaviours. Moreover, the basal ganglia has been implicated in functions such *as prediction and expected utility*, albeit more as an unconscious function, and just as we saw in the aforementioned study by Pessiglione and colleagues (Pessiglione 2008). Finally, as we saw in the chapter about emotions and feelings, and as we will see in the chapter on wanting and liking, regions of the basal ganglia are involved in unconscious emotional responses and even unconscious motivation. Its central positioning in the brain, combined with a missing access by conscious control, makes the basal ganglia the "hidden agenda" of the brain, which allows for many direct actions to be made with little or no conscious control.

CONDITIONING – CLASSICAL AND OPERANT

One of the most well established terms in memory research is *conditioning*, which can be broadly defined as have something that has a significant influence on or determines the manner or outcome of something else. More specifically, conditioning is the process with which organism learns to associate elements both in terms of the world and their inner states, which ultim-

ately can provide increased likelihood of survival.

Today, we are talking of two kinds of conditioning – classical and operant. Classical conditioning is a case of *passive* learning in which we learn to associate two or more things. Operant conditioning can be labeled *active* learning, as it depends on actions on behalf of the organism.

Classical conditioning

How we learn to associate things was more or less accidentally discovered by Ivan Pavlov and Edwin Twitmyer[16] at the end of the 19th century. While doing his work on the function of salivation in mammals, Pavlov noted that salivation did not occur at the time that food was provided – it could occur much earlier, such as the sound of footsteps of an assistant bringing the food. This led Pavlov to the idea that mammals, including ourselves, could learn to associate events (footsteps and food), which would eventually lead us to display a particular behaviour (e.g., salivating) to an otherwise irrelevant stimulus (e.g., footsteps).

In researching this further, he observed that this kind of learning occurred in stages. We start off by the original behavioural response, such as the drooling response to food being presented, which is typically cal-

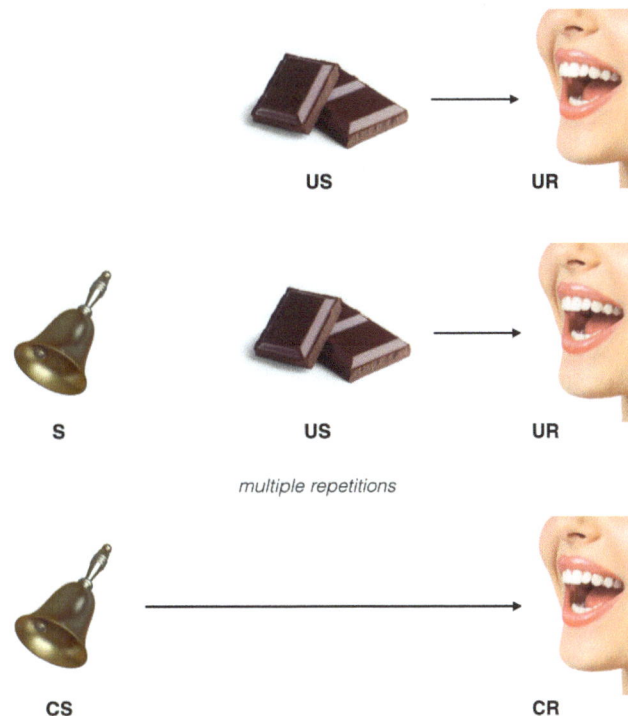

US UR

S US UR

multiple repetitions

CS CR

Figure The process of classical conditioning. An unconditioned stimulus (US) can trigger the unconditioned response (UR), since it is an inborn response tendency. If an otherwise unrelated stimulus (S) is presented with the US repeatedly, it will be associated with the US, and can finally trigger the response by itself. Here, the stimulus has become a conditioned stimulus (CS) and the response to the CS is called a conditioned response (CR). Notably, the example with the bell here can be replaced by a brand, a person, a jingle, or any other item.

led the **unconditioned response (UR).** The food that causes the UR is similarly called the **unconditioned stimulus (US)** since drooling to food does not need any prior learning:

1. A stimulus, let's say a bell, is presented prior to the US. At this stage, there is no relationship between the bell, the food and the drooling response.
2. After repeated trials with the bell preceding the food, the bell can trigger the drooling response, without the food ever being present. At this stage, the bell can act as a trigger to the response, and is called the **conditioned stimulus (CS)** and the accompanying drooling response is called the **conditioned response (CR).**

The really interesting aspect of classical conditioning is that it provides a vehicle for brand communication. Brands that are successfully coupled to positive experiences (e.g., eating chocolate) has the potential to provide positive experiences in themselves. This provides an effect from which brands can introduce novel products, and have a "flying start."

Operant conditioning

Classical conditioning is, if anything, a relatively passive learning experience. That is, it assumes that an organism is passively observing a stimulus and learns to associate it with a particular outcome. However, organisms are in constant flux and movements, and therefore a major learning mechanism should have evolved to deal effectively in learning to associate its behaviours with outcomes.

This was treated by several scholars from the so-called *behaviourist school of psychology,* which assumed that conditioning was such a prominent learning mechanism to animals and humans alike that our stored associations were identical to who we were as persons. For example, John B. Watson, who was the founder of Behaviourism, made the following bold statement in 1930:

> "Give me a dozen healthy infants, well-formed, and my own specified world to bring them up in and I'll guarantee to take any one at random and train him to become any type of specialist I might select--doctor, lawyer, artist, merchant-chief, and, yes, even beggarman and thief, regardless of his talents, penchants, tendencies, abilities, vocations, and race of his ancestors. I am going beyond my facts and I admit it, but so have the advocates of the contrary and they have been doing it for many thousands of years."

Operant conditioning was the active learning mechanism counterpart to classical conditioning, in which one studied how an individual's behaviour was modified by the consequences of its actions. For example, let's assume that a child reaches out for a wasp and gets stung. The child soon learns how to associate its behaviour (reaching out) to the pain, and subsequently avoid repeating the behaviour (and the wasp altogether). The formula for this is relatively simple:

1. Reaching out to object X
2. Experiencing pain (punishment)
3. Learning to associate reaching out for X and pain
4. Next time X is encountered, reaching behaviour is less likely to occur

Please note that in terms of learning, we rarely, if ever, deal with absolutes. Preferences and contingencies can reverse, and although a particular behaviour leads to an adverse effect, it does not necessarily leads to a 100% avoidance of that behaviour. More likely, learning to associate pain with a behaviour is likely to lead to a reduced likelihood of repeated behaviour, but not an absolute either-or change.

Let us also go back to the example of Pessiglione's study (Pessiglione 2008), where we found that even subliminally presented abstract symbols could affect gambling behaviour. In this study, the gamble was the original stimulus, but unconsciously the abstract symbols signalled what was about to happen. In itself this is a case of classical conditioning, but it also requires that the person makes the link and acts upon it. The learning to associate abstract symbols to outcomes, and act accordingly, is a good case of operant conditioning

Brain bases of conditioning

As with learning, many brain functions tend to be associated with conditioning, and probably depends on the type of learning that occurs. For example, studies have demonstrated that the learning to associate a stimulus with reward – which also increases the likelihood of approach behaviours – is associated with the basal ganglia, and notably the ventral striatum (aka nucleus accumbens). For example, Lau and Glimcher (Lau 2008) demonstrated that the striatum was engaged both during pre-choice and post-choice, albeit they showed two distinct network activations. Similarly, O'Doherty and colleagues (Odoherty 2004) found that the engagement of the ventral striatum (aka nucleus accumbens) was more related to outcome prediction, while engagement of the dorsal striatum was related to the maintenance of information about the rewarding outcomes of actions, which would enable better actions to be chosen more frequently.

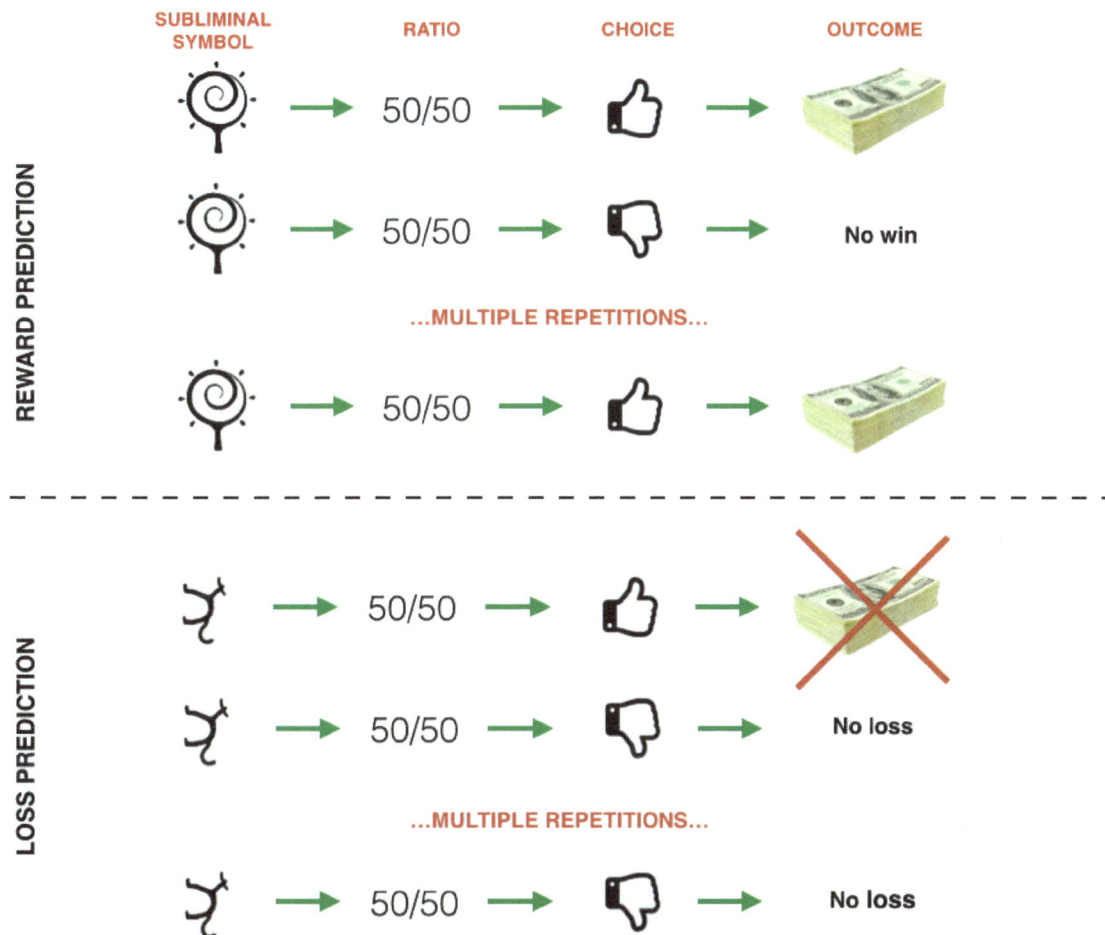

SUBLIMINAL SYMBOL — **RATIO** — **CHOICE** — **OUTCOME**

REWARD PREDICTION

...MULTIPLE REPETITIONS...

LOSS PREDICTION

...MULTIPLE REPETITIONS...

Figure Unconscious operant conditioning as found in the study by Pessiglione and colleagues. In the upper case, a subliminally presented symbol is presented prior to a 50/50 gamble, and regardless of choice, people are more likely to win. If they choose the gamble, they win. If they don't take the gamble, they won't win. After repeats, an unconscious learning takes place, where the brain learns to associate a symbol and action with winning. This increases the likelihood that the abstract cue can trigger risk taking. In the lower case, a subliminally presented abstract symbol predicts monetary loss during the gamble. If a person chooses the gamble, he will lose money. If he avoids the gamble, he also avoids losing the money. After repetitions, the person unconsciously learns to associate the symbol with outcome, and to act accordingly.

Conversely, the amygdala, and its link to the hippocampus, has often been cited as an important structure for the aversive learning of events (Coricelli 2005; Haber 2010; Sehlmeyer 2009; Pribram 1975; Edelson 2011). That is, when we are exposed to something that predicts a negative events, studies have implicated the amygdala, and via its direct links to the hippocampus, as an important network for aversive learning.

However, recent studies have also shown that this view is slightly simplistic. For example, studies have shown that the ventral striatum can be engaged to both expected rewards and expected punishment and pain (Levita 2009). Conversely, the amygdala has been implicated in reward prediction and processing (Murray 2007). These studies show that there is no simple or straightforward relationship between learning and brain regions.

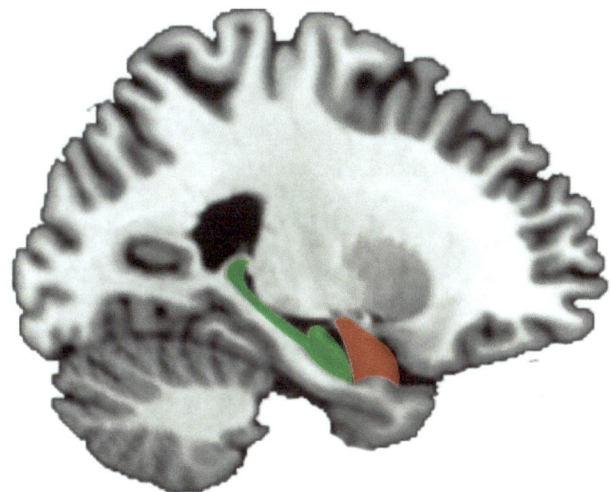

Figure The amygdala (red) and hippocampus (green) are positioned adjacent on the medial side of the temporal lobe and are tightly interconnected, working in unison in the processing of emotional stimuli and events.

Recently, a novel brain structure has come into the attention of neuroscientists. The habenula is a brain structure that has been largely forgotten and misunderstood. The position of this minuscule region is even hard to find on a brain scan, and the use of fMRI frequently counts its activation as mere noise. However, recent studies have shown that the structure is important for reward processing, such as reward prediction errors (BrombergMartin 2011). This is the term for describing the activation that occurs when a reward is either unexpectedly cued, omitted or delivered. Today, much research has focused on the role of the habenula and its role in decision making (Morissette 2008; Matsumoto 2007; Hikosaka 2010; Hong 2008; Haber 2010; Ikemoto 2010; Ullsperger 2003; Hikosaka 2008).

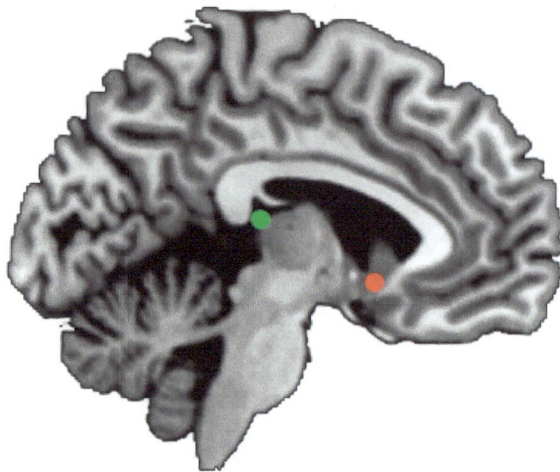

Figure The tentative positions of the habenula (green) and the nucleus accumbens (red).

ASSOCIATIONS AND SUBJECTIVE KNOWLEDGE

Write down anything that comes to mind when you read this word: "Coca-Cola"

You can do the same for anything: brands, products, people, events, words and so on. For each word we tend to have a number of connected thoughts, feelings and impressions. We call these *associations*. What you associate with a particular word, name or item is highly personal, but also tells a story about your relationship to this piece of information.

By just asking people to name their associations – also known as *free recall* – we can assess what their experiences and thoughts are about something. In a study for a famous children's toy manufacturer, we asked participants about their subjective knowledge as well as their free associations and preferences towards different toy brands.

Here, we found several interesting relationships. First, we saw that there is a positive relationship between subjective knowledge and preference. While this may not be that surprising, we nevertheless were surprised to see that there was a large variance in this relationship between the different brands. While some brands showed a steady incline in liking when knowledge increased, some brands had peaks in preference at a certain knowledge level. While our study aims did not allow us to explore the nature of these differences, such insights can provide essential insights and points of departure for further research for both academic researchers as well as brand managers.

Another interesting finding was made when we looked at the relationship between associations and preference. Overall, there was a positive relationship between the number of associations – what we call **associative density** – and preference. That is, there more things that people could mention when given a brand, the more they liked it. Digging more into this relationship, we found that the types of associations were important: positive associations (positive words connected to a brand) was strongly associated with a stronger preference. An even stronger relationship was found between negative associations and preference, although this relationship was opposite: the more negative words people noted in relationship with a brand, the lewes they liked it. Finally, the a higher number of factual associations (e.g., the word "bricks" when given the brand LEGO) was also related to increased preference, albeit not very strong.

The finding of a correlation between associative density and preference does not provide any insights into the causal mechanisms or the direction of the relationship. That is, it may be that the number of associations cause a higher liking, or it may be that a higher liking leads to the generation of more associations, or even that both factors are driven by a third, unknown factor. To study this, my student Helle Shin Andersen and I ran a follow-up study[17], where we manipulated the number of associations that people were given in ads. Two groups received identical ads, but some ads were manipulated to contain more visual associations. Here, we found that more associations was associated with increased brand preference.

What these findings tell us is that the way in which we work with associations and brands can be extremely important as indicators or even effectors of brand

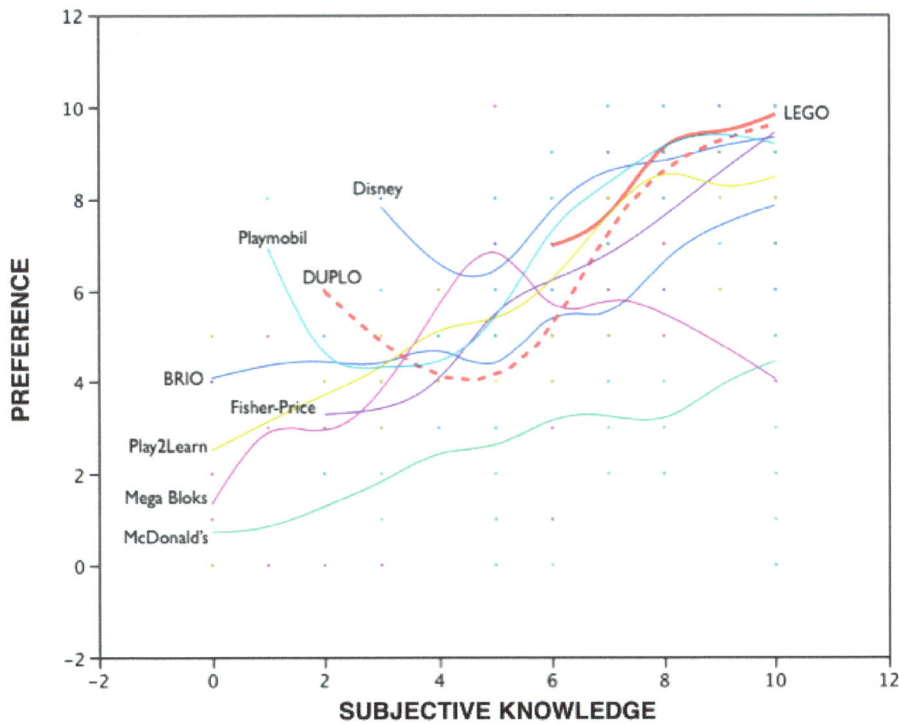

Figure Subjective brand knowledge and preference. As can be seen, there is both an overall positive relationship, but also highly individual characteristics of this relationship. While some brands show the positive trend, others like Mega Blocks, showed a peak preference around knowledge score 5, and then a drop off. Courtesy of Neurons Inc (www.neuronsinc.com)

preference. For brand managers, it is obvious what the priority list should be for working with consumers' associations:

1. Reduce or eliminate the number of negative associations
2. Increase the number of positive associations
3. Increase the number of factual associations

It thus seems that brand managers can definitely use the insights from studying associative density to gauge their consumers' thoughts and feelings, as well as working on changing the ration of positive and factual vs negative associations to their favour. It is interesting to note that factual associations had the least impact on preference, especially in light of what many companies believe that the most important thing about advertising is to inform the customer about the facts about their products. In this sense, tests of associative density can be important neuropsychological tools for consumer researchers.

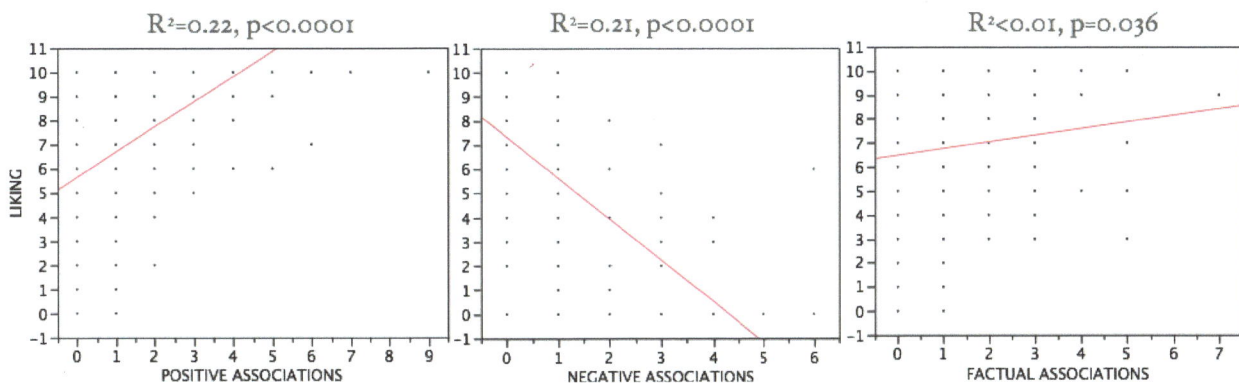

Figure Relationship between number and kind of associations, and preference. There is a strong positive relationship between the number of positive associations and brand liking. For negative associations, there is an even stronger but negative relationship. For factual associations, the relationship is positive and significant, although not particularly strong. Courtesy of Neurons Inc (www.neuronsinc.com)

Associative density and the brain

In a recent study[18], we have tested the brain bases of associative density. In particular, we focused on the role of the hippocampus in rapid encoding of associations when presented with brand-related figures. Here, we used fMRI to assess hippocampal activation while participants were making rapid judgments about whether an object had just been presented or not, a so-called 1-back task. What they saw was a number of different figures, including famous cartoon figures (e.g., Mickey Mouse) and brand mascots (e.g., the Michelin man). After the scan, all participants were given a surprise test of their free associations, subjective knowledge, preference and other questions related to each figure they had seen.

Based on prior research by Howard Eichenbaum (Eichenbaum 2001), a famous memory researcher, it has been suggested that the hippocampus plays a crucial role in rapid coding of associations to something we see or think about. Therefore, our fMRI analysis focused on this region, where we expected that the hippocampus would be more engaged for objects that had more associations, even when taking preference and subjective knowledge into account. Furthermore, based on recent suggestions (Friston 2008) that it is possibly not whole brain structures, but rather a subset of a particular structure that is engaged for these functions, we tested whether the whole hippocampus or just a subset of this structure (so-called

sparse coding) was engaged. Our results showed a staggering effect of sparse *coding* of associative density. This result demonstrates that subregions of the hippocampus encode the number of associations that people have to rapidly presented brand related items. Furthermore, our result also demonstrates that it is possible to predict consumers' associative density from brain scans.

Associative assets and associative goals

Associative density, while a highly interesting concept, remains a very novel concept, both in consumer research as well as in psychology and neuroscience. Therefore, much more research is needed before we have a complete understanding of the phenomenon and how it influences consumers' mind and actions.

That said, the assessment of associations can lead us to many novel insights in terns of consumers' attitudes, thoughts and actions towards brands and products. In particular, we can think of two main ways in which brand association assessment can be used as brand managers and academic scholars alike:

- **Mapping of association density and strengths** – this provides an insight into the number and kinds of associations to the brand and product
  - This also provides the foundation for actionable insights into how brand and product managers can communicate relative to their goals

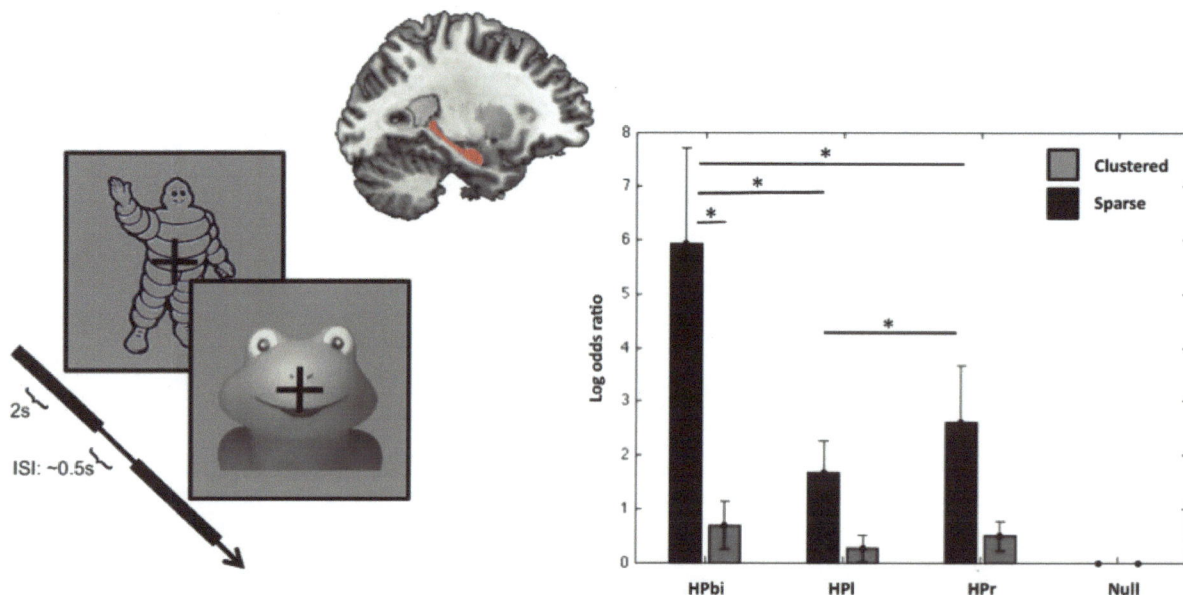

Figure   The effects of associative density on hippocampal responses. LEFT: the experimental design where participants were shown were shown figures for 2 seconds, with an inter-stimulus interval (ISI) of approximately 0.5 seconds. MIDDLE: the hippocampus region, wherein our main analysis was concerned. RIGHT: the effects of sparse vs clustered analyses of activity in different hippocampus subareas (HPbi, HPl, HPr).

- **Identification of unknown and possibly broader associations** – when analysing the associations, are any associations completely surprising and not previously thought about?
  - This can provide insights into how the brand can extend associations and "own" domains for which no links have yet been identified

Below are a few examples of this. From recent studies of different brands, it is obvious that certain associations are more surprising than others, while other associations confirm prior expectations.

Figure Associations to the brand Hershey's, using word cloud for illustrative purposes. Notice how Hershey's seems closely linked to coffee from this test.

In this way, we can talk about brand associations as a strategic goal for a company – the **associative goals** that they have. These are the associations that they want the consumers to make. A chocolate company wants consumers to automatically associate chocolate and certain positive words to their brands and products. Conversely, when provided a particular product type – chocolate – brand managers want their brand to be on the top of the list of the brands that first comes to mind. Working on these associations is one kind of brand management work, and assessing brand associations continuously can be crucial.

However, when measuring consumer associations, we often make some unexpected discoveries. Some brands are associated with words and experi-

Figure Associations to the brand BMW, using word cloud for illustrative purposes. Notice here how words like "team" and "DTM" (Deutsche Tourenwagen Masters) are associated with BMW from this source.

ences that nobody have thought about, or brand associations between a brand and words have a different strength than previously thought. Maybe brand managers at Gillette expect the word "smooth" to be a word that customers are associating with to their product, but that upon measuring no such association is found. Conversely, some words, like "disposable" may be more frequently associated than was expected (or, maybe it comes as a complete surprise). This provides what we can call **associative assets** – the actual associations that customers make to a particular brand or product. When we are surprised by associations, it can provide powerful and actionable insights that brand managers can use to build a new set of associations and communications upon.

Figure Associations to the product and brand Gillette, using word could for illustrative purposes. Words like "version" and "disposable" seem surprising and important, following more obvious associations such as obvious words like "blades" and "razor".

FRAMING AND BRANDING – ASSOCIATIVE SPREAD

The fact that branding works is actually a puzzle to many economists. Why should people value a particular product any differently just because you put a particular label on to it? Why do people favour Coca-Cola over Pepsi cola (or the opposite), why do they favour what they believe is clothing from a high-fashion compared to other brands, and why do they find that a wine they believe is expensive is tasting better than a wine they believe is cheap? And why does this happen even though we cheat them and secretly give them a third brand of cola, self-made garments, or a generic wine?

Framing puzzles us, but we have long realised that it is a foundation of how we work as consumers. If framing did not work, using brands would be meaningless. They would simply not work. They could not imbue products with value, and to consumers all products would be the same. This also illustrates some of the reasons that we have brands at all:

- To stand out from the clutter
- To signal a particular value
- To be recognised
- To be associated with a narrative

In a recent study in my lab[19], we tested how different aspects of branding affected people's product preference. We tested how different aspects – product price, the product's country of origin (CoO) and participants' nationality – could affect the total effect of product preference. In our case we had participants taste wine and rate their preference and willingness to pay for the product. As can be seen below, we made several interesting observations: the price of wine very much depends on where people perceive the wine to come from, combined with their own nationality. Interestingly, all participants were always given the exact same wine!

What this study shows us is that our product preferences are amenable to change by mere information that has really nothing to do with the product. If we believe that the product is French and expensive, we are automatically led to enjoy the product more, and be willing to pay more money for the product – *even though the quality of the product has never changed.*

How on earth can this happen? Let us look at a classic consumer neuroscience study.

Coke and preference

One of the best examples of the effects we observed in the wine tasting study, even has a label: the Pepsi Challenge. A few decades ago, Pepsi had realised that they were not winning the battle on soft drinks with Coca-Cola. However, it was infuriating that this happened in the face of numerous studies showing that consumers preferred their product in blind tasting – that is, when they were blind to which brand they were drinking.

First, although you may think so, you are not particularly good at telling Pepsi and Coca-Cola from each other. Try it yourself: make a blind tasting competition and see how many times you get it right. Chances are that you will be on around 50-50 chance in getting it right.

When Pepsi did this, they also discovered that consumers tended to prefer their product, which was likely due to an incrementally higher sugar level in Pepsi cola (possibly triggering a stronger activation of the nucleus accumbens). However, when they told people the brand they were drinking, things changed: people preferred Coca-Cola, even though they were sometimes secretly given Pepsi cola...

A decade ago, Sam McClure ran a study where he repeated this effect, now while scanning people's brain

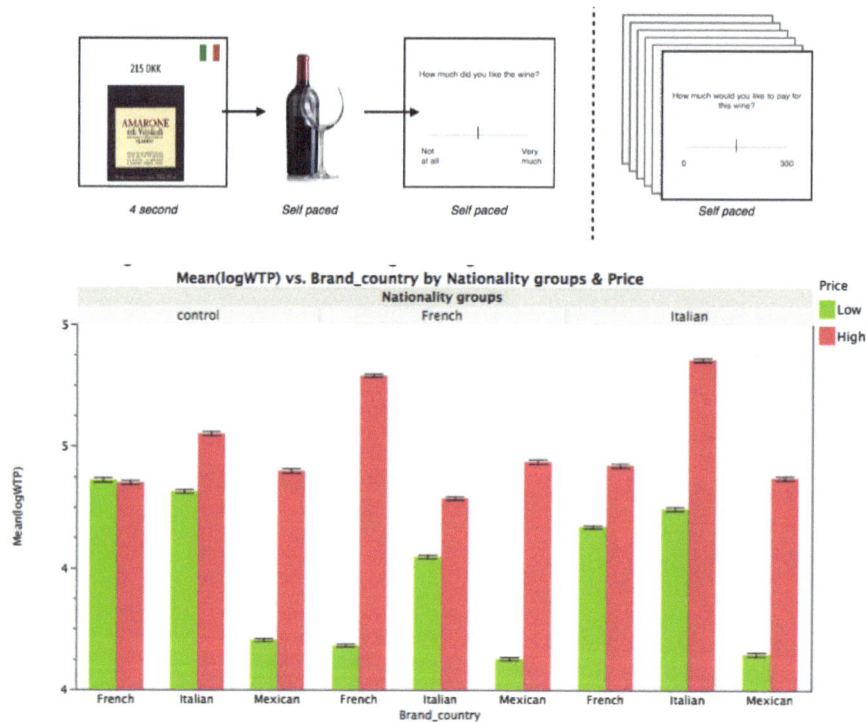

Figurte   Effects on wine tasting preference and willingness to pay. TOP: the experimental paradigm where participants tasted wine and rated it, followed by a post-test choice of their willingness to pay for the product. BOTTOM: the interaction effects of CoO, price and nationality. For example, French participants were much more affected by price of French wine than control participants and Italian participants, while the opposite was the effect for Italians and Italian wine. Interestingly, all participants were strongly affected by the price for Mexican wine.

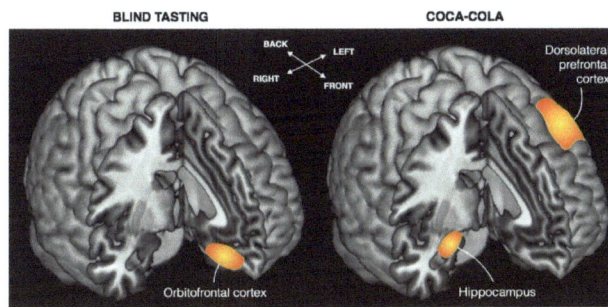

Figure The effects of branding on preference for soft drinks, where stronger product preference was associated with increased activation in the medial orbitofrontal cortex. LEFT: Seeing the Coca-Cola brand prior to tasting the cola produced a strong engagement of the hippocampus and dorsolateral prefrontal cortex. By comparison, seeing the Pepsi cola brand showed no such effect (not shown).

to better understand what actually caused this effect. First, the researchers had participants blind taste cola, and indicate their enjoyment of the drink. Here, he saw that the more people enjoyed a drink, the stronger activity was found in the medial orbitofrontal cortex. As we saw in the chapter of emotions and feelings, this nicely corroborates previous findings.

Next, McClure and his colleagues provided his participants with brands before they received the drink. Besides confirming the long-established effect on preference, McClure also made an interesting observation: when participants believed that they were drinking Coca-Cola, there was a stronger engagement of the hippocampus and dorsolateral prefrontal cortex. As we have seen, these are regions highly implicated in memory functions. However, when he looked at the same effects for Pepsi cola, no such additional activity was found!

This study reveals something crucial about branding: we are affected by brands through triggering of our memory responses, and brands that have a sufficiently strong impact on our experience and choice have this by triggering strong mnemonic (memory) processes in our brains and minds.

The hidden dualism

In an interesting paper in the latest version of Progress in Neurobiology, Yuri I. Arshavsky from UCSD writes about the epistemological dualism that exists in modern neuroscience. Basically, Arshavsky claims that there is a covert dualism in the way that neuroscientists are treating mind-related topics, especially the study of "consciousness". Indeed, as he claims:

> "This covert dualism seems to be rooted in the main paradigm of neuroscience that suggests that cognitive functions, such as language production and comprehension, face recognition, declarative memory, emotions, etc., are performed by neural networks consisting of simple elements."

This might initially sound a bit strange. Is not cognitive functions such as face perception due to operational simple elements? Face perception as such is a combination of many simple processes that operate in unison. So what is Arshavsky proposing? Indeed he suggests the existence of a certain kind of brain cells:

> "(The) performance of cognitive functions is based on complex cooperative activity of "complex" neurons that are carriers of "elementary cognition." The uniqueness of human cognitive functions, which has a genetic basis, is determined by the specificity of genes expressed by these "complex" neurons. The main goal of the review is to show that the identification of the genes implicated in cognitive functions and the understanding of a functional role of their products is a possible way to overcome covert dualism in neuroscience."

So there should exist a subset of neurons that integrate information from a variety of input. This sounds strange, since all neurons integrate inputs from thousands of inputs, many from a large variety of inputs. So what are complex neurons? Here, we are told that:

> "(...) neural networks involved in performing cognitive functions are formed not by simpleneurons whose function is limited to the generation of electrical potentials and transmission of signals to other neurons, but by complex neurons that can be regarded as carriers of "elementary" cognition. The performance of cognitive functions is based on the cooperative activity of this type of complex neurons."

In this way, complex neurons seem to be integrative neurons, i.e. cells that integrate information from a variety of processes. This could include the multi-modal neurons found in the functional sub-structures of the medial temporal lobe, such as the hippocampus, perirhinal, entorhinal and temporopolar cortex. But would it not mean the colour processing nodes in the visual cortex? This leads us back to a basic question: what is a functional unit in the brain. yes, the neuron is a basic building block of information processing in the brain. But what is special about language, memory and so forth in the brain?

It is possible that Arshavsky is not radical enough: what we should seek out is to avoid using generalistic and folk-psychological concepts in the first place. We should possibly not study "language", "memory" or "consciousness", since these concepts will always allude to fundamental assumptions of "language-ness", "memory-ness" and "consciousness-ness", IOW that there is something more to explain after we have found out how the brain produces what we recognize and label a cognitive function.

Maybe neuroscientists are not using a poor strategy after all? Maybe ignoring the past history of philosophy of mind is the best solution. I'm not sure (nor am I sure that I represent Arshavsky's view properly). But how we choose to label a cognitive function depend on our past historical influence and learning, as well as our current approach.

THE NEUROEQUITY BATTERY

Based on what we have learned, it is possible to make out a memory battery that addresses the different aspects of brand memory. I call this the NeuroEquity Battery. The test covers aspects such as memory strength, richness and emotional content, and is sub-divided into the following subtests:

- Top of Mind
- Category cued recall
- Recognition
- Subjective knowledge
- Associative density

Top of Mind

In this test, we ask people to name any brand they can think of. This is a completely open-ended question-naire, but it is important to note down the sequence of the brands as they are named. Typically, you should limit the number of brands to 20 or 30, but it really depends on the scope of your study.

Rating is done by giving the highest score to the brand that is mentioned first, and decreasing the score incrementally for each subsequently named brand. This score will then depend on how many brands you allow. There are other scoring techniques, but for our purposes here, this will suffice.

Category cued recall

This is similar to the Top of mind test, but here, you ask people to name the brands they can think of wi-thin a certain category of brands. For example "Name as many smartphone brands you can think of". This is a so-called *assisted recall*, where we allow ourselves to narrow down the range of brands that are mentioned

Limit yourself to 10 brands per category, which will also make a max score of 10 when you score as above. Depending on the scope of your study, you can also record any errors made, such as when a brand is erroneously mentioned under a wrong category. This could provide interesting insights both in terms of brand management, but also if you are studying a par-ticular kind of consumers.

Recognition

Show a brand – the name and/or logo – and ask whether people recognise the brand. This is often a very simple task, and most people tend to recognise brands when shown.

Subjective knowledge

The recognition task should be automatically ac-companied by a rating of how well a person knows a particular brand. Here, I suggest using an analogue scale, which allows people to freely move between mi-nimum and maximum scores. This can also work on a piece of paper, as you can draw a 10-cm long hori-zontal line and have people make a mark where their rating is, and then apply a ruler to extract the score.

Associative density

Finally, have participants list all kinds of associations, feelings, thoughts and impressions they have to each brand. Here, you list the associations as they come – or have participants write them down themselves.

Rating is done in two stages: first, merely counting the number of associations provides you with the ge-neral associative density score. Second, by rating each association as positive, negative and factual, and then count associations within each kind. This provides you with a more nuanced image of associations. Please note that what you determine to be positive, negative and factual may not always be straightforward, and should be done as consistently as possible.

Collect the scores

Take each of the scores for each brand and join the scores. This provides a total score for each brand. The-re are multiple alternatives to this rating, as you may want to weigh Top of mind scores higher than other scores, or you may like to make a composite "density score" by adding positive and factual associations and subtracting negative associations to provide a total score. My description here is not prescriptive, but pro-vides the skeleton with which you can address brand

and product memory in a more sophisticated manner than a binary yes-no manner.

Notably, the NeuroEquity Battery can be used to much more than addressing brand memory. It can be used to measure what people recall from a movie or commercial, how they link brands to commercials, and much more.

Beyond providing an aggregate score, the Neuro-Equity Battery also allows us to look at different sub-scores. As you can see below, in a recent study we assessed the different scores on brand memory for toy brands by assessing the group's top-of-mind, subjective knowledge and associative density in parents of 4-year old parents:

As the test shows, LEGO and DUPLO scored by far the highest on all tests, and from the associative density test, we can see that both of those brands only have positive and factual associations, and no getative associations. By comparison, McDonald's has many more negative associations, and Mega Blocks have more negative than positive associations. Interestingly, Disney have most factual associations, but also more negative than positive associations. Together, this allows us to draw a picture of brand memory for this group of consumers.

PROSPECTIVE MEMORY

Finally, we should note that memory is not only *retrospective*, that is, not only pointing backwards.

Humans are well known, relative to other animals, for our ability to plan ahead. Prospective memory is another kind of such memories that affects the way in which we behave. Think about what you will do tomorrow or next year. Our ability to "think ahead" is a crucial facet to our minds, and how we act in consumer situations. We plan our purchases ahead, and we might dread the outcome of choices we make in the moment.

In one study, we looked at how the brain makes those plans, albeit on a much shorter time span: how does the brain plan how to solve a task just before executing it? In an fMRI study led by James Rowe (Rowe 2007), participants were asked to solve a task of either remembering the sequence of letters or sequence of positions. The task was signalled prior to the task. Importantly, we compared healthy participants to patients with lesions to their prefrontal cortex.

What we found was that the patients with prefrontal lesions were unable to maintain an initial "preparation" activation between the prefrontal cortex and more posterior brain regions such as the parietal cortex. This finding suggested that the prefrontal cortex is necessary for our ability to maintain a preparation-to-act and thus plan ahead. When planning on how to walk down the store to obtain a particular product, patients with lesions to the prefrontal cortex will be unable to maintain this plan and thus fail to execute it properly.

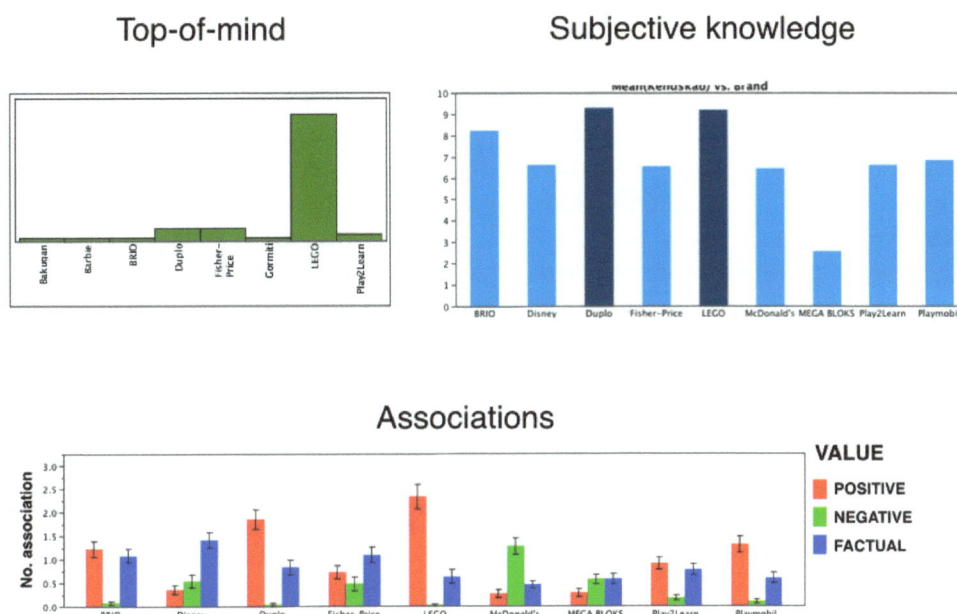

Figure   The effects of the NeuroEquity Battery, here showing the results of subtests on top-of-mind, subjective knowledge and associations. As you can see, LEGO and the related product brand DUPLO scores by far the highest on these scores. Courtesy of Neurons Inc (www. neuronsinc.com)

CHAPTER 8

Wanting, Liking & Deciding

If there is one topic that studies of consumers is concerned with, it is what drives consumer choices. Ranging from our self-reflections and narratives about our desires and motives, to the unconscious drivers of choices, motivation is a key concept in consumer psychology and behaviour, and in marketing.

But first, we need to define what we need by the more general notion of preference, which can be seen as:

- an individual's attitude towards an object, person, company or event, typically reflected in an explicit decision-making process
- evaluative judgment in the sense of liking or disliking an object

Preference can both be seen as the **observed and actual choice behaviours** that a consumer performs, but also a person's **explicit statements** about what he or she prefers.

Do we really need two measures of preference? Is it not sufficient to just ask people, or just observe their behaviours? Are there ways in which consumption behaviour can lead to conflicts between what people do and what they say? Indeed, there are instances where saying and doing are not equal and do not add up! In this chapter, we will focus on this dual nature of consumer motivation. As a result of this chapter, you will hopefully realise that we cannot just suffice with either asking people, or observe their behaviour, to have a complete understanding of consumer behaviour.

Neural predictors of consumer choice

In an article published in 2007 Brian Knutson and colleagues (Knutson 2007) reported the findings from a study of the neural drivers of consumer choice. By first giving money to participants, and then scanning them while they chose whether to buy products or not, the researches were able to tease apart particular neural structures that not only correlated with choice, but even predicted choice. Here, three stages were studied:

1. **Product presentation** – at this stage, only the product was shown. Stronger engagement of the nucleus accumbens (NAcc) was predictive of **increased** likelihood of subsequent choice.
2. **Price viewing** – here, participants were given a suggested product price. Stronger engagement of the insula was predictive of **lower** likelihood of purchase
3. **Decision stage** – at this stage, participants were instructed to make their choice of buying or not buying. Here, the medial PFC was more strongly engaged when participants purchased the product

Even though the NAcc activity was significant for purchased items, it could not predict choice very well. Only about 10% of the variation in purchases could be explained by brain activation. By comparison, self-reports were predictive of about 53% of the variation in purchasing behaviour. Adding the brain data to self reports only improved the predictive ability incrementally. Thus, while the explanatory and predictive power of the NAcc activation was only incremental to only asking people, it is notable that this neural predictor occurred several seconds (8-12 seconds) prior to actual choice. When the researchers asked participants when they felt like making up their minds, they consequently answered they did so during the last stage. This suggests a couple of interesting conclusions:

- NAcc activation is likely a "wanting" response that drives actual choice
- NAcc activation can be used as reliably as self-reports, but at a much earlier stage than self-reports

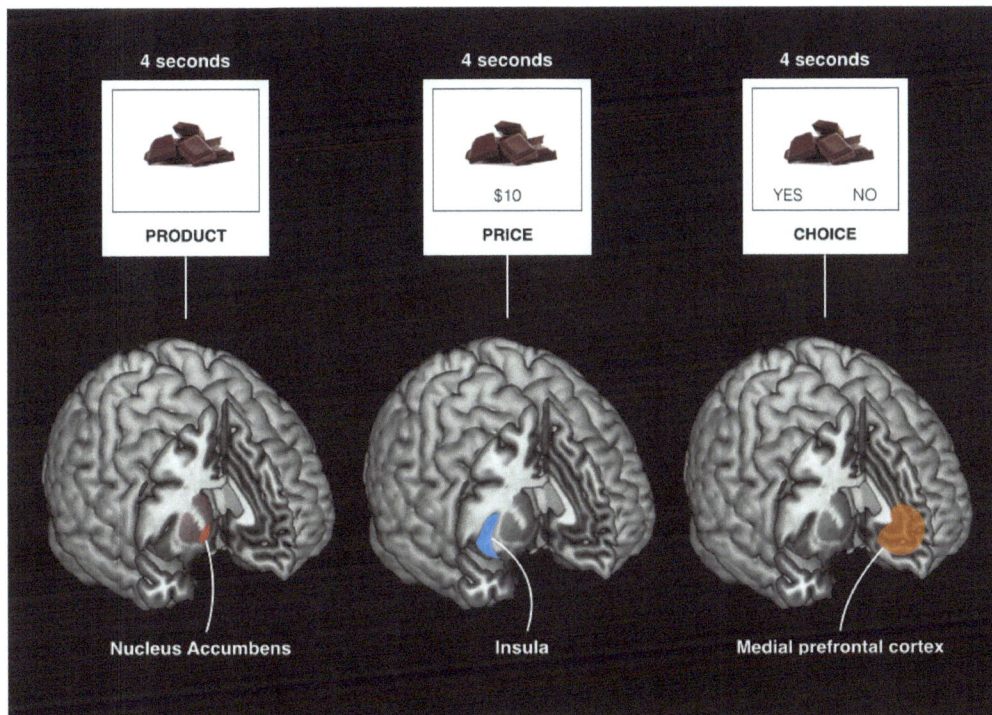

Figure   Neural predictors of purchase. Participants first saw a product for 4 seconds, then saw the product and a price for the product for 4 seconds, and finally had 4 seconds to choose whether to buy or not buy the product for the given price. For each stage, the researchers found 1) stronger engagement of the NAcc to subsequently purchased items; 2) stronger activation of the insula to not chosen items; and 3) stronger engagement of the medial PFC to subsequently chosen items.

- Insula activation is related to reduced purchase willingness, thus suggesting the insula as an unconscious "avoidance" response
- Medial PFC might be more related to a summing of options and choice execution rather than a "wanting" response

A highly related study to this that highlights the medial PFC in "liking" responses was recently reported by Santos and colleagues (Santos 2011). By analysing brain activation before and after a decision was made, they made the interesting observation that activation of the medial PFC occurred *after* the decision was made. That is, brain activation that is usually associated with conscious liking only occurred after a choice, suggesting that it plays little or no causal role in a decision process. As the authors conclude:

> "(...) the results of the present study converge in supporting the notion that the vmPFC may be unimportant in the decision stage concerning brand preference, questioning theories that postulate that the vmPFC is in the origin of brand choice."

Predicting cultural popularity

Another notable study was reported in the Journal of Consumer Psychology by Greg Berns and Sara Moore

(Berns 2012). In a previous study, Berns and colleagues had studied the neural bases of music preference in a group of adolescents (Berns 2010), and demonstrated increased liking for songs that were noted as popular by peers, and an accompanying activation of, among other regions, the caudate nucleus, insula and anterior cingulate cortex.

However, what is notable is the follow-up study. After the original study was completed, Berns incidentally noted that some of the otherwise unknown pieces of music they had played in the first study, had become massive cultural hits. He asked himself if he could indeed predict those hits from his prior study. When analysing both the brain data and the subjective ratings the researchers made a stunning observation: the brain data could predict hits relatively well, but subjective ratings could not. In particular, the ventral striatum was significantly predictive of subsequent sales number at the societal level. As the authors conclude:

> "Our results demonstrate that not only are signals in reward-related regions of the human brain predictive of individual purchase decisions, they are also modestly predictive of population effects. While the nascent field of neuromarketing has made claims to this effect, truly prospective data

has been lacking (...) Surprisingly, our data suggest some validity to these claims."

The revisit to old data by Berns and Moore is highly suggestive that certain aspects of music can reliably trigger emotional responses that not only affect individuals, but do so reliably in many individuals. Perhaps, as many in the music industry suggest, certain aspects of music can reliably lead to hits. If so, the engagement of the ventral striatum/NAcc can be seen as one promising small-sample index for assessing music hits.

A DUAL SYSTEM APPROACH TO MOTIVATION

When we speak of motivation, as with emotions and feelings, we must expect that our choices are not driven by a single, linear and consciously controlled system. Rather, whenever we study choice using behavioural assessment, physiology or neuroimaging, we end up with an emerging picture: our choices are based on at least two processes. Here, we will call them "liking" and "wanting."

A working definition of "liking"

What we like seems relatively straightforward to define. Basically, we can define "liking" as a person's hedonic experience. We assess liking through explicit preference statements, that is, when a person states "I like this brand" and "Oh the taste of that was awful!"

Liking thus seems easy to understand, and in our intuitive sense we feel that what we consciously experience as something we like, is actually what drives our choices. We like something and then we choose it. But as we have already seen in the examples in this and previous chapters, conscious processes are not always "on top" of our choices.

Therefore, we need to ask ourselves: what if our liking is a mere after-effect of our choices? Does liking actually have any effect on our choices? Is liking just sense-making from our unconsciously driven choices? This is one of the recurring discussions that the study of consumer psychology opens, and wherein the application of neuroscience tools and insights provide new insights.

A working definition of "wanting"

In opposition to liking, "wanting" is a different motivational phenomenon. We can define "wanting" as the unconscious approach and avoidance evaluati-

ons related to items, organisms and events. Instead of just asking people – which we simply cannot do with unconscious motivation – we must use a different palette of tools to assess wanting, including changes in:

- **Work and effort** – higher work and effort for desired items
- **Response time** – faster responses to rewards and punishments
- **Eye fixation** – faster first fixation, longer total fixation and number of re-visits for more desired items
- **Behavioural change** – physically approaching vs. avoiding items, leaning forwards vs. backwards
- **Neuroimaging** – stronger engagement of the ventral striatum, amygdala and insula to rewards and punishments, or a higher frontal asymmetry response of the brain to desired items (left frontal cortex more engaged than right frontal cortex, see section on frontal asymmetry later in this chapter)
- **Mental preoccupation** – higher preoccupation relative to desired items
- **Arousal** – higher arousal to rewards and punishments

Wanting is indeed a different kind of species. Notably, these measures employ more sophisticated and nuanced behavioural measures, as well as measures of arousal and all the way to neuroimaging measures.

As we just saw in the study by Berns and Moore (Berns 2012), people's liking responses did not predict cultural popularity of music, but the unconscious neural response did. Indeed, the engagement of the ventral striatum (aka nucleus accumbens) is highly predictive of actual choice and effort.

A prime example to distinguish wanting and liking can be found in the study by Knutson and colleagues that we just treated (Knutson 2007). Here, the researchers found that conscious decision making was related to medial PFC activation just a couple of seconds prior to the actual choice, while several seconds earlier, engagement of the nucleus accumbens during product viewing was equally predictive of choice. Here, medial PFC engagement is an example of a measure of liking, and NAcc activation is a measure of wanting. After all, medial PFC activity was closely related to when participants actively and consciously made a choice. Conversely, since participants explicitly stated that they felt that their decision was made at the last stage, the NAcc activation cannot be directly related to conscious choice, and this suggests it as an index of unconscious motivation, or wanting. This corroborates the aforementioned study by santos and colleagues, who found that activity of the medial PFC occurred after the choice was made.

PREFRONTAL ASYMMETRY AND MOTIVATION

Another index of motivation comes from a broad stroke of research within psychiatry, neurology, psychophysiology and psychology during the past 25 years or so. In a notable research paper, Harmon-Jones, Gable and Peterson (Harmon-Jones 2010) demonstrated how the asymmetric engagement of the frontal parts of the brain are related to motivation. More specifically, this motivation noted as an index of **approach vs avoidance motivation**. That is, whether a person will approach an item or situation, or avoid it. Notably, in the review, the authors draw on sources from psychiatric conditions such as depression, showing that there is a tendency to higher right than left frontal engagement, while in elated states there is a stronger left than right engagement. Similarly, neurological studies of brain injury link right sided lesions to changes towards negative affect, while lesions in the left frontal brain is associated with elation and even mania.

Studies using neuroimaging have also demonstrated that the frontal symmetry is related to moti-

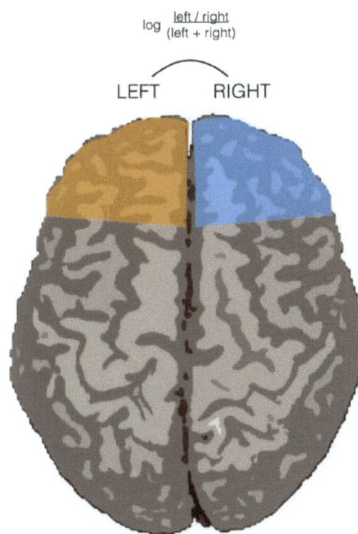

Figure The (pre)frontal asymmetry index. The purpose of the index is to make a relative measure of the engagement of the left versus right frontal cortex. The score tends not to be normally distributed and should therefore be log transformed. This calculation means that higher scores are indicative of approach behaviour, and lower scores are indicative of avoidance motivation. Many studies use the alpha frequency band, in which one should have an inverse logic (alpha is related to de-activation), and therefore multiply by -1 to get the same score directions. Other studies apply the gamma or beta frequency, in which no further operation is needed. For fMRI studies, the same basic calculation also applies, using a Region-of-Interest approach.

vation. In an EEG study by Pizzagalli and colleagues (Pizzagalli 2005) it was found that task-independent measures of asymmetry was highly related to approach behaviours. In another EEG study by Miller and Tomarken (Miller 2001) it was found that the frontal asymmetry varied with monetary incentives. Higher potential monetary gains were associated with stronger left than right activation of the frontal cortex. Finally, in an fMRI study by Berkman and Lieberman (Berkman 2010), it was found that the prefrontal asymmetry was related to action motivation and not stimulus valence as such. That is, they found stronger activation of the left vs right dorsolateral PFC during approach actions, but no such difference for pleasant vs unpleasant stimuli. This confirms that the frontal asymmetry score is not merely a reflection of emotional valence, but rather an index of motivation and ultimately choice.

Interestingly, Ravaja and colleagues employed the frontal asymmetry to consumer choice (Ravaja 2012). Here, they found that the asymmetry index – left vs right engagement of the frontal brain region – during the pre-decision phase was associated with higher likelihood of purchase. Furthermore, higher perceived need for a product, as well as higher perceived product quality, was also related to higher frontal asymmetry and ultimately elevated likelihood of product choice.

Frontal asymmetry and consumer choice

Together, these results suggest that the frontal asymmetry score is a reliable index of motivation, but it is not entirely clear whether it reflects wanting or liking. To abate this, we ran an EEG study[20] where we tested the predictive effects of the frontal asymmetry score on consumers' willingness to pay. In much the same vein as with Knutson's study (Knutson 2007), our participants first saw a product and seconds later made the decision about how much they would be willing to pay (if anything) for the product. When measuring the first 500 milliseconds of product viewing, we found that the asymmetry of the frontal electrodes were significantly predictive of consumers' willingness to pay.

In a follow-up study[21] we performed the same task inside a store environment. Here, participants were given money prior to entering the store and were instructed to purchase items they wanted, or save the money. By using mobile eye-tracking and EEG, we could extract the neural responses while people were looking at particular products or signs, or were in a particular region of the store.

When we analysed the first 500 milliseconds of product viewing, we could reproduce the effects we

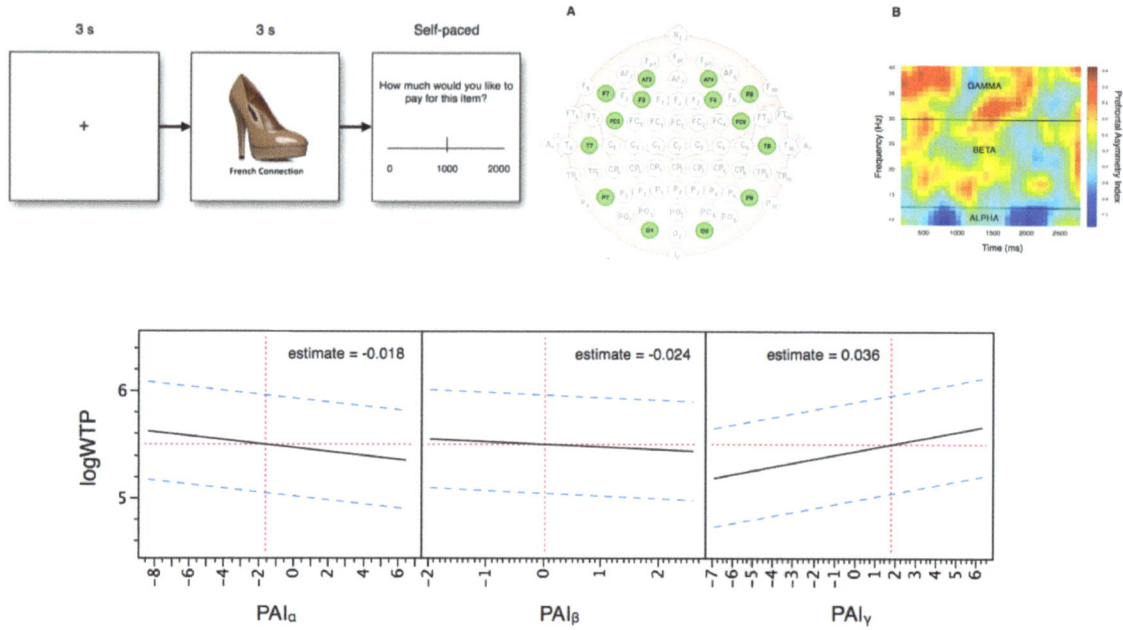

Figure   The prefrontal asymmetry index (PAI) predicts consumers' willingness to pay for a product during the first split second of product viewing. TOP LEFT: Participants first saw a product and seconds later decided how much they were willing to pay for a product. We used a 14-channel low-price EEG system (A) and extracted the raw signals to calculate the frontal asymmetry index for the frequency bands alpha, beta and gamma, projected over time (bottom line indicates milliseconds, warmer colour indicates stronger left than right frontal engagement) (B). BOTTOM: Within each band we found a significant effect for alpha, beta and gamma band scores.

had seen in the lab, as well as the observation made by Knutson and Ravaja and their colleagues (Knutson 2007; Ravaja 2012): a higher motivation score was predictive of increased likelihood of subsequent product purchase. That is, the asymmetry score was significantly, even dramatically, higher when consumers looked at products that they subsequently purchased, compared to when they looked at products they did not buy.

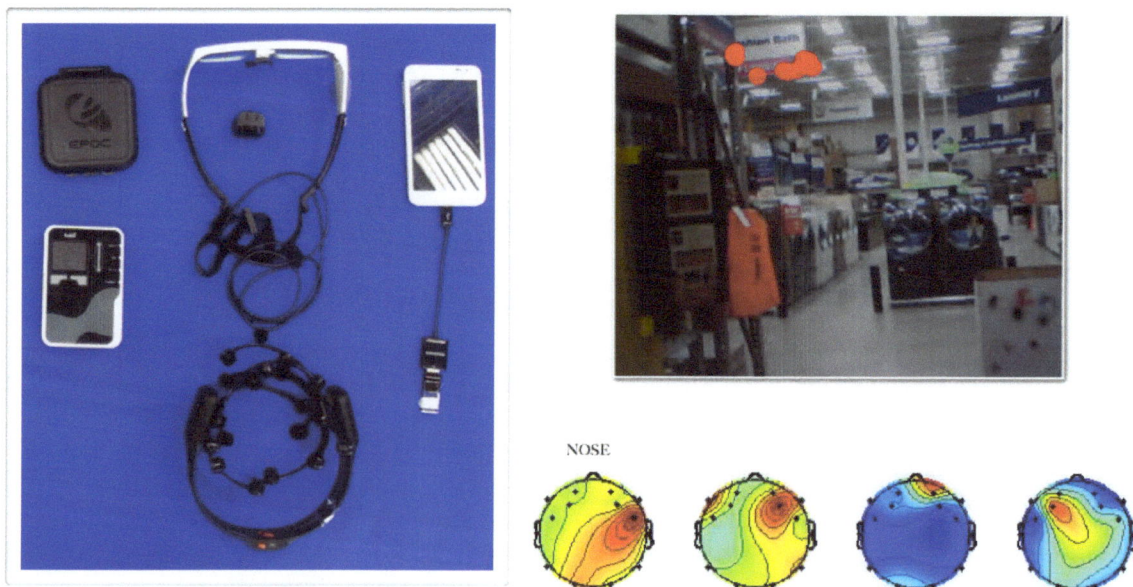

Figure   The mobile data collection setup. LEFT: We used a Tobii Glasses movie eye-tracker, and used the 14-channel Emotiv EPOC headset, which we had hacked and connected wirelessly to a Samsung Galaxy Note 1 smartphone for recording the raw EEG signals. RIGHT TOP: raw eye-tracking signals, showing fixations as red dots overlaid on the video recording from the Tobii Glasses; RIGHT BOTTOM: Distribution of gamma frequency amplitude across the different electrodes, over time. Courtesy of Neurons Inc (www.neuronsinc.com)

Further analyses of the data showed us that we did not even need half a second to make the predictions. Rather, we found that even when consumers had only looked at a product for 100 milliseconds, we could determine whether they would buy the product with an over 70% accuracy. This accuracy increased all the way up to 500 milliseconds, in which it had risen to over 90% accuracy. Mind you, this prediction spanned seconds and even minutes prior to the actual purchase decision.

Today, prefrontal asymmetry has shown many new promises in the way in which we can study and understand consumer thought and behaviour. While there are only few studies looking at the role of frontal asymmetry in consumer choice, it has advantages compared to other motivation metrics, such as nucleus accumbens (NAcc) activation. To summarise some of the notable differences, please consider the following:

- Frontal asymmetry has **direction** – it is related to approach vs avoidance motivation, while NAcc activation is bivalent
- Frontal asymmetry can be studied both with **fMRI and EEG** and other neuroimaging methods, allowing a better ground for scientific validation and cross-comparisons

- The use of EEG allows a **higher temporal resolution** (milliseconds) compared to NAcc measures (seconds), which allows more precise measurements and inferences about motivational effects
- The use of EEG **allows mobile tests,** while NAcc measurements (or any fMRI measure) requires static and highly constrained lab testing.

Where does frontal asymmetry come from?

While it is obvious that we can measure frontal asymmetries using fMRI and EEG, it still leaves out the question about why there is such an asymmetry in the first place. Indeed, it has been somewhat of a puzzle why there would be such gross differences as asymmetric engagement of the brain that is directly related to our behaviour. To be honest, it's been a nuisance to many researchers, including myself: it simply should not be that simple!

Here, an article by Bud Craig (Craig 2005) provides a tentative solution: frontal asymmetries are the result – even magnification – of more basic brain responses that can be traced all the way down to the brainstem. That is, there are asymmetric engagement

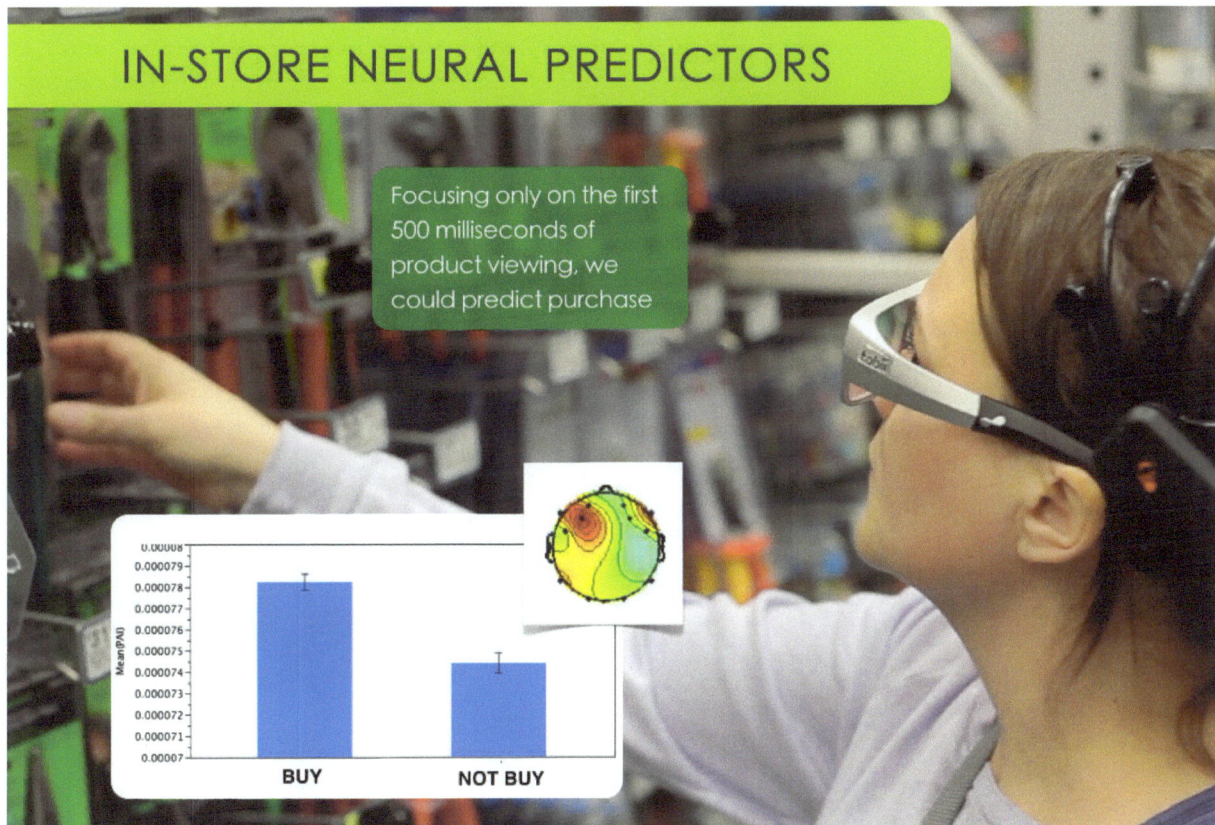

Figure In an in-store study, using only a 14-channel low-price EEG that was connected to a Samsung smartphone, we found that frontal asymmetry during initial product viewing was highly predictive of in-store purchase. Courtesy of Neurons Inc (www.neuronsinc.com)

of the brainstem, in terms of the sympathetic and parasympathetic nervous system, that are reflected and enhanced when we measure at the cortical level. Craig does not address the frontal asymmetries, but focuses a lot on the observed effects of the insula. Therefore, we can at best treat Craig's suggestions as a highly plausible model, but we still need to connect the literatures on frontal and insular cortical activations.

A MODEL OF DECISIONS

When we decide, we rarely decide in a vacuum. Our choices are typically done in competition with other options. According to one such model, the Sequential Choice Model (SCM), an agent's action towards a source of reward reflects the value of the agent's valuation for the reward choice relative to its context (Shapiro 2008). According to the SCM, one quantitative measure of choice strength is response latency, or response time (RT), which is assumed to be negatively related to choice. The implication is that higher valued options should be reflected in shorter RTs, which is also corroborated by empirical evidence (Seed 2011).

Decisions between two or more options will have at least two additional consequences. One dimension concerns the relative valuation of each option. The closer the options are in value, the stronger the response conflict and thus the longer the RT is assumed to be (Braver 2001; Holroyd 2005; Pastötter 2010). From this perspective, the speed of choice can be seen as a reflection of the relative ease of the computation of response options. Recent studies have reported a relationship between the speed of decision-making and the extent of social choice. In a most recent study, Rand, Greene & Nowak used ten different social economic games and demonstrated that fast response times, as reflected in shorter deliberation times, were related to more cooperative behaviors (Rand 2012). Such findings, which are in line with earlier studies linking fair behaviors to shorter response times (Rubinstein 2007, 2008) suggest that prosocial behavior requires less deliberation and as such works as a "default" mode of human social interaction. Other studies contradict these findings, reporting that faster responses were associated with more self-serving choices and slower responses were more associated with prosocial choices (Piovesan 2009). While this discrepancy in results may be due to differences in the nature of the games – such as the use of iteration or one-shot games, and monetary incentives – these studies illustrate the current lack of clear understanding of the relationship between speed, social thinking and social behavior. Accordingly, such differences call for more exploration.

Let's illustrate this with two choice situations that are slightly different. Let's assume that you are low on sugar and in one condition you have the choice between chocolate and rotten fruit. That seems like a relatively straightforward solution: you are more likely to choose chocolate. Moreover, you are more likely to spend less time dwelling over the options, and you are probably going to work harder to obtain the chocolate relative to the rotten fruit.

In a second condition, you have the option between two different kinds of chocolate. Chances are that you will have smaller differences in preference between these options. Consequently, you are likely to spend slightly more time dwelling before you make your choice, and you are more likely to trade one option for the other. That is, you will be likely to sacrifice one solution if the other seems less available.

In this way, we can illustrate how differences in the predicted (or expected) value of something is having a direct impact on our actions. This can be illustrated as with the figure below. If we have choices between two options that are far apart in their expected value (blue bars), then the time we need to choose is shorter; our response (such as grip strength) is stronger; and we are more likely to persevere in trying to obtain the desired item.

Conversely, for choices that are close in there expected value, we will spend more time dwelling and our response time goes up; we tend to show a weaker response (e.g., grip strength); and we will have much less perseverance in trying to obtain a single option (since the other option is available)

Figure Two choice problems. In Problem 1 you have the choice between delicious chocolate or rotten fruits. In Problem 2, you have the choice between two kinds of relatively comparable chocolates. In Problem 1, the choice is relatively simple; in Problem 2 the choice is not equally simple and you are expected to spend more time thinking before making your choice.

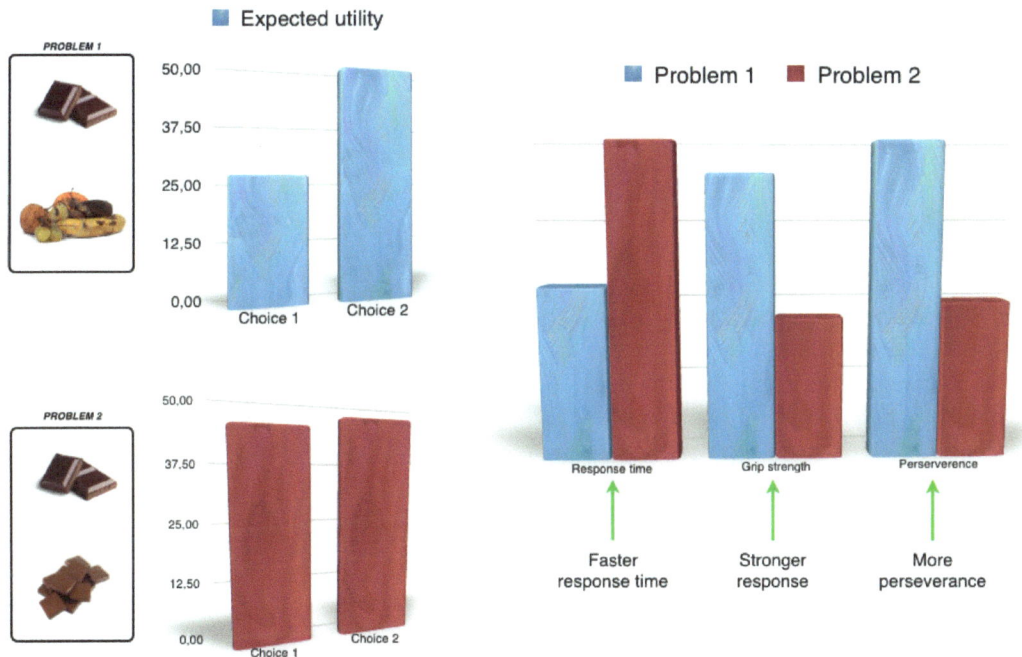

Figure Effects of preference on actions. LEFT: an illustration of the relative expected utility of two options relative to each other. In Problem 1, the difference in expected value (or utility) is large, while in Problem 2, this difference is small. RIGHT: Bar charts showing model predictions of the two problems on different behaviours, such as response time, response strength, and perseverance.

The SCM, which is also highly related to the Diffusion Decision Model (Ratcliff 2008), provides a strong link between the value of options and behavioural responses to obtaining these objects. These models also allows us to make further inferences between the brain bases of choice and our behavioural responses. For example, Mathias Pessiglione and colleagues have previously reported (in another study than we have previously reported in this book) how unconscious priming of monetary value can translate into changes in grip strength (Pessiglione 2007). As the authors show:

"[Even] when subjects cannot report how much money is at stake, they nevertheless deploy more force for higher amounts. (...) These results indicate that motivational processes involved in boosting behavior are qualitatively similar, despite whether subjects are conscious or not of the reward at stake"

To this end, we can say that decision making is the result of not only the single choice we have available, but also the relative context that this choice has to other options.

BRAIN BASES OF LIKING

While wanting seems to be rooted in deep struczzztures of our primordial brains, liking seems to be a more recent evolutionary invention. Our conscious liking of something also corresponds to our flowchart model, where we are calling it "Experienced value". At this time, we have gone through the attention & representation stage, and the predicted value stage, and we are now facing the actual effects of a stimulus. When you are experiencing a delicious meal, listening to soothing music, smelling something rotten, or tasting something sour, we have an explicit and vivid experience of the object and our valuation of it.

Two neural substrates of brain liking

Looking at the brain, we realise that there seems to be at least two regions that are highly associated with the experienced value of something. On the one hand, we see that the **orbitofrontal cortex (OFC)** shows a strong response to liking across different domains. The more we like something, the stronger the activation of this region, especially the medial OFC, while lateral parts seem more related to disliking. Second, but much less known, is the **anterior insula (AI)**. This region is already well known to be associated with feelings of disgust, but many recent studies suggest that it plays a general role in our hedonic experience. Let us look at the evidence

- **Orbitofrontal cortex:** Studies have implicated the OFC as a general "liking spot" of the brain. It receives convergent information from all senses and from cognitive and emotional processes. Therefore, we see that a number of studies demonstrate increased OFC engagement, including:

- Plassmann and colleagues (Plassmann 2010) found that the medial OFC encodes appetitive and aversive preferences and goals at the time of decision making.
- Plassmann and colleagues (Plassmann 2008a) demonstrated how the OFC was implicated in the enjoyment of wine, and how price information affected this enjoyment.
- Kirk and colleagues (Kirk 2009) found that the medial OFC was more engaged when participants enjoyed and reported liking abstract arts, which was further affected by contextual framing.
- Hare and colleagues (Hare 2008) report that the OFC is involved in the computation of goal values and decision values, as opposed to the role of the striatum in prediction error.
- Sescousse, Redouté and Dreher (Sescousse 2010) report that distinct subregions of the OFC compute abstract vs concrete rewards (see the Opinion Piece on the next pages).
- In a notable review of the neuroscientific literature Kringelbach (Kringelbach 2005) points to the OFC as a hedonic hot spot that links reward to hedonic experience.

- **Anterior insula:** Surprisingly often, the AI has been reported in statistical tables but ignored as an activation finding in neuroimaging studies during the focusing of other brain regions such as the OFC. However, careful observation of the results that are reported have implicated the AI in hedonic experience – both positive and negative feelings, including:
  - Chua and colleagues (Chua 2009) found that the AI was strongly engaged during experiences of both regret and disappointment.
  - Liu and colleagues (Liu 2011) found that the AI was more strongly engaged to negative rewards, but also to reward anticipation.
  - In a meta-analysis of neuroimaging studies of aesthetic experiences Brown and colleagues (Brown 2011) identify that the right AI is the most robustly activated region across all aesthetic sensory experiences (seeing, hearing, touch, taste).

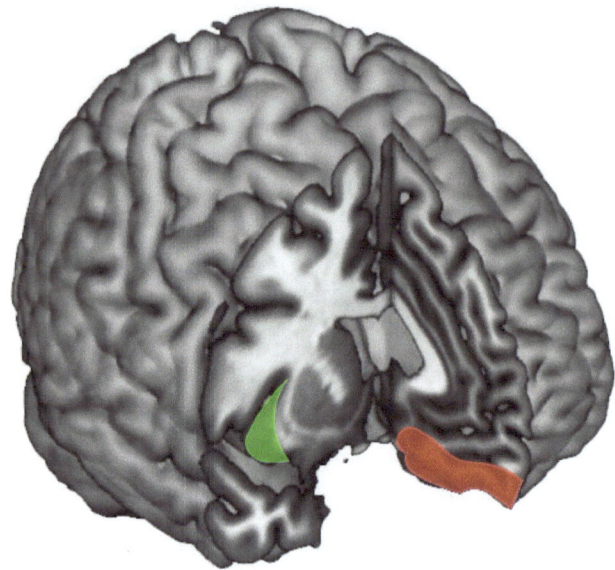

Figure Two brain regions involved in hedonic experience: the anterior insula (green) and the medial orbitofrontal cortex (red)

- Krolak-Salmon and colleagues (KrolakSalmon 2003) identify the AI as a region implicated in processing of disgust, which is further modulated by attention.
- In two notable reviews, Berridge and Kringelbach (Berridge 2013; Kringelbach 2009) implicate the AI as an important structure in both positive and negative hedonic experiences.
- In a Perspective paper, Craig (Craig 2009) suggests that the AI is not only important for the subjective experience of reward and loss, but that it also may be a neural substrate of consciousness.

Together, these studies demonstrate a clear role of both the OFC and the AI in hedonic experiences. As such, they show us that there is no single place in the brain wherein hedonic experiences are represented. Rather, we also see that the OFC and AI are co-activated with many other brain regions, such as the anterior cingulate cortex, medial PFC, and dorsolateral PFC. We cannot yet entirely rule out that deeper structures such as the amygdala or the striatum are involved in hedonic experiences. Indeed, even the visual cortex seems to be engaged when we expect and experience rewards (Serences 2008).

Sex and money: common or different currencies?

How are values computed in the brain? Rewards can be as many things: the expectation when having just ordered your favourite dish; the child's joy at Christmas Eve; the enjoyment of good music or the wonderful taste of strawberries.

But how does the brain process these many different kinds of rewards? Does it treat all types of rewards equally or does the brain distinguish between different kinds of rewards? Rewards can come in many different forms: from sex, social recognition, food when you're hungry, or money. But it is still an open question whether the brain processes such rewards in different ways, or whether there is a "common currency" in the brain for all types of rewards.

Guillaume Sescousse and his colleagues in Lyon recently reported a study on how the brain reacts differently to money and sex. A group of men were scanned with functional MRI. While being tested, subjects played a game in which they sometimes reviewed a reward. The reward could be money or it could be the sight of a lightly dressed woman. So there were two types of rewards. Money can be said to be an indirect reward, and the sexual images can be seen as more immediately rewarding (at least for most heterosexual men). But how did the brain process these rewards?

The researchers found that there were unique activations for both sex and money, but that there were also overlapping regions of activity. On one hand, for both types of reward was a general activation of what we often refer to as the brain's reward system (ventral striatum, anterior insula, anterior cingulate cortex and midbrain; see figure 1). The brain thus uses the some structures to respond to both types of reward.

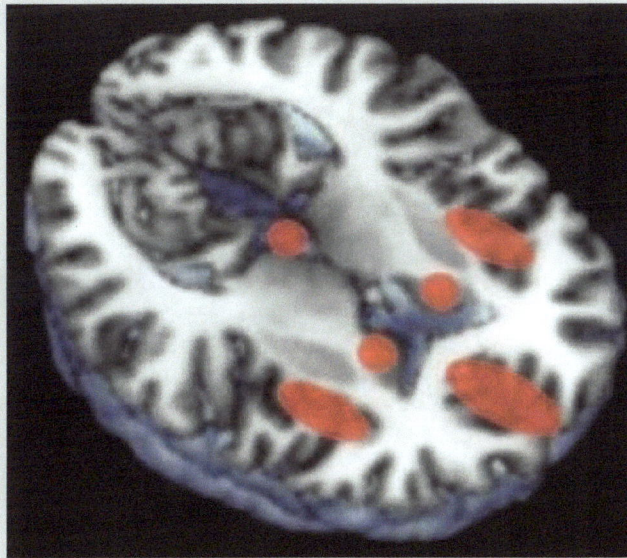

Figure   Regions of common activations

But there were also specific activations for erotic pictures and money. And this difference was primarily made in the brain's prefrontal cortex, especially the orbitofrontal cortex (OFC). Here, it was found that monetary rewards engaged more anterior OFC regions, while erotic images activated more posterior OFC regions.

This could suggest that the brain also treats the two types of reward differently. The crux of this paper, however, is how one explains the difference. As noted, the researchers used two different kinds of reward, but they differ in several ways which I will try to summarise here:

- **Direct vs indirect**
  - Money is indirectly rewarding, because money can not be 'consumed' in itself. They are rewarding to the extent they could be exchanged for other things. Erotic images are in themselves directly rewarding. Not because they symbolise sex, or the possibility of sex, but because they have an immediate rewarding effects.
- **Abstraction level**
  - Another option is to say that erotic pictures and money differ in their level of abstraction: Erotic images are concrete, while money is an abstract reward.
- **Time interval**
  - A final possibility is that there are differences in the time interval: Erotic images are immediately rewarding, while the money can only be converted into real value after a while (for example, after scanning, or after a few days where you spend the money). We already know that the frontopolar regions of the brain is among the regions that are most developed in humans compared to other primates, and is linked to our unique ability to think about the future, i.e. prospective memory and planning, and through this to use complex abstractions for rewards, including money.

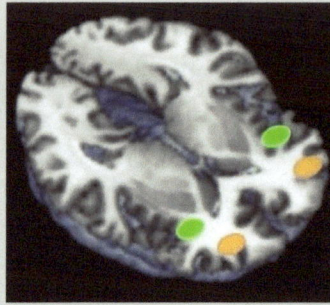

Figure   Regions of distinct activations: orange = monetary rewards, green = sexual rewards

What the exact cause of this common currency as well as the separation between money and erotic pictures is still unclear and warrants further studies (which I am currently undertaking). The essential addition of this study is the separation between the posterior and anterior parts of the OFC in processing different kinds of rewards. By showing common and distinct regions, this study may resolve some of the ongoing debates in the decision neuroscience / neuroeconomics literature. But as always found in science, this study generates more questions than it resolves, and we can only hope that future studies can add to this knowledge.

MEASURING PREFERENCES

As a consequence of this, we now can point to and reiterate the different ways in which preferences can be measured. Here, we should focus on two aspects of preference: wanting as well as liking. That is, both the unconscious and conscious aspects of our motivation.

For conscious preference, and thereby **liking,** we can use overt behaviours such as

- **Preference ratings** – surveys, interviews, focus groups
- **Willingness to buy and Willingness to Pay** – subjective estimates of people's willingness to either buy a product, or specific estimates of how much they expect to be willing to pay for a product

Conversely, for estimates of unconscious preference, or wanting, we can use other measures that should be controlled for being present when people report having no particular preference:

- **Response times** – faster response times are indicative of "easier" choices
- **Effort** – how much effort, strength and perseverance is put into obtaining an item
- **Mental preoccupation** – how much people report thinking about something
- **Attentional focusing** – eye-tracking analyses of number and duration of fixations
- **Neuroimaging measures** – prefrontal asymmetry and ventral striatum engagement

Together, these measures provide a reasonable estimate of unconscious and conscious motivation. Of course, more options are available, but here, these options are plenty to obtain a proper image of consumer motivation and preference.

Can context affect preference?

While work by people such as Diederik Stapel has been deemed fraudulent, research into the effects of contexts on thinking and behaviour has almost come to a halt. There has been much skepticism towards whether there are any effects at all. Could it still be the case that context can affect our emotional responses and judgments of specific items? If so, what are the causal mechanisms?

A recent publication from my lab has just come out in the Journal of Neuroscience, Psychology and Economics, and is entitled "*Effects of Perceptual Uncertainty on Arousal and Preference Across Different Visual Domains*". The paper describes an interesting finding: simple, unpredictable sounds sequences can affect judgment of simultaneously presented visual materials such as brand logos and abstract art.

The work was inspired by a recent study by Herry et al (2007) which had demonstrated that unpredictable sounds lead to an increased engagement of the amygdala and increased avoidance behaviour, both in rodents and humans. The question we were interested in was whether such effects also would hold for more complex stimuli, such as cultural artefacts, including brands.

In the study, we showed either a previously unknown brand logo (either a beer, financial, cosmetic or electronic brand), or a piece of unknown abstract art. One second prior to the image, subjects heard a simple sound sequence that was either in a predictable or an unpredictable manner. Here are two examples:

Here is the predictable sound sequence: 1000

And here is an example of an unpredictable sound sequence: 1021

The sound started one second before the image and lasted throughout the whole 3-second image presentation. After this, subjects were asked to judge the image.

The results showed that items that were associated with unpredictable sounds received significantly lower ratings than those that were presented with unpredictable sounds.

While this is an interesting effect in itself, we went a step further: we tested whether unpredictable sounds were associated with increased arousal. By measuring pupil dilation using high-resolution eye-tracking, we found that indeed, unpredictable sounds were associated with increased arousal, and related to lower ratings.

Interestingly, this negative relationship between arousal seemed to be related to the first second, that is, when only the sound was present. Once the image got on, there was a positive relationship between pupil dilation response and preference.

This study suggests that contextual "noise" and unpredictability can affect first impressions and judgments of novel stimuli, and that it can do so across different visual domains. Thus, one take-home message would be to avoid contextual unpredictability if you are in need of good first impressions – and who's not in need of that!

CAUSAL EFFECTS OF CONSCIOUSNESS

As we have seen, hedonic experiences are rooted in particular brain responses such as the medial orbitofrontal cortex (mOFC) and more generally the ventromedial prefrontal cortex (vmPFC). However, as we have also seen, choices can be set off completely unconsciously, and this raises the question about whether the responses of the mOFC or vmPFC can play any causal role in choices at all. Put differently, are our experiences of making a choice only an after-the-fact response of the brain, and are we in fact not consciously in control?

In the aforementioned study by Santos and colleagues (Santos 2011) the researchers wanted to test whether the vmPFC was engaged before or after a choice was being made. By setting up their experiment in a way that allowed them to analyse this difference, the authors made an interesting observation: the vmPFC was activated after a choice had occurred, not before the choice was made. In other words, the vmPFC was not engaged during the pre-decision phase, and consequently cannot be said to play any causal role in the actual decision making

As we have already seen, both the study by Knutson and colleagues (Knutson 2007), as well as Pessiglione and colleagues (Pessiglione 2008), unconscious motivation is rooted in regions such as the nucleus accumbens.

This suggests that our conscious experience of making a choice may indeed be an after-the-fact matter of filling in the gaps. Our sense of autonomy may be an illusion created by the brain, just as we have seen that visual illusions are created by the brain. This suggests two broad suggestions

- We cannot trust conscious reports – they are, after all, only after-the-fact retrospective accounts of a choice that has already happened
- We should seek to broaden the use and validation of unconscious measures, such as neuroimaging, physiology and behavioural responses that the participant is not privy to

Besides these consequences, there are a host of other issues that are raised as a consequence of these findings. For example, if customer choice is not conscious, and conscious introspection cannot reveal the root causes of consumer choice, where does this leave us when it comes to making ads or any kind of communication to consumers? After all, this also suggests that commercials may primarily work unconsciously, and our conscious control is only a minor process with little or no true impact on our choices.

To better understand this, we also need to focus on more deliberate, conscious decision making. We feel like being in control over our actions and choices, so what does the scientific literature say about this? Here, we turn to the science of executive functions.

A RECIPE FOR CONFLICT

Indeed, having two (or more) sources of motivations is a recipe for conflict. If you have an unconscious motivation system that is driven by one thing and another system motivated towards something else, chances are that you will be having a conflict. Several studies have show that a number of measures can detect conflict. Most notably, we see that **response time** increases linearly with the degree of decision conflict (Dyer 1973;

Eriksen 1995).

Take for example the Stroop task. The task is simple: say, as fast as you can, the colour with which the words are written in. Do not read the words themselves:

RED BLUE GREEN YELLOW

BLUE GREEN RED YELLOW

RED BLUE YELLOW GREEN

The first line was pretty easy to solve: there was a congruence between what the colour was and what the words said. But in the next two lines, you probably made many mistakes and your time to solve the task increased dramatically. You could probably also feel the conflict – the intention to say something that needed to be stopped and replaced by a correct response.

What you did not observe was what your brain did at this moment: if we scanned your brain using fMRI (or high-resolution EEG or MEG), we would see an increased engagement of the medial prefrtonal cortex, in particular the anterior cingulate cortex (ACC). Several studies have pointed to this region as a region that detects, mediates and solves conflicts.

For example, Joshua Greene and colleagues (Greene 2004a) wanted to study the neural bases of social conflict, and scanned participants using fMRI while they solved different kinds and degrees of moral dilemmas. Here, they found that the ACC, together

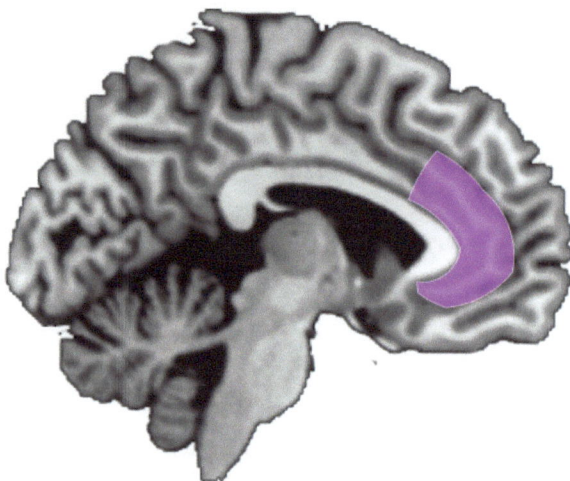

Figure The Anterior Cingulate Cortex lies on the medial side of each hemisphere, and is often counted as part of the ventromedial prefrontal cortex (vmPFC). The medial view of the brain is seen from the right side, where the nose is to your right.

with the dorsolateral prefrontal cortex (dlPFC) was more strongly engaged for hard compared to easy moral dilemmas. Indeed, they found that the ACC is related to the detection of conflict as well as the recruitment of prefrontal brain regions in solving the conflict. This builds on prior research in which they had also found a similar role for the ACC and dlPFC in non-social conflicts, where the ACC is involved in the detection of conflict and the dlPFC is involved in the resolution of conflicts (Botvinick 1999; VanVeen 2001; Botvinick 2004).

We are also different in the degree with which we can cope with and resolve conflicts. Some people are good – fast and precise – at resolving conflicts while other may be slower and/or poorer. Studies have pointed to a significant role of the ACC in our individual differences in resolving such conflicts (Haas 2007; Brown 2008; Botvinick 1999). For example, in a study by Deppe and colleagues (Deppe 2007), it was found that ACC activity could predict individual differences in the effects of branding. In their study, fMRI was employed to study brain responses when participants were exposed to advertisements together with one of four possible newspaper brand logos. The newspaper brands had different cultural values, and were thus believed to influence ad liking. What the researchers found was that people's ratings of ad liking were significantly affected by the newspaper brand they were accompanied with. Most notably, the researchers found that the degree of activation in the ACC predicted how affected an individual was. This suggests that the region of the ACC engaged must play a role in functions beyond mere conflict monitoring, but also an effect wherein contextual information can affect perception and action.

Indeed the findings reported by Deppe and colleagues is highly relevant to branding: we are all influenced by brands in our perception of products, goods and services. But we also know that some people are more affected than others. Beyond this *inter-individual* difference between people, we can also suspect that there is an *intra-individual* variance, in which the same person can be highly influenced by brands within one domain, but not so much within another domain. While this is still an unexplored area even in the behavioural sciences, we may contend that both inter- and intra-individual variance has a foundation in the operations of the ACC.

Political opinions: fast herd or slow ideology?

Sometimes politicians make claims that may seem at odds with the ideological background they represent. Would you agree with your party if they presented a statement that went against their ideological foundation? Put this way, you would probably not, right?

But in fact, you'd very likely do so. It's been known for a while that in the face of conflicting information about political ideology and group belonging, people tend to follow their group and dispose of their ideology (at least for a while). Group think comes first.

But how does this happen? Does knowing the party provenance of a statement lead to faster or slower responses? Does knowing that statement X comes from party B make it less or more easy to make up your mind?

In a recent study, we have approached this problem, in which the literature has suggested two opposing proposals. On the one hand, knowing the political party behind a statement could trigger some heuristic that makes decisions easier. On the other hand, knowing the sender could trigger a more complex weighing of the opportunities, leading to an overall more difficult decision process (even though it would only take an additional few milliseconds). In two related studies, we used subjects response time to assess the level of conflict, adhering to prior studies (e.g. this one, PDF)

The article, published in the journal Political Behavior, bears the saying title "*Motivated Reasoning and Political Parties: Evidence for Increased Processing in the Face of Party Cues*". The abstract reads as follows:

Extant research in political science has demonstrated that citizens' opinions on policies are influenced by their attachment to the party sponsoring them. At the same time, little evidence exists illuminating the psychological processes through which such party cues are filtered. From the psychological literature on source cues, we derive two possible hypotheses: (1) party cues activate heuristic processing aimed at minimizing the processing effort during opinion formation, and (2) party cues activate group motivational processes that compel citizens to support 15 the position of their party. As part of the latter processes, the presence of party cues would make individuals engage in effortful motivated reasoning to produce arguments for the correctness of their party's position. Following psychological research, we use response latency to measure processing effort and, in support of the motivated reasoning hypothesis, demonstrate that across student and nationally representative samples, the presence of party cues increases processing effort.

The PDF is available from this page..

EXECUTIVE FUNCTIONS

If you did the Stroop task in the previous section of this chapter, you probably realised that you cannot be completely in control. Some processes are just so automatic and direct that once they are learned – such as reading – it is virtually impossible to undo them. The problems you faced with the task suggest that there is at least some kind of inhibition process that our minds can do (at least attempt to do). But once we look at this, we find that the functions that we often call **executive functions** can be divided into at least three different kinds. Here, we will refer to these as initiation, inhibition, and modulation.

To better illustrate the differences between these functions, I will use a case from my previous clinical work. Among many patients, some are better remembered than others, and some better fit the need for examples than others. In this case, let me introduce my patient, whom I will call H.O., who I treated at a rehabilitation facility in Copenhagen[22]. Prior to his injury H.O. was a family father who saw to his work as an engineer, and with great performance both in terms of his work achievement and financial success. He was respected and liked among his colleagues, and had a normal family life with his wife and son.

However, one crucial day, while working on his house, H.O. fell down from a ladder and his head struck the concrete, producing a major brain injury. Upon neurological exam, it was obvious that H.O. had substantial (and somewhat diffuse) injury to his frontal lobes, but his brain was otherwise relatively intact. Upon neuropsychological exams, it was clear that he had several issues with impulse control and other decision-making disorders. Here, we will take these in turn:

Initiating an action

Let us start by side-stepping the clinical story first: imagine that you are planning to go out the door and buy some candy. The planning stage works fine: you can both think about and explicitly state how you can get to the store and buy some candy, and possibly the kinds of candy you are going to buy. Now let's say that somehow, you never manage to get out of the door. You simply can't get yourself out of the chair.

This phenomenon is well known to neurologists and neuropsychologists when they test patients with injury or disorder to the prefrontal cortex. These patients may be very good at explaining what they are about to do and how to do it, but then nothing happens. It is as if these people simply lack a "go button" in their mind. Indeed, due to the disordered prefrontal cortex, we can actually say that they lack the ability to convert motivation to action.

This is exactly one of the problems that H.O. had. During a clinical session, upon my question he would describe his plans for what he would do after our session, such as going down to the local kiosk and buying a pack of cigarettes (he had started smoking after his injury). He could describe in detail how he would stand up, open the door, walk outside and towards the kiosk, and in detail how he would select the cigarette package and pay for it. However, when the clinical session was over, nothing happened! He did not get up, did not show any initiative to walk out of my office. When I asked him again what he was about to do, he would repeat his intentions to get up, walk out of the office etc. But he would not move. There was obviously some kind of *disconnection between his conscious intentions and his ability to initiate the actual movements*. Strange as it may seem, our ability to talk about our motives and acting upon those motives are two different functions of the brain.

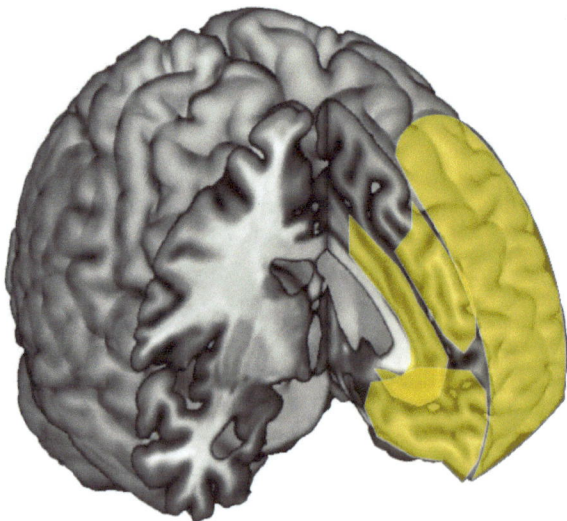

Figure The prefrontal cortex encompasses a number of brain regions, including the anterior cingulate cortex, the orbitofrontal cortex, the medial prefrontal cortex, dorsolateral prefrontal cortex, and polar prefrontal cortex. Our view of the role that different regions play in executive functions is still incomplete. My patient, H.O., had severe and widespread lesions to the prefrontal cortex on both sides of the brain (bilaterally).

Thus, one important function of the prefrontal cortex is to initiate conscious, controlled actions. This is very different from the unconscious, direct and impulse driven behaviours we see from deeper brain structures, such as the ventral striatum and amygdala. These are regions, unlike the prefrontal cortex, that are both fast and automatically lead to actions.

In consumer behaviour, executive control in this sense is important for all aspects of planning our behaviours. This ranges from our planning of our next trip to the grocery store to planning our wedding: our ability to plan and then execute our plan is a crucial aspect of many consumer behaviours.

Inhibiting an impulse

We all know the feeling of stopping an urge. It can range from hindering oneself from eating an extra piece of cake, to avoiding blurting out something that is socially inappropriate. Multiple neuroimaging studies have demonstrated a role for the frontal lobe in so-called response inhibition (Luna 2004; Liddle 2001; Garavan 2002; Rubia 2003; Horn 2003; Aron 2004). Interestingly, with respect to the left-right lateralisation of approach vs avoidance motivation we have described earlier in this chapter, many studies implicate the *right* inferior prefrontal cortex in successful response inhibition.

In the case with my patient H.O., response inhibition was also part of his clinical repertoire of dysfunctions. During clinical sessions, or just walking outside, he would suddenly blurt out with nonsensical words, or start mimicking other's behaviour. During the clinical session, he would sometimes mirror my own movements. If he heard another patient saying or shouting something outside, he would automatically say or shout exactly the same. H.O's control over certain impulses was completely abolished.

The symptoms displayed by H.O. are often falling under the heading of *frontal lobe syndrome and disinhibition syndrome,* which is the clinical state of lack of impulse control or attentional control (but also related to lack of initiative) (Webster 1997; Shulman 1997; Brutkowski 1963; Starkstein 1997). The insight from these conditions is that someone can in fact have an intact initiative, but a failing impulse control, and vice versa.

In consumer behaviour, impulse inhibition can be seen in multiple domains, but let us focus on two phenomena:

- **Media multitasking** – in several studies, multitasking between different media have been linked to poor executive control. For example, in a study by Ophir, Nass and Wagner (Ophir 2009), the degree of media multitasking was found to be negatively correlated with formal measures of impulse control. That is, the more people reported multitasking between smartphones, computers, TV and more, the *poorer* they performed on standardised tests of impulse control and executive control. In my own lab, we have demonstrated the same effects in tweens (10-12 year olds), especially in boys (unpublished data).

- **Eating disorders** – In specific consumer disorders, such as obesity, studies have suggested links to cognitive dysfunction, especially a poorer impulse control (Cserjési 2009). This suggests that besides other clinical symptoms, such as depression, obesity may in part be driven by a poor impulse control.

Modulating a behaviour

Finally, sometimes our urges and impulses fail to be blocked completely, but we can still modulate them in such a way that they are more appropriate. In a social context, we may alter our urge to say something into something more mundane, a direct question becomes a more indirect insinuation. In these and many other situations, we are masters at modulating our behaviours in such a way that they take the physical and social context into account.

While there is no single brain substrate for behavioural modulation, we can say that modulation implies at least some degree of inhibition. We need to inhibit the straightforward and original behavioural impulse, and replace it with something more mundane.

This is best exemplified by by patient, H.O., who had lost the ability to take the physical and especially social context into account. During clinical sessions, he would sometimes run out of things to say, but would still continue to speak, often rambling about and use completely meaningless words and utterances, such as "Yes, I like being here...dumbdidum...dubidubidu... mailbox...lallalla..." but without ever seeing the need to modify these utterances. In a supervised conversation with his wife, he could use explicit description of his sexual fantasies with his wife, in front of everybody else in the room.

Behavioural modification is also an essential part of consumer behaviour. Some instances involve altering an impulse to purchase something to, say, buying less than originally planned; or buying something less expensive than the most desired item. Still, our

understanding of behavioural modification is less understood than straightforward impulse inhibition, and more research is needed to link the neuroscience and psychology of behavioural modulation with consumer behaviour.

TAKING RISKS

When making decisions, we tend to weigh the pros and cons of each choice – the potential benefits relative to their potential costs. Overall, we do not tend to align our decisions to what can be called formally rational choices. That is, while a choice option can be said to have an optimal solution, humans rarely make those optimal solutions. Rather, we are too risk averse to make the best possible choices.

Let us take an example: flip a coin 10 times but before beginning please decide whether you would accept or reject each of the following scenarios:

A: Heads = you win \$5, tails = you lose \$1
B: Heads = you win \$10, tails = you lose \$5
C: Heads = you win \$7, tails = you lose \$5

Would you accept the first bet? Chances are that yes indeed you would. If you have a 50/50 chance of winning \$5 and losing \$1, it makes perfect sense to accept this bet. It even feels right to take the bet. Mathematically, we can express this as the following expected outcomes over ten trials:

1. Winning: $10 \times 0.5 \times \$5$ = winning \$25 on average
2. Losing: $10 \times 0.5 \times \$1$ = losing \$5 on average
3. Net over ten trials = \$25 – \$1 = average win of \$24

This calculus makes it quite simple: you should definitely take the bet.

How, then, about the other bets? Would yo take bet B? How about bet C? Your gut feeling probably says... nah! Those bets are too risky. They don't *feel right*. But if you calculate the expected outcome of either of those, you will see that formally speaking, you would gain money. Even the worst bet, C, has more likelihood of producing a gain than loss to you:

1. Winning: $10 \times 0.5 \times \$7$ = winning \$35 on average
2. Losing: $10 \times 0.5 \times \$5$ = losing \$25 on average
3. Net over ten trials = \$35 – \$25 = average win of \$10

So from a purely mathematical perspective, even this bet looks good, and you should accept it. Yet, most people tend to reject the bet of winning \$7 vs losing \$5. It simply does not feel right, and it feels risky.

Prospect theory

This phenomenon has been nicely described by behavioural economists for decades, also running under the heading of Prospect Theory (Kahneman 2003; Kahneman 1979; Tversky 1992; Tversky 1986). This theory makes two particular claims:

1. people value gains and losses differently, and
2. people will base their decisions on perceived gains rather than perceived losses.

The first claim is related to our examples above: we simply tend to weigh our potential losses more than potential gains. In Prospect Theory, it has been well established that people tend to accept bets when the potential gains are about twice of what the potential loss is, something we also know as the 2:1 ratio. This function is also expressed as the mathematical expression lambda (using the Greek symbol λ). In a recent study in my lab[23], we have found that there are significant individual differences in people's lambda score, as shown in the figure below.

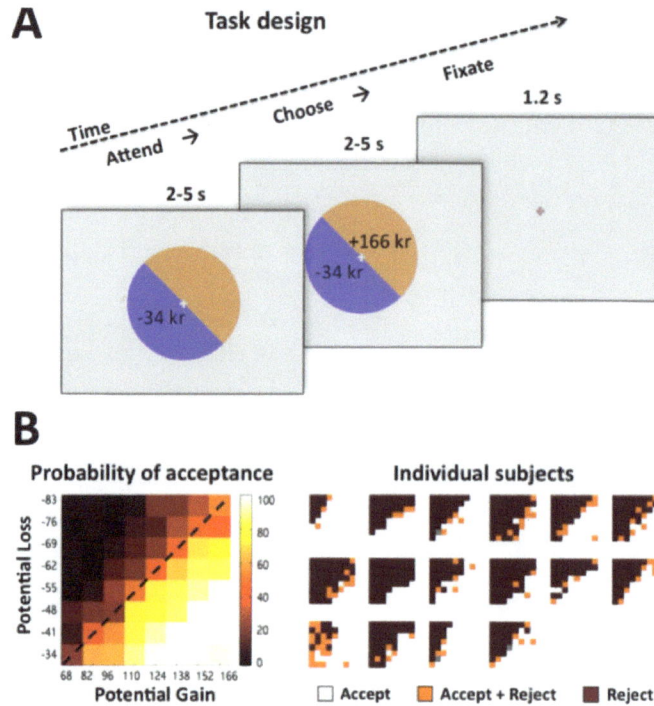

Figure Risk taking and loss aversion. In the task (A), participants made judgments on whether to accept or reject bets with losses (blue) or wins (orange) money (Danish Kroner). While across the group, subjects showed the 2:1 lambda ratio as predicted by Prospect Theory (B, left) there was also a substantial individual difference in individuals' risk willingness (right). Here, the heat map shows accepted bets as white, and rejected bets as brown.

Furthermore, we also found[24] that the anterior insula is involved in computing the magnitude of potential losses, thus providing more evidence to the view of a role for this region in value-based decision making. Notably, we also found[25] that certain structures, such as the amygdala, showed an increased engagement to extreme ratios – both positive and negative. That is, the amygdala signalled the "easiness" of either accepting or rejecting a bet when it was either extremely attractive or unattractive.

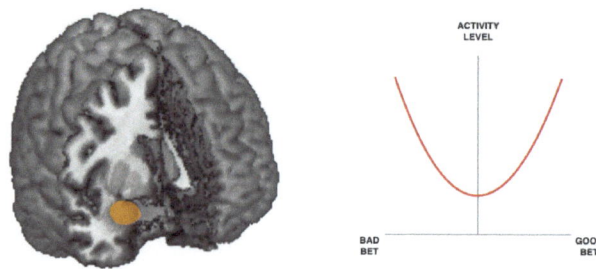

Figure Engagement of the amygdala during betting: the activity of the amygdala was high for both the best and the worst bets, and lower for bets around the person's lambda cutoff. This suggests that the amygdala is related to signalling the decision value – the approach vs avoidance value – of each betting choice.

Contextual cues and risk

As we have seen, we deviate from optimal and "rational" choices, and as seen earlier, our preferences and choices are malleable by anything ranging from single words to brands. The very same applies to our risk taking: during risk taking tasks, we can be affected by contextual cues such as what we call the task.

Indeed, in a study by Benedetto De Martino and his colleagues (DeMartino 2006) participants were undergoing fMRI while they made risky decisions – making the choice between either receiving a fixed amount or taking a risky gamble. By simply changing a single word in the risk taking task – framing the fixed amount as something they could either "keep" or "lose" – De Martino found that participants' choices were significantly affected. In the "keep" frame, they tended to be risk averse and accept the fixed option, while in the "lose" frame they tended to be less loss averse, and even risk seeking, tending to accept the gambles.

When scanning the participants' brains, the researchers found that the amygdala was the mediating factor underlying the framing effects. Furthermore, individual differences in framing effects was related to changes in activation of the medial and orbitofrontal cortex. Together, these results suggest that framing effects can alter our risk behaviours by affecting emotional processes.

In another task, research in my own lab[26] has recent demonstrated that positive reinforcement can affect our decision making substantially. In the study we had two groups of young men make several kinds of decision making tasks. One group received positive factual and social feedback and acknowledgements from the experimenter (such as "you performed that task very well" and "I am impressed, I have not seen this score level before"). The second group only received non-emotional factual feedback. What we found was that the positive reinforcement group made significantly more errors, and when studying their behaviour more carefully, we saw that their decision making process had also switched towards a faster response time. In this way, we found that positive reinforcement made young men make faster but poorer decisions.

INTER-TEMPORAL CHOICE

When it comes to making decisions, we often face choices between something immediate and something delayed. Take, for example, whether to eat a chocolate bar now or stay slim; whether to brush your teeth now to save your teeth in the long run, and whether to exercise now in order to stay healthy in the long run. Many choices involve decisions that differ in terms of their temporal dimension. This is often referred to as *inter-temporal* choices, which can be more formally defined as the the relative value people assign to two or more choices at different points in time.

If you are given the option of receiving $100 now or $120 in a month, what would you choose? Many people would choose the immediate option, been though the monthly interest rate of choosing the second option is 20%, and by far the best economic decision. When given the choice between $100 in a year or $120 in a year and a month, most people would accept the second option, although the interest rate is the same as in the first example. While our choices deviate from "rational" and optimal decisions in this way, we also see that our preferences are unstable with respect to the temporal dimension of our choices.

Another dual system model: beta and delta systems

Researchers have studied inter-temporal choices both in psychological and behavioural terms, and more recently in terms of the underlying neurobiology. For example, in a study by McClure and colleagues (McClure 2004), participants were scanned using fMRI while they made choices between immediate and delayed rewards. Here, the researchers identified two distinct neural systems that were engaged during selection of immediate and delayed rewards, respectively.

In choosing immediate rewards, participants showed stronger engagement of structures related to

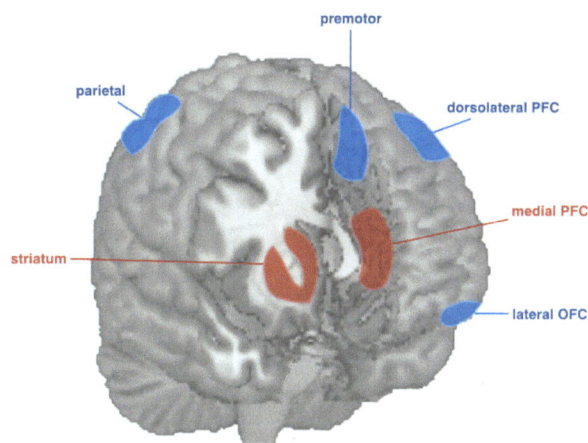

Figure Two decision systems in the human brain. Choosing immediate rewards is associated with the beta system (red), and choosing delayed rewards is associated with the delta system (blue).

the dopamine system, including the ventral striatum, medial PFC, medial OFC and posterior cingulate cortex. By contrast, when participants selected the delayed reward, they showed an increased engagement of more cortical regions, including the dorsolateral PFC, parietal cortex, premotor area and lateral OFC.

The researchers call the two systems the beta and delta system, respectively. The beta system is more related to activation of the limbic structures such as the striatum, while the delta system is related to prefrontal regions and associated with higher cognitive processes.

The beta versus delta system finding is highly reminiscent of the findings made in other value-based models, such as models of the wanting vs liking system. To our knowledge, no direct comparison has been made to uncover the relationship between models of intertemporal choice and wanting–liking, but it is likely that the models are overlapping, converging on a common view that wanting driven choices of immediate gratification are driven by deep brain structures such as the amygdala and ventral striatum, while more controlled and conscious choices are related to more cortical regions such as the prefrontal and parietal cortices.

Serotonin and value-based decision making

Can antidepressive medicine alter your decision behaviour? A recent paper in Science now demonstrates that alterations in subjects' serotonin levels leads to significant changes in their decision making behaviour. In the study, subjects were set to play the Ultimatum Game repeatedly. Subjects had to do the task two times at two different days, and at one of the days they were administered an acute tryptophan depletion (ATD), i.e., their serotonin levels would drop for a period of time. The design was double-blind and placebo controlled.

The Ultimatum Game is an experimental economics game in which two players interact to decide how to divide a sum of money that is given to them. The first player proposes how to divide the sum between themselves, and the second player can either accept or reject this proposal. If the second player rejects, neither player receives anything. If the second player accepts, the money is split according to the proposal. The game is played only once, and anonymously, so that reciprocation is not an issue.

What the researchers found was that the ATD led subjects to reject more offers, but only unfair offers. That is, ATD did not interact with offer size per se, and there was no change in mood, fairness judgement, basic reward processing or response inhibition. So serotonin seemed to affect aversive reactions to unfair offers.

The study is a nice illustration of how we now are learning to induce alterations in preferences and decision making. Along with other studies using, e.g., oxytocin to increase trust in economic games (see also my previous post about this experiment), one may expect that increasing the serotonin level may actually make subjects less responsive to unfair offers.

This knowledge is also important to learn more about, as it poses a wide range of ethical problems. If our preferences and decisions are really influenced by these stimuli, can this be abused? It should be mentioned that many of these substances are not necessarily detected (oxytocin is odourless), so we may be influenced without our consent or knowledge. The wide applicances could include casinos, stores (e.g. for expensive cars), dating agencies and so on. If we did not accept subliminal messages in ads, how can we accept this?.

CHAPTER 9

Consumer Aberrations

When we are studying consumer choice, one of the important insights is the study of aberrant behaviours. There are at least two reasons for doing so. First, a better understanding of when consumption behaviours go wrong may help us better detect, diagnose, understand and treat such problems. Second, the study of abnormal behaviours can allow us a better understanding of "healthy" consumer choices.

While this chapter is concerned with aberrant behaviours, what does it really mean to show a "healthy" consumption behaviour? After all, we have demonstrated earlier that consumer behaviour deviates from rational and optimal choice, and often occurs without a person being privy to it. What, then, constitutes "normal" behaviour?

To better understand this, we should focus on some instances where we are affected. As we already saw earlier:

- We can be affected by framing, names, brands, price and country of origin
- We show individual differences in how "frameable" we are – which seems to be driven by certain brain regions such as the ACC
- Our sense of control can be affected by outcomes, and we "reconstruct" our sense of control, even though it is correct or wrong

Let us now illustrate this with a study where we studied the effects of positive reinforcement on decision making. Usually, and often in lieu of "positive psychology", we assume that positive feedback makes people perform better.

Making good or bad decisions are due to both inter-individual but also intra-individual effects. The **inter-individual** effects are those that occur between people, and may be caused by inherent or acquired causes. On the other hand, **intra-individual** effects are changes that can occur within a single person between two or more points in time. For example, a person undergoing positive reinforcement as in the above study, or changes over time associated with ageing.

COMPULSIVE BUYING DISORDER

Compulsive buying disorder (CBD) is a condition characterised by an obsession with shopping and a chronic, repetitive purchasing behavior that has adverse consequences both for the sufferer and their social surroundings. Today, the prevalence of CBD is estimated to be somewhere between 5% and 7% (Koran 2006; Mueller 2010a), and it is often reported that CBD is significantly more prevalent in women than men, although other studies report relatively equal distributions (Mueller 2010a). While CBD has not been formally classified as a psychiatric disorder in diagnostic manuals such as DSM-IV-R, it shares characteristics of other clinical disorders such as pathological gambling, including impulsivity, dys-executive functions and mood disorders (DiNicola 2010; DeRuiter 2009; VanHolst 2010). As with pathological gambling (Kessler 2008), CBD has also been reported to be characterized by comorbidities including Axis I disorders such as mood and especially anxiety disorders (Mueller 2010), lending further support to a possible common foundation between pathological gambling and CBD.

However, little is known about the neurobiological and neurophysiological underpinnings of CBD (Lejoyeux 2010). On the one hand, studies have suggested that CBD is related to lower executive control and self-control (Claes 2010), while others have reported emotion-related issues, such as elevated levels of anxiety, depression and other mood disorders (Black 2001; Mueller 2010; Mueller 2010a). Thus, to what extent CBD is caused by a failing executive system, altered emotion responses or both, still remains unknown. Moreover, advances in understanding drug addiction have suggested that addiction behaviours could be the result of a "failing willpower" (Bechara 2005), in which sufferers may well generate emotional responses and have executive functions comparable to healthy subjects, but instead display an inability to let emotions guide behavior in value-based decision making. Taken together, these suggest at least three different, yet not mutually exclusive, causal mechanisms underlying CBD.

The aim of a recent study in my lab[27] was to provide a better insight into the basic neuropsychological and neurophysiological foundations of CBD. Let us have a closer look at this work.

Hypotheses

Based on the prior literature, three specific hypotheses were formed. The hypotheses pertained to 1) executive control, 2) emotional responses per se, and 3) the effect of emotional on decision making. Following the literature on impulse control disorders (Regard 2003; VanHolst 2010; Lejoyeux 2010; Black 2010), CBD could be related to dys-executive function and a general lack of impulse control, and thus we would expect to find a negative relationship between CBS score and performance on formal cognitive tests for executive control. Thus, we first expected that high scorers on the CBS would score lower on tests of executive control, compared to low-scorers on the CBS.

However, as other research has suggested (Grant 2006; Regard 2003; Joutsa 2011) compulsive buying disorder could also be related to increased emotional arousal, signifying a stronger drive to obtain goods and items (Bechara 2005; Berridge 1998a; Berridge 2009; Berridge 1996). Thus, a second prediction was that high scorers on the CBS would show stronger emotional arousal than low scorers on the CBS.

Finally, following recent work suggesting that behavioural addictions may be the effect of a "failing willpower" to resist specific stimuli (Bechara 2005), one possibility would be that CBD would be associated with an inability to translate emotional responses to guide actual choice behavior. That is, that sufferers can indeed generate emotional responses comparable or even stronger than healthy subjects, but that these responses do not have any real effect on actual choice behavior.

None of the three hypotheses are mutually exclusive, but could rather integrate into an understanding of CBD as a condition resulting from a collection of heterogenous dysfunctions.

Methods

We recruited women in the range from healthy and non-compulsive, through compensatory buyers, and to compulsive buyers, according to a standardised classification procedure. By testing them on neuropsychological tests of executive function and using high-resolution eye-tracking to assess emotional responses, we sought to determine the psychological mechanisms underlying CBD.

To this end, sixty-three women (age range 19-51, mean/std = 26.5±7.5, all right handed) were recruited from the Copenhagen and Frederiksberg regions, Denmark, using both online (www.forsoegsperson.dk and www.videnskab.dk) and direct recruitment procedures. At enrolment, subjects read and signed an informed consent, and were initially informed and trained with the experimental procedure. They also filled out questionnaires relating to their overall health state, including the Compulsive Buying Scale (Faber 1992a). This CBS score was used as the independent variable for all analyses in this study.

All stimuli were presented on a screen running with a 1920x1200 pixel resolution. Subjects were placed at an approximate distance of 60 cm from the screen. The test consisted of three phases: i) an ISI phase which displayed a fixation cross for 3 seconds; ii) product image phase in which a product from one of four categories (purses, clothes, women's shoes, and fast-moving consumer goods) were shown for three seconds; and iii) a rating phase, in which subjects were asked to report how much they would like to pay for the product they had just seen, by using an on-screen visual analogue scale ranging from zero to 2.000 Danish Kroner (≈$330). The experimental design is illustrated in Figure 1. To increase the external validity of the test, subjects were instructed that the choices from two of the subjects from the cohort would be randomly selected and given 1.500 DKK each, and that five of each subject's choices would be randomly selected, and the product receiving the highest bid of those five would be realised. Should the highest bid not amount to 1.500 DKK, they would be paid the remaining amount in cash. This meant that subjects were motivated to optimise their product choices, which allowed us to better estimate the actual WTP, instead of subjective estimates of WTP.

Results

Willingness To Pay

We first analysed whether compulsive buyers showed an altered willingness to pay (WTP) for products compared to healthy controls by running an independent samples samples t-test. To test for further linear effects of CBS score and WTP we ran a linear regression analysis with WTP as the dependent variable and using CBS as the independent variable.

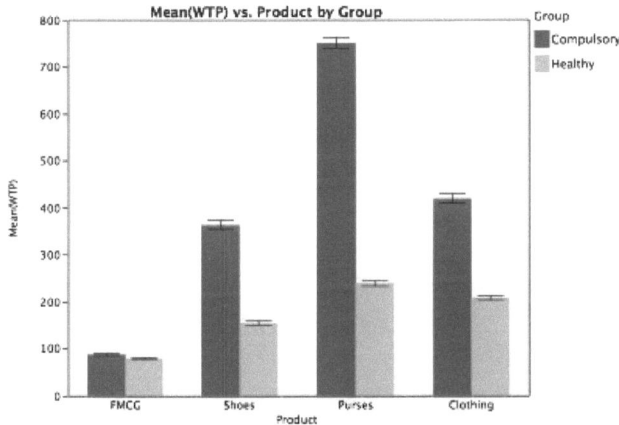

Figure Compulsive buying was associated with substantially higher WTP for relevant products, but not for MFCG

Here, we found a significant group effect, in that CB subjects displayed a disproportionately higher WTP (mean±SEM = 402.22±1.99 DKK) than controls (171.94±1.4, T=-81.6, p<0.0001). Further post-hoc testing showed that this difference was more pronounced for certain products, such as shoes, clothing and especially purses, but not for fast moving consumer goods (FMCG).

Executive functions

Second, to compare whether CBD was related to dysfunctional executive processes, we analysed the relationship between CBD and test results. Here, the neuropsychological tests were first analysed using direct comparison between the CBD group and the healthy volunteers, using an independent samples t-test. To further learn about the relationship between CBS scores and executive functions, we then ran a linear regression analysis on each test by using the test score as the dependent variable and with CBS score as the independent variable.

We found no group differences on the executive tests (Table 1), and only the Stroop test response time is faster in the CBD group compared to healthy controls.

TEST	T	KS	p
Eriksen Flanker Test		0,12	0,731
Stroop Test			
- normal	-2,59		0.0078*
- interference	-1,33		0.0965(*)
- difference	0,28		0,389
Visual reaction		0,17	0,348
Go/No-go			
- result		0,12	0,771
- time		0,17	0,289
- combined		0,12	0,71

Figure * = significance at p<0.05 level; (*)= trend significance at p<0.1

Emotional responses

We then turned to the effect of CBD on emotional responses. Here, we analysed the relationship between the CBD and emotional arousal during product viewing by comparing emotional responses in the CBD group to the healthy subjects using an independent samples t-test.

Here, we found that CBD subjects had a lower pupil dilation responses (3.23±0.01) than controls (3.26±0.01, T=6.03, p<0.0001).

From emotion to decision

Finally, we wanted to test whether compulsive buying was associated with an alteration in the impact that emotions could have on WTP choice. Here, we ran a linear regression analysis using WTP as the dependent variable and using pupil dilation, group and the interaction between pupil dilation and group as the independent variables.

Here, we found that the effects of pupil dilation on WTP was significantly different between the groups. While controls shower little relationship between pupil dilation and WTP, CBD subjects showed a strong relationship.

Discussion

Taken together, these results provide novel and compelling insights into the mechanisms of compulsive

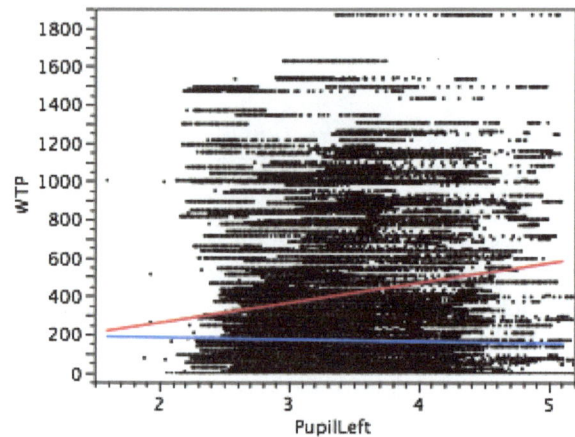

Figuer   Relationship between pupil dilation response and WTP for CBD (red) and controls (blue)

buying disorder. Notably, our data suggest that CBD is not due to a difference in executive functions, but rather a specifically stronger effect of arousal in making WTP choices. Notably, CBD subjects did not show a general stronger emotional response, but still showed a stronger effect of such responses on WTP choices.

Together, these results suggest that CBD may be erroneously linked to impulse control disorders. Rather, it seems that CBD is due to a stronger influence of emotions on decision-making. One possibility is that CBD should rather be classified as a behavioural addiction, in which urges and impulses have a stronger impact on decisions.

On the limits of human reasoning in environmental issues

Can we ever understand the environmental crisis? Can our limited minds ever get around to do "the right thing"?

I am currently appearing the Huffington Post, inan interview on the human mind and the environmental crisis. From a recent report(PDF), it is suggested that humans lack the true ability to both comprehend the abstract features and overall picture of global warming. In addition, we seem to lack the ability to plan ahead properly and cat "rationally". In particular, three factors come into play:

- Climate change has a very long time-scale – we are talking decades and centuries when it comes to alterations and consequences caused by global warming.
- Climate change is abstract and indefinite – we do not know the exact consequences and extent.
- Climate change has a very high level of complexity – we are faced with endless lines of statistics and scientific data documenting the reality of climate change.

I think the interview went well, and that I'm properly cited. However I have a concern regarding the Huff Post. It is well known to be featuring articles of lesser quality, and to lend voices to pseudoscientific claims. Indeed, it is frequently featured on the top ten list of skeptical sites' most criticised sources.

So, although I do think that this particular interview went well, my concern is that this could be used as a way to boost credibility for the HP, giving all the more prominence and credit to their other and non-science based articles. I've tried to ask some of the more prominent skeptics whether the HP is a good or a bad idea, but no responses thus far.

UNDERSTANDING PATHOLOGICAL GAMBLING

Pathological gambling is a disorder characterised by an irresistible urge to engage in monetary gambling despite harmful consequences. Its prevalence in the general population reaches 1-2% in Western societies, making pathological gambling a serious public and personal health problem. Pathological gambling has recently been classified as a behavioural addiction, and shares many core symptoms with drug addictions such as withdrawal, tolerance, and high preoccupation.

Pathological gamblers show several abnormalities when making gambling decisions. For example, pathological gambling is associated with a range of cognitive distortions. These are inappropriate beliefs about randomness, personal control, luck or talent that nurture an expectation to win. One example is "loss chasing" where the experience of repeated losses may prompt pathological gamblers to continue with gambling because they think that the lost money can be "won back". This belief of being able to win back previous losses can lead to intensified betting with higher stakes, causing substantial losses and ensuing financial, social and legal problems.

Pathological gambling is also associated with altered reward processing. A seminal functional neuroimaging study on pathological gambling reported an attenuated responsiveness of the ventral striatum to wins as opposed to losses (Reuter 2005). Only healthy controls displayed additional activation in the ventromedial and ventrolateral prefrontal cortex. Pathological gamblers showed a reduced ability to learn from aversive auditory events relative to healthy controls and enhanced response perseveration in a probabilistic reversal-learning task. In agreement with a reduced sensitivity to reward and punishment, response preservation was associated with hypo activation of the ventrolateral prefrontal cortex when money was gained and lost.

Abnormal dopamine function?

Although functional neuroimaging studies in pathological gambling have consistently found abnormal activation of dopaminergic regions implicated in reward learning, motivation and emotion, the direction of these effects appears to be highly dependent on relatively subtle differences in task design (LimbrickOldfield 2013). While some studies have found a hypo-activation of the so-called mesolimbic reward pathway, other studies have found a hyper-activation in the same areas. This opposing pattern of activation changes is mirrored in two main accounts of the aetiology of pathological gambling, the reward deficiency theory and the sensitisation theory.

The **reward deficiency theory,** which is adapted from the addiction literature (Blum 2000; Comings 2000), predicts a hypo-sensitive reward system due to a dysfunctional dopamine D2 receptor found in substance addicts and gamblers. A lower dopaminergic tone in the brain would push pathological gamblers to seek higher rewards, in order to reach the threshold at which a "reward cascade" is initiated in the brain. This theory accounts for decreased activation of the ventral striatum, ventromedial and ventrolateral prefrontal cortex by rewarding outcomes. Pathological gamblers also showed decreased activation of ventral striatum, ventromedial prefrontal cortex, and insula during the prospect and anticipation of both gains and losses with the activity in the ventral striatum being correlated inversely with levels of impulsivity. The reward deficiency theory predicts a global reduction in dopamine response to all stimuli, regardless of type. At variance with this prediction, a recent study found an influence of the type of reward (Sescousse 2013): while primary (i.e., erotic) stimuli presented during the anticipation period lead to lower activation of ventral striatum compared in pathological gamblers relative to healthy controls, no between-group difference emerged for financial rewards.

The **sensitisation theory** of pathological gambling states that pathological gambling results from a strong motivational bias towards the object of addiction (Robinson 2000; Robinson 2001; Robinson 1993; Berridge 1998; Berridge 1998a; Robinson 2008) leading to a hyper-sensitivity in dopaminergic regions. The motivation to gamble is triggered by gambling cues in the environment, which would override the incentive value of alternative sources of reward. This deficit would develop through repeated exposure to the combination of gambling cues and the pleasure or excitement related to the action of gambling. Cue-elicited drug-seeking behaviour in rats is believed to rely on habitual learning in the dorsal striatum. The dorsal striatum has been shown to exhibit hyper-sensitivity towards anticipation of rewarding gambling stimuli in pathological gamblers (van Holst et al., 2012). This theory is also supported by studies showing enhanced responses of the dopaminergic system to gambling cues.

The results reviewed above underscore that the precise neural mechanisms underlying pathological gambling deficits remain to be clarified.

Studying reward and punishment in gamblers

Here, one can ask: do pathological gamblers have problems with balancing possible gains against possible losses during choice behaviour? To address this issue, we recently performed a functional magnetic resonance imaging (fMRI) study [28] during a mixed gamble task in which pathological gamblers and matched controls had to accept or reject gambles on the basis of their relative gain-loss ratio. Our study design allowed us to address whether pathological gamblers balance positive and negative values during choice evaluations differently from controls, and whether the integration of gain-loss ratios in gambling decisions was associated with abnormal activity in brain regions involved in value based decisions.

A second focus of the study was to assess differences in decision bias towards loss aversion. This decision bias reflects the fact that people are more sensitive to losses than to gains when balancing potential gains and losses. For example, people typically reject fifty-fifty gambles unless they can win around twice as much as they can lose (Tom 2007a). While it is known that pathological gamblers exhibit distortions from normal decision-making, such as in the phenomena of loss chasing, it remains unclear whether they differ from healthy individuals in terms of loss aversion.

The task

During the fMRI session, participants performed a gambling task, which required them to accept or reject mixed gain-loss gambles with an equal probability of winning or losing. Each gamble trial offered different gain-loss ratios. On each trial, subjects were presented with a pie chart with either a potential gain amount or a potential loss amount, according to main condition. After a varying display time (2-5s), the second amount of the mixed gamble was presented and subjects used right index and middle fingers to either accept or reject the bet by pressing a button.

The findings

In analysing the fMRI data, we found that pathological gamblers displayed a dorsal prefrontal-striatal network with higher neural sensitivity to the most appetitive and aversive gain-loss ratios compared to healthy matched controls. The stronger tuning of dorsal prefrontal-striatal areas to extreme gain-loss ratios indicates that pathological gamblers put more weight on the extremes of the decision frame offered by the gambling task. Importantly, this u-shaped neural response to gambling ratios was not observed in control subjects, suggesting that this specific hyper-sensitivity to extreme ratios constitutes a neural signature of pathological gambling.

The u-shaped tuning of neural activity to most aversive and most appetitive gambles was not expressed in core regions of the reward network, such as ventral striatum or orbitofrontal cortex, but in a bilateral dorsal prefrontal-striatal associative network, including the dlPFC and the caudate nucleus. Based on this, we infer that in pathological gamblers, the dorsal prefrontal-striatal associative network weights these extreme gain-loss ratios more strongly when making gambling decisions.

Pathological gamblers revealed a particular pattern in relation to loss aversion. In gamblers, the relation between loss aversive behaviour and neural activity to gamble ratios revealed only a non-significant trend in the amygdala. Instead it was the dlPFC where the degree of loss aversion during gambling increased the u-shaped tuning of dlPFC activity to extreme gambles. This effect was significantly stronger for gamblers compared to control subjects. Interestingly, this effect peaked at the same location in dlPFC where we found the stronger hyper-sensitivity to extreme ratios relative to healthy controls. This indicates that in pathological gamblers, the individual degree of loss aversion is not reflected by traditional emotion and reward prediction areas such as the amygdala and the ventral striatum, but instead in the dlPFC, a cortical area subserving executive control functions such as working memory, task switching, and representing action-outcome contingencies. This suggests that, in this population, the DLPFC has replaced the amygdala as key region for biasing loss aversive behaviours.

In sum, we show that a dorsal cortico-striatal network involved in action-outcome contingencies and sense of agency expresses a hyper-sensitivity to extreme gain-loss ratios in pathological gamblers. The u-shaped response profile in DLPFC and precuneus was related to the individual degree of loss aversion during gambling task and severity of pathological gambling, respectively. These results stimulate future research to extend the focus of neuroimaging from the core reward system to dorsal cortico-striatal networks in pathological gambling.

BRAIN LESION AND CONSUMER CHOICE

What happens to consumer behaviour when someone is brain injured or has a brain disorder? While this is something that has yet been studied specifically in the domain of consumer neuroscience, even consumer psychology. There is a vast neuropsychological literature on the effects of brain lesions on everyday behaviours (called ADL = Activities of Daily Living), but the connection has, to my knowledge, not yet been done in the field of consumer neuroscience.

Below is therefore a tentative list of specific lesions and associated and expected effects they can have on consumption behaviour. The list is highly tentative and needs to be filled out as more studies come in that can clarify and specify these relationships.

LESION REGION	EXAMPLE DISORDER	SYMPTOMS
Prefrontal cortex	Dysexecutive syndrome	Failure to initiate action Impulse control disorder Failure to modulate behaviours Lack of enjoyment
Parietal cortex	Unilateral neglect	Failure to know product actions Spatial and navigation problems
Temporal cortex	Visual agnosia	Failure to recognise brands and products, as well as people Associative learning errors and limitations
Occipital cortex	Cortical blindness	Lack of sight, pronounced blind spots, otherwise intact mind
Basal ganglia	Ahedonia, Parkinson's	Lack of motivation and urges Need for extraordinary stimulation
Amygdala	Ahedonia, ablation of emotions	Lack of emotional tone and responses Lack of emotional learning
Hippocampus	Alzheimer's, amnesia	Failing declarative memory, including brand knowledge Failing meta-memory, i.e, failure to estimate one's own memory performance
Anterior Cingulate cortex		Failure to detect, mediate and solve decision conflicts Effects on fremeability (unknown)

CHAPTER 10

Epilogue

Where is neuromarketing and consumer neuroscience heading? Today, we see that several developments shape the way neuromarketing is changing almost on a daily basis:

- **Technology** – advances in technology, ranging from mobile apps to advances in measurements, allow cheaper, faster and more reliable assessments of consumption behaviours and their underlying causes. Today, dry electrodes can be connected to smartphone apps; fMRI machines gain substantially better temporal resolution; and solutions become abundant around the world and not only inside major labs. I believe we are facing a "Do-It Yourself" trend that can bring consumer neuroscience from being a highly expensive and slow method, to something that is becoming available to any researcher across the planet.

- **Science** – Our knowledge of the brain and how its structures and processes drive consumer behaviour is ever increasing. As our understanding grows, and our statistical and processing methods improve, solutions for assessing unconscious processes in consumer choice will improve substantially.

- **Awareness** – The application of neuroscience to the study of consumer behaviour is, despite it's decade-old track record, still considered novel and unknown to many. That is, today we can see that many professionals – academic and industrial researchers alike – are well acquainted with the possibility of neuroscience in their field, few know where to start, and many see many hurdles before fully incorporating such methods and insights. The challenge today is to provide better insights, better communication and solutions that provide a better return of investment to all researchers alike.

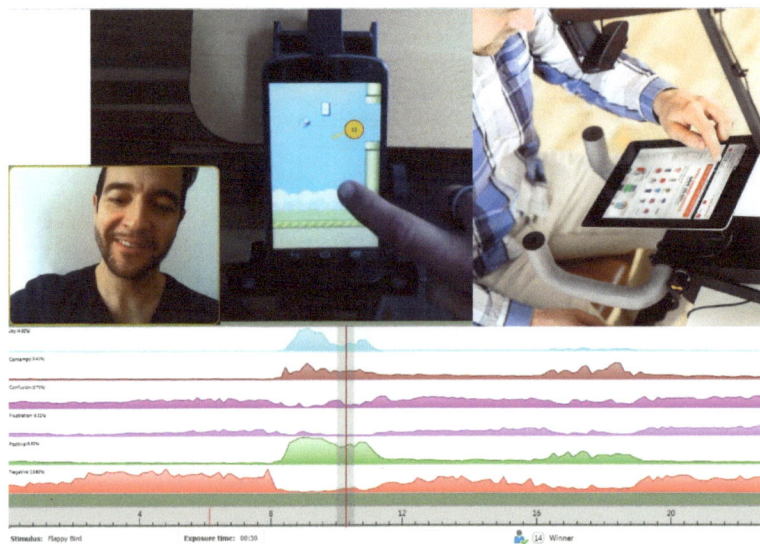

Figure Consumer neuroscience is moving out of the lab and into the real world. Today, it is already possible to assess customer responses to what is otherwise challenging situations, such as mobile apps on smartphones and tablets. Today, it is possible to assess both eye-tracking, brain responses and facial expressions in such condition, and with further advances in this domain, we see that research on consumer preference and behaviour is at the tipping point of becoming highly available, scalable and valid. Courtesy of iMotions (www.imotionsglobal.com)

We round off by pointing to a couple of topics related to the neuroethics of neuromarketing, and what lies beyond. Although this book only scratches the surface of how neuroscience can improve our understanding of consumers, the disciplines holds both promise and challenges to academia, industry and society as a whole.

NEUROETHICS OF NEUROMARKETING

What can advertisers learn from modern neuroscience[29]? Any online query will indicate that despite it's recent emergence, the term "neuromarketing" is already abundantly used. Popular science books and branding gurus alike suggest that "going neuro" is indeed the future of advertising, and such accounts are supported by academic approaches. But how do we know? Wherein lies the added value of neuroscience in advertising and marketing? Rather than being blinded by the flashing brain pictures shown in abundance, we need to focus on added values.

Today, neuromarketing has already demonstrated novel insights into consumers' minds and actions, even to the stage that it raises ethical concerns on behalf of the consumer. The methods of neuromarketing – ranging from measures of facial expression, skin conductance and pupil dilation, to state-of-the-art measures of brain activation – all bear on a common theme: they provide measures of mental processes that the consumer is not privy to, or willing to convey. Hence, these measures have already demonstrated an added value to the academic and commercial market researcher in their understanding of the consumer.

Several trends can be identified in which it is suggested, but not necessarily expected, that neuroscience can affect advertising in the foreseeable future. While marketing research has already seen a flourishing of tools for the assessment of consumer perception and choice, the impact for advertising has been less prominent. Indeed, one may claim that it is highly potent to employ the insights from neuroscience – a biological understanding of human behavior – in advertising. In this paper, I present a few notable insights in neuroscience, and discuss how they can influence the future of advertising.

Three neuroethical concerns

The rapid growth of neuromarketing and consumer neuroscience has led to considerable debate in the media and other forums. Today, we can think of three mail ethical concerns that apply when we talk about neuromarketing:

1. **Neuromarketing makes communication too effective**. This raises two concerns. On the one hand, there will be a need for protection of various parties who may be harmed or exploited by the research, marketing, and deployment of neuromarketing. A second concern is a protection of consumer autonomy if neuromarketing reaches a critical level of effectiveness (stealth neuromarketing). The suggestion and answer to this challenge is for academics and companies using neuromarketing techniques to adopt a code of ethics, to ensure beneficent and non-harmful use of the technology in consideration of both categories of ethics concerns.

2. **Consumer privacy is compromised**. Here, the concern concerns consumers' awareness, consent, and understanding to what may be viewed as invasion of their privacy rights. The solution to these problems is to better inform, provide better "escape clauses" for neuromarketing and consumer neuroscience research, and to apply better systems for regulating practices for consumer research

3. **Overselling and underdelivering**. This concern focuses on the use and presentation of neuroscience findings that is not warranted, which can be further subdivided into three subdomains:

- **Bad experimental design** – A lack of control over experimental variables, leading to a lower validity of the study and the results.
- **Non-validated methods** – Ignoring the need for independent demonstration and validation, benchmark tests.
- **Poor data preprocessing and analysis** – Increased likelihood of false negatives and false positives.
- **Misrepresentation, over-interpretation of results** – Including "reverse inference".

Avoiding doing bad

Knowing how the brain works reflects a way to understand how our minds work and, by extension, how consumers 'tick'. In doing so, we can identify several cases in which neuroscience and related sciences can help advertisers avoid making bad decisions. As the famous quote from John Wanamaker goes: "Half of my marketing budget is misspent--I just don't know which half..." With recent insights from neuroscience, it may be possible to avoid at least some of the inefficient part of the marketing budget.

One such example is the case of inattentional blindness. A number of studies have demonstrated that even when information is presented at plain sight, it can fail to be processed. Notably, in older healthy people, the inattentional blindness effect is even larger. What drives such effects? Just the sheer complexity of the visual display, or that specific attractors from visual or other senses, can lead subjects to be inattentive to otherwise salient information. Knowing exactly the mechanisms and premises for this effect can have a tremendous impact on how communication is shaped, and what the target group is. Neuromarketing insights like these can have significant impacts on how we choose to communicate to specific groups of consumers, and in avoiding bad designs that fall prey to inattentional blindness and other communication maladies. In addition, neuromarketing provides the tools for determining exactly where the cognitive and emotional traps reside. For instance, complexity and effort have significant effects on pupil dilation. Thus, knowing complexity responses up front during beta-testing of advertisements can provide an important tool for knowing when complexity level exceeds a critical threshold and information gets lost on consumers.

Today, neuromarketing is only emerging as a new player on the business arena, and it has barely passed the tests of validation and reliability. With increasing insights from neuroscience being transferred into advertising, neuromarketing can gain an additional footing in affecting the way in which we communicate to consumers.

Motivating forces

Models of consumer motivation – what drives consumers to seek out and obtain information and products – traditionally describe such processes as overt, conscious and controlled. In this perspective, consumer choice is the process of a deliberate, controlled and conscious decision making process. However, recent studies have demonstrated that consumer choices can be significantly affected by unconscious stimuli and processes. Similar findings have been demonstrated in a variety of situations related to consumption choice and monetary decision making. These findings suggest that unconscious processes both can affect choice behavior, but also demonstrate that specific brain regions are related to such effects. Indeed, it has been demonstrated that some of these deep brain structures, including the striatum, most likely operate unconsciously. This implies that the generation and maintenance of motivated consumer choice can operate altogether at an unconscious level. Indeed, it has been suggested that two motivational systems operate simultaneously in our minds: an overt and conscious "liking" system, and motivational "wanting" system, often operating under the limen of consciousness. Such a distinction both parallels accounts in behavioural economics and have distinct foundations in the brain.

Knowing how to employ this and other sources of knowledge from neuroscience will be one of the challenges for advertising in the near future. Indeed, it can be contended that consumer assessment is only one half of what neuromarketing is about. The other half, barely used in today's business, consists of the more tedious work of becoming updated on the science of how consumers feel and think. During the past couple of decades, cognitive neuroscience has provided dramatic changes in our understanding of human perception, attention, consciousness, memory, emotions, preference formation, decision making and social behavior. Becoming up to date on this knowledge will provide advertisers with new skills and tools for communicating with consumers in their struggle to obtain attention, preference, choice and satisfaction for their products and services. It will provide a crucial next step in the toolbox of advertising.

Patenting (part of) the brain?

I have recently become aware of the news that a company has patented regional brain responses to "appeal" and "engagement".

Through the scarcity of the material presented, it is really hard to get an idea of what the patent really entails. But from the sound of it, we are suggested that the patent is about the responses of particular brain regions, and that their responses predict consumer engagement and product/information appeal. If this is the case, it is very disturbing!

What's more disturbing, is the note about what brain regions we are looking at. The regions implied are the temporal and frontal gyri of the brain. OK, so WHICH frontal gyrus are we talking about? The superior frontal gyrus, the medial frontal gyrus, the inferior frontal gyrus? If the patent says only "frontal gyrus", we're talking pretty much the entire frontal cortex! It's exactly the same thing with the temporal gyrus: do they mean superior, inferior, lateral? If this is the state of patenting, I'm on my way to the patent office to submit a patent for all activations in the brain's gyrus (patent 1) and the brain's sulcus (patent 2)...

The way I see it, the patent should only be possible if it has a VERY detailed description of:

1. the stimuli being used (specific images used, stimulus duration, stimulus order, ISI, jittering etc.)
2. which MRI scanner specs is used (e.g. a Siemens Trio 3T with 8-channel head coil)

3. the EPI sequence is being used
4. preprocessing procedures
5. 1st and 2nd order analysis (PDF) protocol, and
6. statistical threshold and method (PDF) for selection of a priori brain region (regions of interest, ROI, protocol).

This would mean that others who use the exact same stimuli and precisely the same procedure should pay the patent owner. But any other scanning protocol, analysis method, use of other stimuli etc etc would be an abuse of the patenting system and should be discarded immediately. You cannot and should never be able to patent the response of a particular brain region – it is the worst case of inverse inference that I would have heard about!

The way I see it, the patent should only be possible if it has a VERY detailed description of:

1. the stimuli being used (specific images used, stimulus duration, stimulus order, ISI, jittering etc.)
2. which MRI machine is being used (e.g. a Siemens Trio 3T with 8-channel head coil)
3. what EPI sequence is being used
4. preprocessing procedures
5. 1st and 2nd order analysis protocol
6. statistical threshold and method for selection of a priori brain region (regions of interest, ROI, protocol).

This would mean that others who use the exact same stimuli and precisely the same procedure should pay the patent owner. But any other scanning protocol, analysis method, use of other stimuli etc etc would be an abuse of the patenting system and should be discarded immediately. You cannot and should never be able to patent the response of a particular brain region – it is the worst case of reverse inference that I would have heard about!

Dodging the one-sided approach to neuromarketing

It's all in the media, ethics discussions and blogosphere. Neuromarketing is a bad thing. It holds all the promises of using technology to improve marketing efforts, use covert persuasion and stealth marketing. As such, neuromarketing companies are providing the tools for companies to persuade consumers to choose their product.

Just as you can see neuromarketing companies popping up with the same message: we can predict liking, attention, emotion, consciousness, purchase intent...(use your own word), there is an equal share of people voicing their fear and skepticism agains this kind of neuromarketing. Suffice to say, most of those critics take the neuromarketing hype for granted.

Luckily, there are more sober treatments of neuromarketing. Take, for example, a listing of seven sins of neuromarketing. Indeed, the main problem with neuromarketing is that it is still untested! There is still no shared validation of the different methods, tools and analyses. As opposed to tests for the effectiveness of medication, or even the predictive power of neuropsychological tests, there are clear effect measures that one needs to address. If your medication does not work better than a placebo treatment, you're not allowed to sell it. If your predictive test for Alzheimer's Disease shows a low positive or negative predictive value, it won't be used, let alone published.

None of this is yet happening in commercial neuromarketing. This is the real problem with the current debate on neuromarketing: companies claim effective measures, but users are not able to check the validity of such claims. On the other hand, the gross overstatements of neuromarketing tool efficacy, combined with the lack of proper information, leads to criticism, fear and skepticism.

But this is a wholly one-sided approach to neuromarketing. Just because these approaches are the loudest, they need not be the most representative. Here, I provide two alternative accounts of neuromarketing. Indeed, they may seem so different, that even researchers consider changing the name of such approaches to "consumer neuroscience", "decision neuroscience" and the like. But at it's core, it is still neuromarketing.

Academic neuromarketing

A different take than the "commercial neuromarketing" industry is stressing the scholarly values that neuromarketing can provide. At its base, marketing itself is often defined as "the activity, set of institutions, and processes for creating, communicating, delivering, and exchanging offerings that have value for customers, clients, partners, and society at large". Moreover, marketing science is the attempt to understand the processes of such efforts, including how consumers attend to, respond to and are influenced by communication efforts.

For scientists in the cognitive neurosciences, this provides a wonderful model behaviour for research. Since we all make consumer choices each day, it is a behaviour that is easily understood during testing, and tests can be designed in such ways that subjects engage without much training, yet the test remains sufficiently controlled for empirical assessment. This is important when doing neuroimaging research, where the low signal to noise ration most often requires multiple repetitions of the same kind of behaviour.

In this way, neuromarketing is an academic approach to better understand how information leads to changes in attention, emotional responses, preference formation, choices and learning. While most discussions of neuromarketing is unidirectional – neuroscience tools being used to test marketing efforts – this take suggests that neuroscientists should cherish the golden opportunity to use marketing and consumer actions to better understand the human mind.

De-biasing aberrant behaviours

Much of the (neuro)ethics debate on neuromarketing suggests that it can affect consumer decisions significantly, possibly without subjective access to such effects. Prominent neuroaethicists such as Judy Illes have debated the effects of "stealth marketing," and this is indeed a valid concern.

However, much of this debate still seems to miss the major insights made through behavioural economics and neuro-economics: subjects deviate from rational choices, are affected by information (overt as well as covert), and most decisions can be traced back to unconscious antecedents. Indeed, if our "normal" choices are based on processes in our minds that we are not ourselves privy to, why is "stealth marketing" a problem? This claim raises the bar for which consumer decisions can be considered valid, to a level that no normal decisions can be found. If the bases of your behaviours are not the result of conscious and deliberate processes, how can you expect to make rules and laws prohibiting communication efforts affecting your unconscious?

Put differently: should we ban facial expression, intonation and gestures? Maybe brands should be banned altogether, because the value they signify are not readily available to our awareness, yet affect our behaviour significantly.

Better than focusing on these pointless and anachronistic assumptions, is to use neuromarketing for the sake of improving the life of people. In the study of marketing effects and consumer behaviour, we learn how we are affected by information, how biological mechanisms change our motivations, and how our choices are biased and affected "naturally," and without our awareness of such effects.

In particular, we also get to know how aberrant consumer decisions occur, ranging from impulse control disorders and behavioural addictions such as "shopaholism" and pathological/problem gambling, to other unhealthy behaviours such as obesity.

In this way, neuromarketing can be a window to new insights into many conditions. It is worth considering two approaches here:

Debiasing decisions – Given the insights provided by behavioural economics and neuroeconomics, we have come to realise that our decisions are not rational or optimal. Our choices are suboptimal and there is much room for improvement. With the knowledge gained from neuromarketing and related disciplines, we can employ it to improve conditions. Consumers can be trained in making better decisions, better at controlling the decision making process and having an improved insights into factors affecting one's choices. Based on this, de-biasing decisions is a means to improve overall wealth for each individual, and to take better control of one's own mind and behaviour. If is anywhere the famous dictum dictum "know thyself" would make sense, this is it.

Improving health – Better insights into the factors involved in aberrant consumer behaviours will have three main effects, driven by the increased knowledge into its mechanisms, causes and effects:

1. **Improve detection** – we will be better at picking up, defining and predicting aberrant behaviours, even before they manifest clinically
2. **Nuanced aetiology** – our knowledge of a specific phenomenon will possibly provide sub-categories of a particular disease (e.g., there is a difference between being a pathological gambler playing slot machines, poker or online games)
3. **Better treatment** – early detection and intervention and improved understanding leads to increased probability of improved and more specific treatment efforts

In sum, neuromarketing should be treated as a much more heterogenous concept. It does not – at least should not – belong to commercial parties alone. Just as marketing and medicine both host academic and commercial approaches, neuromarketing should have an equal share of both. Only through this realisation we will be able to grab the problem with over-commercialised neuromarketing by the root.

CHAPTER 11

References

REFERENCES

Abraham & von Cramon 'Reality = Relevance? Insights from Spontaneous Modulations of the Brain's Default Network when Telling Apart Reality from Fiction' PLOS 2009 Vol 4 e741.

Adcock, R. A., Thangavel, A., Whitfield-Gabrieli, S., Knutson, B., & Gabrieli, J. D. (2006). Reward-motivated learning: Mesolimbic activation precedes memory formation. Neuron, 50(3), 507-17. doi:10.1016/j.neuron.2006.03.036

Alexander, W. H., & Brown, J. W. (2011). Medial prefrontal cortex as an action-outcome predictor. Nat Neurosci, 14(10), 1338-44. doi:10.1038/nn.2921

Areni, C. S., & Kim, D. (1993). The influence of background music on shopping behavior: Classical versus top-forty music in a wine store. Advances in Consumer Research, 20(1), 336-340.

Aron, A. R., Robbins, T. W., & Poldrack, R. A. (2004). Inhibition and the right inferior frontal cortex. Trends in Cognitive Sciences, 8(4), 170-177.

Baars, B. J. (1993). A cognitive theory of consciousness. Cambridge University Press.

Baars, B. J. (1997). In the theater of consciousness: The workspace of the mind. Oxford University Press.

Baars, B. J. (2002). The conscious access hypothesis: Origins and recent evidence. Trends in Cognitive Sciences, 6(1), 47-52.

Baars, B. J., Franklin, S., & Ramsøy, T. Z. (2013). Global workspace dynamics: Cortical "binding and propagation" enables conscious contents. Front Psychol, 4, 200. doi:10.3389/fpsyg.2013.00200

Baars, B. J., Ramsøy, T. Z., & Laureys, S. (2003). Brain, conscious experience and the observing self. Trends in Neurosciences, 26(12), 671-675. doi:10.1016/j.tins.2003.09.015

Banks, S. J., Eddy, K. T., Angstadt, M., Nathan, P. J., & Phan, K. L. (2007). Amygdala-frontal connectivity during emotion regulation. Soc Cogn Affect Neurosci, 2(4), 303-12. doi:10.1093/scan/nsm029

Barbano, M. F., & Cador, M. (2007). Opioids for hedonic experience and dopamine to get ready for it. Psychopharmacology, 191(3), 497-506.

Barbeau, E., Wendling, F., Régis, J., Duncan, R., Poncet, M., Chauvel, P., & Bartolomei, F. (2005). Recollection of vivid memories after perirhinal region stimulations: Synchronization in the theta range of spatially distributed brain areas. Neuropsychologia, 43(9), 1329-37. doi:10.1016/j.neuropsychologia.2004.11.025

Barraza, J. A., & Zak, P. J. (2009). Empathy toward strangers triggers oxytocin release and subsequent generosity. Ann N Y Acad Sci, 1167, 182-9. doi:10.1111/j.1749-6632.2009.04504.x

Battaglia, F. P., Benchenane, K., Sirota, A., Pennartz, C. M., & Wiener, S. I. (2011). The hippocampus: Hub of brain network communication for memory. Trends in Cognitive Sciences, 15(7), 310-8. doi:10.1016/j.tics.2011.05.008

Bechara, A. (2005). Decision making, impulse control and loss of willpower to resist drugs: A neurocognitive perspective. Nat Neurosci, 8(11), 1458-63. doi:10.1038/nn1584

Bechara, A., Damasio, H., & Damasio, A. R. (2000a). Emotion, decision making and the orbitofrontal cortex. Cerebral Cortex (New York, N.Y. : 1991), 10(3), 295-307.

Bechara, A., Damasio, H., & Damasio, A. R. (2000b). Emotion, decision making and the orbitofrontal cortex. Cerebral Cortex, 10(3), 295-307.

Bechara, A., Damasio, H., Damasio, A. R., & Lee, G. P. (1999). Different contributions of the human amygdala and ventromedial prefrontal cortex to decision-making. J Neurosci, 19(13), 5473-81.

Bechara, A., Damasio, H., Tranel, D., & Damasio, A. R. (1997). Deciding advantageously before knowing the advantageous strategy. Science, 275(5304), 1293-5.

Bechara, A., Tranel, D., & Damasio, H. (2000). Characterization of the decision-making deficit of patients with ventromedial prefrontal cortex lesions. Brain, 123 (Pt 11), 2189-202.

Bennett, C. M., Miller, M. B., & Wolford, G. L. (2009). Neural correlates of interspecies perspective taking in the post-mortem atlantic salmon: An argument for multiple comparisons correction. NeuroImage, 47(1), S125.

Berger, T. W., Alger, B., & Thompson, R. F. (1976). Neuronal substrate of classical conditioning in the hippocampus. Science, 192(4238), 483-485.

Berkman, E. T., & Lieberman, M. D. (2010). Approaching the bad and avoiding the good: Lateral prefrontal cortical asymmetry distinguishes between action and valence. Journal of Cognitive Neuroscience, 22(9), 1970-1979.

Berns, G. S., & Moore, S. E. (2012). A neural predictor of cultural popularity. Journal of Consumer Psychology, 22(1), 154-160. doi:10.1016/j.jcps.2011.05.001

Berns, G. S., Capra, C. M., Moore, S., & Noussair, C. (2010). Neural mechanisms of the influence of popularity on adolescent ratings of music. NeuroImage, 49(3), 2687-96. doi:10.1016/j.neuroimage.2009.10.070

Berridge, K. C. (1996). Food reward: Brain substrates of wanting and liking. Neuroscience & Biobehavioral Reviews, 20(1), 1-25.

Berridge, K. C., & Kringelbach, M. L. (2013). Neuroscience of affect: Brain mechanisms of pleasure and displeasure. Current Opinion in Neurobiology, 23(3), 294-303. doi:10.1016/j.conb.2013.01.017

Berridge, K. C., & Robinson, T. E. (1998a). What is the role of dopamine in reward: Hedonic impact, reward learning, or incentive salience? Brain Research Reviews, 28(3), 309-369.

Berridge, K. C., & Robinson, T. E. (1998b). What is the role of dopamine in reward: Hedonic impact, reward learning, or incentive salience? Brain Research Reviews, 28(3), 309-369.

Berridge, K. C., Robinson, T. E., & Aldridge, J. W. (2009). Dissecting components of reward: 'Liking', 'wanting', and learning. Curr Opin Pharmacol, 9(1), 65-73. doi:10.1016/j.coph.2008.12.014

Black, D. W. (2001). Compulsive buying disorder: Definition, assessment, epidemiology and clinical management. CNS Drugs, 15(1), 17-27.

Black, D. W., Shaw, M., & Blum, N. (2010). Pathological gambling and compulsive buying: Do they fall within an obsessive-compulsive spectrum? Dialogues Clin Neurosci, 12(2), 175-85.

Blackford, J. U., Buckholtz, J. W., Avery, S. N., & Zald, D. H. (2010). A unique role for the human amygdala in novelty detection. NeuroImage, 50(3), 1188-1193.

Block, N. (1996). How not to find the neural correlate of consciousness. Royal Institute of Philosophy Lectures.

Block, N. (2005). Two neural correlates of consciousness. Trends in Cognitive Sciences, 9(2), 46-52.

Blum, K., Braverman, E. R., Holder, J. M., Lubar, J. F., Monastra, V. J., Miller, D., . . . Comings, D. E. (2000). The reward deficiency syndrome: A biogenetic model for the diagnosis and treatment of impulsive, addictive and compulsive behaviors. Journal of Psychoactive Drugs, 32(sup1), 1-112.

Blumenfeld, H. (2005). Consciousness and epilepsy: Why are patients with absence seizures absent? Progress in Brain Research, 150, 271-603.

Blumenfeld, H., & Taylor, J. (2003). Why do seizures cause loss of consciousness? The Neuroscientist, 9(5), 301-310.

Boettiger, C. A., Mitchell, J. M., Tavares, V. C., Robertson, M., Joslyn, G., D'Esposito, M., & Fields, H. L. (2007). Immediate reward bias in humans: Fronto-parietal networks and a role for the catechol-o-methyltransfe-

rase 158val/val genotype. The Journal of Neuroscience, 27(52), 14383-14391.

Botvinick, M., Nystrom, L. E., Fissell, K., Carter, C. S., & Cohen, J. D. (1999). Conflict monitoring versus selection-for-action in anterior cingulate cortex. Nature, 402(6758), 179-181.

Botvinick, M. M., Cohen, J. D., & Carter, C. S. (2004). Conflict monitoring and anterior cingulate cortex: An update. Trends in Cognitive Sciences, 8(12), 539-46. doi:10.1016/j.tics.2004.10.003

Bradley, M. M., Miccoli, L., Escrig, M. A., & Lang, P. J. (2008). The pupil as a measure of emotional arousal and autonomic activation. Psychophysiology, 45(4), 602-7. doi:10.1111/j.1469-8986.2008.00654.x

Braver, T. S., Barch, D. M., Gray, J. R., Molfese, D. L., & Snyder, A. (2001). Anterior cingulate cortex and response conflict: Effects of frequency, inhibition and errors. Cerebral Cortex (New York, N.Y. : 1991), 11(9), 825-36.

Bromberg-Martin, E. S., & Hikosaka, O. (2011). Lateral habenula neurons signal errors in the prediction of reward information. Nat Neurosci, 14(9), 1209-16. doi:10.1038/nn.2902

Brown, J. W., & Braver, T. S. (2008). A computational model of risk, conflict, and individual difference effects in the anterior cingulate cortex. Brain Research, 1202, 99-108.

Brown, S., Gao, X., Tisdelle, L., Eickhoff, S. B., & Liotti, M. (2011). Naturalizing aesthetics: Brain areas for aesthetic appraisal across sensory modalities. NeuroImage, 58(1), 250-8. doi:10.1016/j.neuroimage.2011.06.012

Brutkowski, S., & Dabrowska, J. (1963). Disinhibition after prefrontal lesions as a function of duration of intertrial intervals. Science, 139(3554), 505-506.

Bryant, R. A., Hung, L., Guastella, A. J., & Mitchell, P. B. (2012). Oxytocin as a moderator of hypnotizability. Psychoneuroendocrinology, 37(1), 162-6. doi:10.1016/j.psyneuen.2011.05.010

Burns, L. H., Annett, L., Kelly, A. E., Everitt, B. J., & Robbins, T. W. (1996). Effects of lesions to amygdala, ventral subiculum, medial prefrontal cortex, and nucleus accumbens on the reaction to novelty: Implications for limbic striatal interactions. Behavioral Neuroscience, 110(1), 60.

Canessa, N., Crespi, C., Motterlini, M., Baud-Bovy, G., Chierchia, G., Pantaleo, G., . . . Cappa, S. F. (2013). The functional and structural neural basis of individual differences in loss aversion. J Neurosci, 33(36), 14307-17. doi:10.1523/JNEUROSCI.0497-13.2013

Chartrand, T., Huber, J., Shiv, B., & Tanner, R. (2008). Nonconscious goals and consumer choice. Journal of Consumer Research, 35(2), 189-201. doi:10.1086/588685

Chelnokova, O., Laeng, B., Eikemo, M., Riegels, J., Løseth, G., Maurud, H., . . . Leknes, S. (2014). Rewards of beauty: The opioid system mediates social motivation in humans. Mol Psychiatry. doi:10.1038/mp.2014.1

Chen, A. C., Feng, W., Zhao, H., Yin, Y., & Wang, P. (2008). EEG default mode network in the human brain: Spectral regional field powers. NeuroImage, 41(2), 561-74. doi:10.1016/j.neuroimage.2007.12.064

Christensen, M. S., Ramsøy, T. Z., Lund, T. E., Madsen, K. H., & Rowe, J. B. (2006). An fmri study of the neural correlates of graded visual perception. NeuroImage, 31(4), 1711-25. doi:10.1016/j.neuroimage.2006.02.023

Chua, H. F., Gonzalez, R., Taylor, S. F., Welsh, R. C., & Liberzon, I. (2009). Decision-related loss: Regret and disappointment. NeuroImage, 47(4), 2031-40. doi:10.1016/j.neuroimage.2009.06.006

Chun, M. M., & Phelps, E. A. (1999). Memory deficits for implicit contextual information in amnesic subjects with hippocampal damage. Nature Neuroscience, 2(9), 844-847.

Churchland, P. S., Koch, C., & Sejnowski, T. J. (1993). What is computational neuroscience? In Computational neuroscience (pp. 46-55).

Claes, L., Bijttebier, P., Eynde, F. V. D., Mitchell, J. E., Faber, R., Zwaan, M. D., & Mueller, A. (2010). Emotional reactivity and self-regulation in relation to compulsive buying. Personality and Individual Differences, 49(5), 526-530. doi:10.1016/j.paid.2010.05.020

Coan, J. A., & Allen, J. J. (2003). Frontal EEG asymmetry and the behavioral activation and inhibition systems. Psychophysiology, 40(1), 106-14.

Cohen, M. A., & Dennett, D. C. (2011). Consciousness cannot be separated from function. Trends in Cognitive Sciences, 15(8), 358-64. doi:10.1016/j.tics.2011.06.008

Comings, D. E., & Blum, K. (2000). Reward deficiency syndrome: Genetic aspects of behavioral disorders. Progress in Brain Research, 126, 325-341.

Coricelli, G., Critchley, H. D., Joffily, M., O'Doherty, J. P., Sirigu, A., & Dolan, R. J. (2005). Regret and its avoidance: A neuroimaging study of choice behavior. Nat Neurosci, 8(9), 1255-62. doi:10.1038/nn1514

Craig, A. D. (2005). Forebrain emotional asymmetry: A neuroanatomical basis? Trends in Cognitive Sciences, 9(12), 566-71. doi:10.1016/j.tics.2005.10.005

Craig, A. D. (2009). How do you feel--now? The anterior insula and human awareness. Nat Rev Neurosci, 10(1), 59-70. doi:10.1038/nrn2555

Cryer, D., & Burchinal, M. (1997). Parents as child care consumers. Early Childhood Research Quarterly, 12(1), 35-58.

Damasio, A. R. (2005). Descartes' error : Emotion, reason, and the human brain. London; New York: Penguin. Retrieved from Amazon.

Dang-Vu, T. T., Desseilles, M., Laureys, S., Degueldre, C., Perrin, F., Phillips, C., . . . Peigneux, P. (2005). Cerebral correlates of delta waves during non-rem sleep revisited. NeuroImage, 28(1), 14-21.

David, N., Newen, A., & Vogeley, K. (2008). The "sense of agency" and its underlying cognitive and neural mechanisms. Conscious Cogn, 17(2), 523-34. doi:10.1016/j.concog.2008.03.004

Davidson, R. J. (1988). EEG measures of cerebral asymmetry: Conceptual and methodological issues. International Journal of Neuroscience, 39(1-2), 71-89.

Davidson, R. J. (1992). Anterior cerebral asymmetry and the nature of emotion. Brain and Cognition, 20(1), 125-151.

Davidson, R. J. (2004). What does the prefrontal cortex "do" in affect: Perspectives on frontal EEG asymmetry research. Biol Psychol, 67(1-2), 219-33. doi:10.1016/j.biopsycho.2004.03.008

Decety, J., Jackson, P. L., Sommerville, J. A., Chaminade, T., & Meltzoff, A. N. (2004). The neural bases of cooperation and competition: An fmri investigation. NeuroImage, 23(2), 744-51. doi:10.1016/j.neuroimage.2004.05.025

Dehaene, S., & Naccache, L. (2001). Towards a cognitive neuroscience of consciousness: Basic evidence and a workspace framework. Cognition, 79(1), 1-37.

Dehaene, S., Changeux, J. P., Naccache, L., Sackur, J., & Sergent, C. (2006). Conscious, preconscious, and subliminal processing: A testable taxonomy. Trends in Cognitive Sciences, 10(5), 204-11. doi:10.1016/j.tics.2006.03.007

Demany, L., Semal, C., Cazalets, J. R., & Pressnitzer, D. (2010). Fundamental differences in change detection between vision and audition. Exp Brain Res, 203(2), 261-70. doi:10.1007/s00221-010-2226-2

Demany, L., Trost, W., Serman, M., & Semal, C. (2008). Auditory change detection: Simple sounds are not memorized better than complex sounds. Psychological Science, 19(1), 85-91. doi:10.1111/j.1467-9280.2008.02050.x

Deppe, M., Schwindt, W., Pieper, A., Kugel, H., Plassmann, H., Kenning, P., . . . Ringelstein, E. B. (2007). Anterior cingulate reflects susceptibility to framing during attractiveness evaluation. Neuroreport, 18(11), 1119-23. doi:10.1097/WNR.0b013e3282202c61

Di Muro, F., & Murray, K. B. (2012). An arousal regulation explanation of mood effects on consumer choice. Journal of Consumer Research, 39(3), 574-584. doi:10.1086/664040

Di Nicola, M., Tedeschi, D., Mazza, M., Martinotti, G., Harnic, D., Catalano, V., . . . Janiri, L. (2010). Behavioural addictions in bipolar disorder patients: Role of impulsivity and personality dimensions. J Affect Disord, 125(1-3), 82-8. doi:10.1016/j.jad.2009.12.016

Dolcos, F., LaBar, K. S., & Cabeza, R. (2004). Dissociable effects of arousal and valence on prefrontal activity indexing emotional evaluation and subsequent memory: An event-related fmri study. NeuroImage, 23(1), 64-74. doi:10.1016/j.neuroimage.2004.05.015

Dorfman, J., Shames, V. A., & Kihlstrom, J. F. (1996). Intuition, incubation, and insight: Implicit cognition in problem solving. Implicit Cognition, 257-296.

Dreher, J. C. (2007). Sensitivity of the brain to loss aversion during risky gambles. Trends in Cognitive Sciences, 11(7), 270-2. doi:10.1016/j.tics.2007.05.006

Dunn, B. D., Billotti, D., Murphy, V., & Dalgleish, T. (2009). The consequences of effortful emotion regulation when processing distressing material: A comparison of suppression and acceptance. Behav Res Ther, 47(9), 761-73. doi:10.1016/j.brat.2009.05.007

Dyer, F. N. (1973). The stroop phenomenon and its use in the stlldy of perceptual, cognitive, and response processes. Memory & Cognition, 1(2), 106-120.

Edelson, M., Sharot, T., Dolan, R. J., & Dudai, Y. (2011). Following the crowd: Brain substrates of long-term memory conformity. Science (New York, N.Y.), 333(6038), 108-11. doi:10.1126/science.1203557

Eichenbaum, H., & Cohen, N. J. (2001). From conditioning to conscious recollection: Memory systems of the brain. Oxford University Press.

Ekman, P. (2009). Lie catching and microexpressions. The Philosophy of Deception, 118-133.

Eriksen, C. W. (1995). The flankers task and response competition: A useful tool for investigating a variety of cognitive problems. Visual Cognition, 2(2-3), 101-118.

Esposito, F., Bertolino, A., Scarabino, T., Latorre, V., Blasi, G., Popolizio, T., . . . Di Salle, F. (2006). Independent component model of the default-mode brain function: Assessing the impact of active thinking. Brain Research Bulletin, 70(4), 263-269.

Faber, R. J., & O'Guinn, T. C. (1992). A clinical screener for compulsive buying. Journal of Consumer Research, 459-469.

Falkenbach, K., Schaab, G., Pfau, O., Ryfa, M., & Birkan, B. (2013). Mere exposure effect.

Farage, M. A., Osborn, T. W., & MacLean, A. B. (2008). Cognitive, sensory, and emotional changes associated with the menstrual cycle: A review. Arch Gynecol Obstet, 278(4), 299-307. doi:10.1007/s00404-008-0708-2

Farrer, C., Franck, N., Georgieff, N., Frith, C. D., Decety, J., & Jeannerod, M. (2003). Modulating the experience of agency: A positron emission tomography study. NeuroImage, 18(2), 324-333. Retrieved from http://www.hubmed.org/display.cgi?uids=12595186

Farrer, C., Frey, S. H., Van Horn, J. D., Tunik, E., Turk, D., Inati, S., & Grafton, S. T. (2008). The angular gyrus computes action awareness representations. Cerebral Cortex (New York, N.Y. : 1991), 18(2), 254-261. Retrieved from http://www.hubmed.org/display.cgi?uids=17490989

Fitzsimons, G., Chartrand, T., & Fitzsimons, G. (2008). Automatic effects of brand exposure on motivated behavior: How apple makes you "think different". Journal of Consumer Research, 35(1), 21-35. doi:10.1086/527269

Fitzsimons, G. J., Hutchinson, J. W., Williams, P., Alba, J. W., Chartrand, T. L., Huber, J., . . . Russo, J. E. (2002). Nonconscious influences on consumer choice. Marketing Letters, 13(3), 269-279.

Fox, P. T., Mintun, M. A., Raichle, M. E., Miezin, F. M., Allman, J. M., & Van Essen, D. C. (1986). Mapping human visual cortex with positron emission tomography.

Fox, P. T., Mintun, M. A., Reiman, E. M., & Raichle, M. E. (1988). Enhanced detection of focal brain responses using intersubject averaging and change-distribution analysis of subtracted PET images. Journal of Cerebral Blood Flow & Metabolism, 8(5), 642-653.

Friese, M., Hofmann, W., & Wänke, M. (2008). When impulses take over: Moderated predictive validity of explicit and implicit attitude measures in predicting food choice and consumption behaviour. Br J Soc Psychol, 47(Pt 3), 397-419. doi:10.1348/014466607X241540

Friston, K., Chu, C., Mourão-Miranda, J., Hulme, O., Rees, G., Penny, W., & Ashburner, J. (2008). Bayesian decoding of brain images. NeuroImage, 39(1), 181-205. doi:10.1016/j.neuroimage.2007.08.013

Garavan, H. (2002). Dissociable executive functions in the dynamic control of behavior: Inhibition, error detection, and correction. NeuroImage, 17(4), 1820-1829. doi:10.1006/nimg.2002.1326

Goldin, P. R., McRae, K., Ramel, W., & Gross, J. J. (2008). The neural bases of emotion regulation: Reappraisal and suppression of negative emotion. Biol Psychiatry, 63(6), 577-86. doi:10.1016/j.biopsych.2007.05.031

Goodale, M. A., & Milner, A. D. (1992). Separate visual pathways for perception and action. Trends in Neurosciences, 15(1), 20-25.

Gorn, G., Pham, M. T., & Sin, L. Y. (2001). When arousal influences ad evaluation and valence does not (and vice versa). Journal of Consumer Psychology, 11(1), 43-55. Retrieved from WorldCat.

Grant, J. E., Brewer, J. A., & Potenza, M. N. (2006). The neurobiology of substance and behavioral addictions (2006). CNS Spectr, 11(12), 924-930.

Graybiel, A. M. (2008). Habits, rituals, and the evaluative brain. Annual Review of Neuroscience, 31, 359-87. doi:10.1146/annurev.neuro.29.051605.112851

Greene, J. D., Nystrom, L. E., Engell, A. D., Darley, J. M., & Cohen, J. D. (2004). The neural bases of cognitive conflict and control in moral judgment. Neuron, 44(2), 389-400. doi:10.1016/j.neuron.2004.09.027

Greicius, M. D., Krasnow, B., Reiss, A. L., & Menon, V. (2003). Functional connectivity in the resting brain: A network analysis of the default mode hypothesis. Proceedings of the National Academy of Sciences, 100(1), 253-258.

Groeppel-Klein, A. (2005). Arousal and consumer in-store behavior. Brain Res Bull, 67(5), 428-37. doi:10.1016/j.brainresbull.2005.06.012

Gusnard, D. A., Akbudak, E., Shulman, G. L., & Raichle, M. E. (2001). Medial prefrontal cortex and self-referential mental activity: Relation to a default mode of brain function. Proceedings of the National Academy of Sciences, 98(7), 4259-4264.

Haas, B. W., Omura, K., Constable, R. T., & Canli, T. (2007). Emotional conflict and neuroticism: Personality-dependent activation in the amygdala and subgenual anterior cingulate. Behavioral Neuroscience, 121(2), 249.

Haber, S. N., & Knutson, B. (2010). The reward circuit: Linking primate anatomy and human imaging. Neuropsychopharmacology, 35(1), 4-26. doi:10.1038/npp.2009.129

Hamann, S. (2012). Mapping discrete and dimensional emotions onto the brain: Controversies and consensus. Trends in Cognitive Sciences. doi:10.1016/j.tics.2012.07.006

Hare, T. A., Camerer, C. F., Knoepfle, D. T., & Rangel, A. (2010). Value computations in ventral medial prefrontal cortex during charitable decision making incorporate input from regions involved in social cognition. J Neurosci, 30(2), 583-90. doi:10.1523/JNEUROSCI.4089-09.2010

Hare, T. A., O'Doherty, J., Camerer, C. F., Schultz, W., & Rangel, A. (2008). Dissociating the role of the orbitofrontal cortex and the striatum in the computation of goal values and prediction errors. J Neurosci, 28(22), 5623-30. doi:10.1523/JNEUROSCI.1309-08.2008

Hariri, A. R., Mattay, V. S., Tessitore, A., Kolachana, B., Fera, F., Goldman, D., . . . Weinberger, D. R. (2002). Serotonin transporter genetic variation and the response of the human amygdala. Science (New York, N.Y.), 297(5580), 400-403. Retrieved from JSTOR.org: http://www.jstor.org/stable/3077181

Harlé, K. M., Chang, L. J., van 't Wout, M., & Sanfey, A. G. (2012). The neural mechanisms of affect infusion in social economic decision-making: A mediating role of the anterior insula. NeuroImage, 61(1), 32-40. doi:10.1016/j.neuroimage.2012.02.027

Harmon-Jones, E., Gable, P. A., & Peterson, C. K. (2010). The role of asymmetric frontal cortical activity in emotion-related phenomena: A review and update. Biol Psychol, 84(3), 451-62. doi:10.1016/j.biopsycho.2009.08.010

Harrison, B. J., Pujol, J., López-Solà, M., Hernández-Ribas, R., Deus, J., Ortiz, H., . . . Cardoner, N. (2008). Consistency and functional specialization in the default mode brain network. Proceedings of the National Academy of Sciences, 105(28), 9781-9786.

Havlicek, J., Roberts, S. C., & Flegr, J. (2005). Women's preference for dominant male odour: Effects of menstrual cycle and relationship status. Biol Lett, 1(3), 256-9. doi:10.1098/rsbl.2005.0332

He, B. J., & Raichle, M. E. (2009). The fmri signal, slow cortical potential and consciousness. Trends in Cognitive Sciences, 13(7), 302-9. doi:10.1016/j.tics.2009.04.004

Heath, R. (2012). Seducing the subconscious: The psychology of emotional influence in advertising. John Wiley & Sons.

Henke, K. (2010). A model for memory systems based on processing modes rather than consciousness. Nature Reviews Neuroscience, 11(7), 523-532. doi:10.1038/nrn2850

Herry, C., Bach, D. R., Esposito, F., Di Salle, F., Perrig, W. J., Scheffler, K., . . . Seifritz, E. (2007). Processing of temporal unpredictability in human and animal amygdala. J Neurosci, 27(22), 5958-66. doi:10.1523/JNEUROSCI.5218-06.2007

Hewig, J., Hagemann, D., Seifert, J., Naumann, E., & Bartussek, D. (2006). The relation of cortical activity and BIS/BAS on the trait level. Biol Psychol, 71(1), 42-53. doi:10.1016/j.biopsycho.2005.01.006

Hikosaka, O. (2010). The habenula: From stress evasion to value-based decision-making. Nature Reviews Neuroscience, 11(7), 503-513.

Hikosaka, O., Sesack, S. R., Lecourtier, L., & Shepard, P. D. (2008). Habenula: Crossroad between the basal ganglia and the limbic system. The Journal of Neuroscience, 28(46), 11825-11829.

Holroyd, C. B., Yeung, N., Coles, M. G., & Cohen, J. D. (2005). A mechanism for error detection in speeded response time tasks. J Exp Psychol Gen, 134(2), 163-91. doi:10.1037/0096-3445.134.2.163

Hong, S., & Hikosaka, O. (2008). The globus pallidus sends reward-related signals to the lateral habenula. Neuron, 60(4), 720-729.

Horn, N. R., Dolan, M., Elliott, R., Deakin, J. F. W., & Woodruff, P. W. R. (2003). Response inhibition and impulsivity: An fmri study. Neuropsychologia, 41(14), 1959-1966.

Hubert, M., & Kenning, P. (2008). A current overview of consumer neuroscience. Journal of Consumer Behaviour, 7(4-5), 272-292. doi:10.1002/cb.251

Ikemoto, S. (2010). Brain reward circuitry beyond the mesolimbic dopamine system: A neurobiological theory. Neurosci Biobehav Rev, 35(2), 129-50. doi:10.1016/j.neubiorev.2010.02.001

Isen, A. M., & Means, B. (1983). The influence of positive affect on decision-making strategy. Social Cognition, 2(1), 18-31.

Itti, L., & Koch, C. (2000). A saliency-based search mechanism for overt and covert shifts of visual attention. Vision Research, 40(10-12), 1489-1506.

Jacoby, J., & Kyner, D. B. (1973). Brand loyalty vs. Repeat purchasing behavior. Journal of Marketing Research, 1-9.

James, W. (2008). The stream of consciousness. Psychology.

Jensen, O., & Tesche, C. D. (2002). Frontal theta activity in humans increases with memory load in a working memory task. European Journal of Neuroscience, 15(8), 1395-1399.

Jones, B. C., DeBruine, L. M., Perrett, D. I., Little, A. C., Feinberg, D. R., & Law Smith, M. J. (2008). Effects of menstrual cycle phase on face preferences. Arch Sex Behav, 37(1), 78-84. doi:10.1007/s10508-007-9268-y

Jones, B. C., Penton-Voak, I. S., & Perrett, D. I. (n.d.). Evidence for menstrual cycle shifts in women⊠s preferences for masculinity: A response to harris (in press)⊠menstrual cycle and facial preferences reconsidered⊠.

Joutsa, J., Saunavaara, J., Parkkola, R., Niemelä, S., & Kaasinen, V. (2011). Extensive abnormality of brain white matter integrity in pathological gambling. Psychiatry Res, 194(3), 340-6. doi:10.1016/j.pscychresns.2011.08.001

Kahneman, D. (2003a). Maps of bounded rationality: Psychology for behavioral economics. The American Economic Review, 93(5), 1449-1475. Retrieved from JSTOR.org: http://www.jstor.org/stable/3132137

Kahneman, D. (2003b). A perspective on judgment and choice: Mapping bounded rationality. Am Psychol, 58(9), 697-720. doi:10.1037/0003-066X.58.9.697

Kahneman, D., & Tversky, A. (1979). Prospect theory: An analysis of decision under risk. Econometrica, 47(2), 263-292. Retrieved from JSTOR.org: http://www.jstor.org/stable/1914185

Kemp AH, Gray MA, Eide P, Silberstein RB, Nathan PJ. Steady-state visually evoked potential topography during processing of emotional valence in healthy subjects. Neuroimage. 2002; 17:1684-92.

Kessler, R. C., Hwang, I., LaBrie, R., Petukhova, M., Sampson, N. A., Winters, K. C., & Shaffer, H. J. (2008). DSM-IV pathological gambling in the national comorbidity survey replication. Psychol Med, 38(9), 1351-60. doi:10.1017/S0033291708002900

Kircher, T. T. J., & Leube, D. T. (2003). Self-consciousness, self-agency, and schizophrenia. Consciousness and Cognition, 12(4), 656-669.

Kirk, U., Skov, M., Hulme, O., Christensen, M. S., & Zeki, S. (2009). Modulation of aesthetic value by semantic context: An fmri study. NeuroImage, 44(3), 1125-32. doi:10.1016/j.neuroimage.2008.10.009

Knutson, B., Adams, C. M., Fong, G. W., & Hommer, D. (2001). Anticipation of increasing monetary reward selectively recruits nucleus accumbens. J Neurosci, 21(16), RC159.

Knutson, B., Rick, S., Wimmer, G. E., Prelec, D., & Loewenstein, G. (2007). Neural predictors of purchases. Neuron, 53(1), 147-56. doi:10.1016/j.neuron.2006.11.010

Koran, L., Faber, R., Aboujaoude, E., Large, M., & Serpe, R. (2006). Estimated prevalence of compulsive buying behavior in the united states. American Journal of Psychiatry, 163(10), 1806-1812.

Kouider, S., & Dehaene, S. (2007). Levels of processing during non-conscious perception: A critical review of visual masking. Philosophical Transactions of the Royal Society B: Biological Sciences, 362(1481), 857-875.

Kousta, S. T., Vinson, D. P., & Vigliocco, G. (2009). Emotion words, regardless of polarity, have a processing advantage over neutral words. Cognition, 112(3), 473-81. doi:10.1016/j.cognition.2009.06.007

Kringelbach, M. L. (2005). The human orbitofrontal cortex: Linking reward to hedonic experience. Nature Reviews Neuroscience, 6(9), 691-702.

Kringelbach, M. L., & Berridge, K. C. (2009). Towards a functional neuroanatomy of pleasure and happiness. Trends in Cognitive Sciences, 13(11), 479-87. doi:10.1016/j.tics.2009.08.006

Krolak-Salmon, P., Hénaff, M. -A., Isnard, J., Tallon-Baudry, C., Guénot, M., Vighetto, A., . . . Mauguiere, F. (2003). An attention modulated response to disgust in human ventral anterior insula. Annals of Neurology, 53(4), 446-453.

Kron, A., Goldstein, A., Lee, D. H., Gardhouse, K., & Anderson, A. K. (2013). How are you feeling? Revisiting the quantification of emotional qualia. Psychol Sci, 24(8), 1503-11. doi:10.1177/0956797613475456

Kuhnen, C. M., & Knutson, B. (2005). The neural basis of financial risk taking. Neuron, 47(5), 763-70. doi:10.1016/j.neuron.2005.08.008

Kumaran, D., & Maguire, E. A. (2005). The human hippocampus: Cognitive maps or relational memory? J Neurosci, 25(31), 7254-9. doi:10.1523/JNEUROSCI.1103-05.2005

Kuraoka, K., & Nakamura, K. (2007). Responses of single neurons in monkey amygdala to facial and vocal emotions. J Neurophysiol, 97(2), 1379-87. doi:10.1152/jn.00464.2006

Kühn, S., & Gallinat, J. (2012). The neural correlates of subjective pleasantness. NeuroImage, 61(1), 289-94. doi:10.1016/j.neuroimage.2012.02.065

Lang, P. J., & Bradley, M. M. (2010). Emotion and the motivational brain. Biol Psychol, 84(3), 437-50. doi:10.1016/j.biopsycho.2009.10.007

Lau, B., & Glimcher, P. W. (2008). Value representations in the primate striatum during matching behavior. Neuron, 58(3), 451-63. doi:10.1016/j.neuron.2008.02.021

Laufs, H., Kleinschmidt, A., Beyerle, A., Eger, E., Salek-Haddadi, A., Preibisch, C., & Krakow, K. (2003). EEG-correlated fmri of human alpha activity. NeuroImage, 19(4), 1463-1476.

Lauritzen, M. (2001). Relationship of spikes, synaptic activity, and local changes of cerebral blood flow. Journal of Cerebral Blood Flow & Metabolism, 21(12), 1367-1383.

LeDoux, J. (1998). The emotional brain: The mysterious underpinnings of emotional life. Simon and Schuster.

Lejoyeux, M., & Weinstein, A. (2010). Compulsive buying. Am J Drug Alcohol Abuse, 36(5), 248-53. doi:10.3109/00952990.2010.493590

Levita, L., Hare, T. A., Voss, H. U., Glover, G., Ballon, D. J., & Casey, B. J. (2009). The bivalent side of the nucleus accumbens. NeuroImage, 44(3), 1178-87. doi:10.1016/j.neuroimage.2008.09.039

Liang, M., Mouraux, A., & Iannetti, G. D. (2013). Bypassing primary sensory cortices--a direct thalamocortical pathway for transmitting salient sensory information. Cerebral Cortex (New York, N.Y. : 1991), 23(1), 1-11. doi:10.1093/cercor/bhr363

Liddle, P. F., Kiehl, K. A., & Smith, A. M. (2001). Event-related fmri study of response inhibition. Human Brain Mapping, 12(2), 100-109.

Limbrick-Oldfield, E. H., van Holst, R. J., & Clark, L. (2013). Fronto-striatal dysregulation in drug addiction and pathological gambling: Consistent inconsistencies? NeuroImage: Clinical, 2, 385-393.

Lindström, M. (2010). Buy ology: Truth and lies about why we buy. Random House LLC.

Lindström, M. (2011). Brandwashed: Tricks companies use to manipulate our minds and persuade us to buy. Random House LLC.

Liu, X., Hairston, J., Schrier, M., & Fan, J. (2011). Common and distinct networks underlying reward valence and processing stages: A meta-analysis of functional neuroimaging studies. Neurosci Biobehav Rev, 35(5), 1219-36. doi:10.1016/j.neubiorev.2010.12.012

Logothetis, N. K., Pauls, J., Augath, M., Trinath, T., & Oeltermann, A. (2001). Neurophysiological investigation of the basis of the fmri signal. Nature, 412(6843), 150-157.

Luna, B., & Sweeney, J. A. (2004). The emergence of collaborative brain function: FMRI studies of the development of response inhibition. Annals of the New York Academy of Sciences, 1021(1), 296-309.

MacLean, P. D. (1970). The triune brain, emotion, and scientific bias. The Neurosciences: Second Study Program, 336-349.

MacLean, P. D. (1990). The triune brain in evolution: Role in paleocerebral functions. Springer.

MacLean, P. D., & Kral, V. A. (1973). A triune concept of the brain and behaviour. Published for the Ontario Mental Health Foundation by University of Toronto Press.

Madsen, K. S., Baaré, W. F., Vestergaard, M., Skimminge, A., Ejersbo, L. R., Ramsøy, T. Z., . . . Jernigan, T. L. (2010). Response inhibition is associated with white matter microstructure in children. Neuropsychologia, 48(4), 854-62. doi:10.1016/j.neuropsychologia.2009.11.001

Maguire, E. A., Gadian, D. G., Johnsrude, I. S., Good, C. D., Ashburner, J., Frackowiak, R. S., & Frith, C. D. (2000). Navigation-related structural change in the hippocampi of taxi drivers. Proceedings of the National Academy of Sciences, 97(8), 4398-4403.

Maguire, E. A., Woollett, K., & Spiers, H. J. (2006). London taxi drivers and bus drivers: A structural MRI and neuropsychological analysis. Hippocampus, 16(12), 1091-101. doi:10.1002/hipo.20233

Mangan, B. (1993). Taking phenomenology seriously: The" fringe" and its implications for cognitive research. Consciousness and Cognition, 2(2), 89-108.

De Martino, B., Kumaran, D., Seymour, B., & Dolan, R. J. (2006). Frames, biases, and rational decision-making in the human brain. Science (New York, N.Y.), 313(5787), 684-7. doi:10.1126/science.1128356

Matsumoto, D., & Hwang, H. S. (2011). Evidence for training the ability to read microexpressions of emotion. Motivation and Emotion, 35(2), 181-191.

Matsumoto, M., & Hikosaka, O. (2007). Lateral habenula as a source of negative reward signals in dopamine neurons. Nature, 447(7148), 1111-1115.

Mayer, J. D., Gaschke, Y. N., Braverman, D. L., & Evans, T. W. (1992). Mood-congruent judgment is a general effect. Journal of Personality and Social Psychology, 63(1), 119-132. doi:10.1037/0022-3514.63.1.119

McClure, S. M., Laibson, D. I., Loewenstein, G., & Cohen, J. D. (2004). Separate neural systems value immediate and delayed monetary rewards. Science (New York, N.Y.), 306(5695), 503-7. doi:10.1126/science.1100907

McClure, S. M., Li, J., Tomlin, D., Cypert, K. S., Montague, L. M., & Montague, P. R. (2004). Neural correlates of behavioral preference for culturally familiar drinks. Neuron, 44(2), 379-87. doi:10.1016/j.neuron.2004.09.019

McConnell, A. R., & Leibold, J. M. (2001). Relations among the implicit association test, discriminatory behavior, and explicit measures of racial attitudes. Journal of Experimental Social Psychology, 37(5), 435-442.

Meloy, M. (2000). Mood-driven distortion of product information. Journal of Consumer Research, 27(3), 345-359. Retrieved from JSTOR.org: http://www.jstor.org/stable/10.1086/317589

Mickley Steinmetz, K. R., Addis, D. R., & Kensinger, E. A. (2010). The effect of arousal on the emotional memory network depends on valence. NeuroImage, 53(1), 318-24. doi:10.1016/j.neuroimage.2010.06.015

Miller, A., & Tomarken, A. J. (2001). Task-dependent changes in frontal brain asymmetry: Effects of incentive cues, outcome expectancies, and motor responses. Psychophysiology, 38(3), 500-11.

Milner, A. D., & Goodale, M. A. (2008). Two visual systems re-viewed. Neuropsychologia, 46(3), 774-785.

Milner, A. D., Goodale, M. A., & Vingrys, A. J. (2006). The visual brain in action (Vol. 2). Oxford University Press Oxford.

Milosavljevic, M., Navalpakkam, V., Koch, C., & Rangel, A. (2012). Relative visual saliency differences induce sizable bias in consumer choice. Journal of Consumer Psychology, 22(1), 67-74. doi:10.1016/j.jcps.2011.10.002

Monahan, J. L., Murphy, S. T., & Zajonc, R. B. (2000). Subliminal mere exposure: Specific, general, and diffuse effects. Psychol Sci, 11(6), 462-6.

Moore, J. W., Wegner, D. M., & Haggard, P. (2009). Modulating the sense of agency with external cues. Consciousness and Cognition, 18(4), 1056-1064.

Morissette, M. -C., & Boye, S. M. (2008). Electrolytic lesions of the habenula attenuate brain stimulation reward. Behavioural Brain Research, 187(1), 17-26.

Morrin, M., & Ratneshwar, S. (2003). Does it make sense to use scents to enhance brand memory? Journal of Marketing Research, 40(1), 10-25. Retrieved from JSTOR.org: http://www.jstor.org/stable/30038832

Mueller, A., Mitchell, J. E., Black, D. W., Crosby, R. D., Berg, K., & de Zwaan, M. (2010a). Latent profile analysis and comorbidity in a sample of individuals with compulsive buying disorder. Psychiatry Res, 178(2), 348-53. doi:10.1016/j.psychres.2010.04.021

Mueller, A., Mitchell, J. E., Crosby, R. D., Gefeller, O., Faber, R. J., Martin, A., . . . de Zwaan, M. (2010b). Estimated prevalence of compulsive buying in germany and its association with sociodemographic characteristics and depressive symptoms. Psychiatry Res, 180(2-3), 137-42. doi:10.1016/j.psychres.2009.12.001

Murray, E. A. (2007). The amygdala, reward and emotion. Trends in Cognitive Sciences, 11(11), 489-97. doi:10.1016/j.tics.2007.08.013

Nasrallah, M., Carmel, D., & Lavie, N. (2009). Murder, she wrote: Enhanced sensitivity to negative word valence. Emotion (Washington, D.C.), 9(5), 609-18. doi:10.1037/a0016305

Nitschke, J. B., Sarinopoulos, I., Mackiewicz, K. L., Schaefer, H. S., & Davidson, R. J. (2006). Functional neuroanatomy of aversion and its anticipation. NeuroImage, 29(1), 106-16. doi:10.1016/j.neuroimage.2005.06.068

North, A. C., Hargreaves, D. J., & McKendrick, J. (1997). In-store music affects product choice. Nature.

Northoff, G., Grimm, S., Boeker, H., Schmidt, C., Bermpohl, F., Heinzel, A., . . . Boesiger, P. (2006). Affective judgment and beneficial decision making: Ventromedial prefrontal activity correlates with performance in the iowa gambling task. Hum Brain Mapp, 27(7), 572-87. doi:10.1002/hbm.20202

O'Doherty, J., Dayan, P., Schultz, J., Deichmann, R., Friston, K., & Dolan, R. J. (2004, April). Dissociable roles of ventral and dorsal striatum in instrumental conditioning. Science (New York, N.Y.), 304(5669), 452-454. doi:10.1126/science.1094285

Ohme, R., Reykowska, D., Wiener, D., & Choromanska, A. (2010). Application of frontal EEG asymmetry to advertising research. Journal of Economic Psychology, 31(5), 785-793. doi:10.1016/j.joep.2010.03.008

Olofsson, J. K., Nordin, S., Sequeira, H., & Polich, J. (2008). Affective picture processing: An integrative review of ERP findings. Biol Psychol, 77(3), 247-65. doi:10.1016/j.biopsycho.2007.11.006

Onton, J., Delorme, A., & Makeig, S. (2005). Frontal midline EEG dynamics during working memory. NeuroImage, 27(2), 341-56. doi:10.1016/j.neuroimage.2005.04.014

Ophir, E., Nass, C., & Wagner, A. D. (2009). Cognitive control in media multitaskers. Proceedings of the National Academy of Sciences, 106(37), 15583-15587.

Overgaard, M., Rote, J., Mouridsen, K., & Ramsøy, T. Z. (2006). Is conscious perception gradual or dichotomous? A comparison of report methodologies during a visual task. Conscious Cogn, 15(4), 700-8. doi:10.1016/j.concog.2006.04.002

Pastötter, B., Hanslmayr, S., & Bäuml, K. H. (2010). Conflict processing in the anterior cingulate cortex constrains response priming. NeuroImage, 50(4), 1599-605. doi:10.1016/j.neuroimage.2010.01.095

Peeters, G., & Czapinski, J. (1990). Positive-negative asymmetry in evaluations: The distinction between affective and informational negativity effects. European Review of Social Psychology, 1(1), 33-60.

Pessiglione, M., Petrovic, P., Daunizeau, J., Palminteri, S., Dolan, R. J., & Frith, C. D. (2008). Subliminal instrumental conditioning demonstrated in the human brain. Neuron, 59(4), 561-7. doi:10.1016/j.neuron.2008.07.005

Pessiglione, M., Schmidt, L., Draganski, B., Kalisch, R., Lau, H., Dolan, R. J., & Frith, C. D. (2007). How the brain translates money into force: A neuroimaging study of subliminal motivation. Science (New York, N.Y.), 316(5826), 904-6. doi:10.1126/science.1140459

Pfister, T., Li, X., Zhao, G., & Pietikainen, M. (2011). Recognising spontaneous facial micro-expressions. In Computer vision (ICCV), 2011 IEEE international conference on (pp. 1449-1456).

Pieters, R. (2008). A review of eye-tracking research in marketing. Review of Marketing Research, 4, 123-147.

Pinto, Y., van der Leij, A. R., Sligte, I. G., Lamme, V. A., & Scholte, H. S. (2013). Bottom-up and top-down attention are independent. J Vis, 13(3), 16. doi:10.1167/13.3.16

Piovesan, M., & Wengström, E. (2009). Fast or fair? A study of response times. Economics Letters, 105(2), 193-196. doi:10.1016/j.econlet.2009.07.017

Pizzagalli, D. A., Sherwood, R. J., Henriques, J. B., & Davidson, R. J. (2005). Frontal brain asymmetry and reward responsiveness: A source-localization study. Psychol Sci, 16(10), 805-13. doi:10.1111/j.1467-9280.2005.01618.x

Plassmann, H., O'Doherty, J., Shiv, B., & Rangel, A. (2008). Marketing actions can modulate neural representations of experienced pleasantness. Proceedings of the National Academy of Sciences of the United States of America, 105(3), 1050-4. doi:10.1073/pnas.0706929105

Plassmann, H., O'Doherty, J. P., & Rangel, A. (2010). Appetitive and aversive goal values are encoded in the medial orbitofrontal cortex at the time of decision making. J Neurosci, 30(32), 10799-808. doi:10.1523/JNEUROSCI.0788-10.2010

Plassmann, H., Ramsøy, T. Z., & Milosavljevic, M. (2012). Branding the brain: A critical review and outlook. Journal of Consumer Psychology, 22(1), 18-36. doi:10.1016/j.jcps.2011.11.010

Preuschoff, K., Quartz, S. R., & Bossaerts, P. (2008). Human insula activation reflects risk prediction errors as well as risk. J Neurosci, 28(11), 2745-52. doi:10.1523/JNEUROSCI.4286-07.2008

Pribram, K. H., & McGuinness, D. (1975). Arousal, activation, and effort in the control of attention. Psychol Rev, 82(2), 116-49.

Price, C. J., & Friston, K. J. (2005). Functional ontologies for cognition: The systematic definition of structure and function. Cognitive Neuropsychology, 22(3-4), 262-275.

Quiroga, R. Q., Reddy, L., Kreiman, G., Koch, C., & Fried, I. (2005). Invariant visual representation by single neurons in the human brain. Nature, 435(7045), 1102-1107.

Raghunathan, R., & Pham, M. T. (1999). All negative moods are not equal: Motivational influences of anxiety and sadness on decision making. Organizational Behavior and Human Decision Processes, 79, 56-77.

Raichle, M. E. (2006). The brain's dark energy. SCIENCE-NEW YORK THEN WASHINGTON-, 314(5803), 1249.

Raichle, M. E. (2010). Two views of brain function. Trends in Cognitive Sciences, 14(4), 180-90. doi:10.1016/j.tics.2010.01.008

Raichle, M. E., & Gusnard, D. A. (2002). Appraising the brain's energy budget. Proceedings of the National Academy of Sciences, 99(16), 10237-10239.

Raichle, M. E., & Snyder, A. Z. (2007). A default mode of brain function: A brief history of an evolving idea. NeuroImage, 37(4), 1083-1090.

Raichle, M. E., MacLeod, A. M., Snyder, A. Z., Powers, W. J., Gusnard, D. A., & Shulman, G. L. (2001). A default mode of brain function. Proceedings of the National Academy of Sciences, 98(2), 676-682.

Ramsøy, T. Z., & Overgaard, M. (2004). Introspection and subliminal perception. Phenomenology and the Cognitive Sciences, 3(1), 1-23.

Ramsøy, T. Z., & Skov, M. (2010). How genes make up your mind: Individual biological differences and value-based decisions. Journal of Economic Psychology, 31(5), 818-831. doi:10.1016/j.joep.2010.03.003

Ramsøy, T. Z., & Skov, M. (2014). Brand preference affects the threshold for perceptual awareness. Journal of Consumer Behaviour, 13(1), 1-8. doi:10.1002/cb.1451

Ramsøy, T. Z., Friis-Olivarius, M., Jacobsen, C., Jensen, S. B., & Skov, M. (2012). Effects of perceptual uncertainty on arousal and preference across different visual domains. Journal of Neuroscience, Psychology, and Economics, 5(4), 212-226. doi:10.1037/a0030198

Ramsøy, T. Z., Liptrot, M. G., Skimminge, A., Lund, T. E., Sidaros, K., Christensen, M. S., ... Jernigan, T. L. (2009). Regional activation of the human medial temporal lobe during intentional encoding of objects and positions. NeuroImage, 47(4), 1863-72. doi:10.1016/j.neuroimage.2009.03.082

Rand, D. G., Greene, J. D., & Nowak, M. A. (2012). Spontaneous giving and calculated greed. Nature, 489(7416), 427-430. doi:10.1038/nature11467

Rao, H., Korczykowski, M., Pluta, J., Hoang, A., & Detre, J. A. (2008). Neural correlates of voluntary and involuntary risk taking in the human brain: An fmri study of the balloon analog risk task (BART). NeuroImage, 42(2), 902-10. doi:10.1016/j.neuroimage.2008.05.046

Ratcliff, R., & McKoon, G. (2008). The diffusion decision model: Theory and data for two-choice decision tasks. Neural Computation, 20(4), 873-922.

Ravaja, N., Somervuori, O., & Salminen, M. (2012). Predicting purchase decision: The role of hemispheric asymmetry over the frontal cortex. Journal of Neuroscience, Psychology, and Economics. doi:10.1037/a0029949

Regard, M., Knoch, D., Gütling, E., & Landis, T. (2003). Brain damage and addictive behavior: A neuropsychological and electroencephalogram investigation with pathologic gamblers. Cogn Behav Neurol, 16(1), 47-53.

Reuter, J., Raedler, T., Rose, M., Hand, I., Gläscher, J., & Büchel, C. (2005). Pathological gambling is linked to reduced activation of the mesolimbic reward system. Nat Neurosci, 8(2), 147-8. doi:10.1038/nn1378

Robinson, T. E., & Berridge, K. C. (1993). The neural basis of drug craving: An incentive-sensitization theory of addiction. Brain Research Reviews, 18(3), 247-291.

Robinson, T. E., & Berridge, K. C. (2000). The psychology and neurobiology of addiction: An incentive--sensitization view. Addiction, 95(8s2), 91-117.

Robinson, T. E., & Berridge, K. C. (2001). Incentive-sensitization and addiction. Addiction, 96(1), 103-114.

Robinson, T. E., & Berridge, K. C. (2008). The incentive sensitization theory of addiction: Some current issues. Philosophical Transactions of the Royal Society B: Biological Sciences, 363(1507), 3137-3146.

Rolls, E. T., & Deco, G. (2002). Computational neuroscience of vision. Oxford university press.

Rossiter, J. R., Silberstein, R. B., Harris, P. G., Nield, G. Brain-imaging detection of visual scene encoding in long-term memory for TV commercials. Journal of Advertising Research. 2001; 41: 13-21.

Rowe, J. B., Sakai, K., Lund, T. E., Ramsøy, T., Christensen, M. S., Baare, W. F., . . . Passingham, R. E. (2007). Is the prefrontal cortex necessary for establishing cognitive sets? J Neurosci, 27(48), 13303-10. doi:10.1523/JNEUROSCI.2349-07.2007

Rubia, K., Smith, A. B., Brammer, M. J., & Taylor, E. (2003). Right inferior prefrontal cortex mediates response inhibition while mesial prefrontal cortex is responsible for error detection. NeuroImage, 20(1), 351-358.

Rubinstein, A. (2007). Instinctive and cognitive reasoning: A study of response times*. The Economic Journal, 117(523), 1243-1259.

Rubinstein, A. (2008). COMMENTS ON NEUROECONOMICS. Economics and Philosophy, 24(03), 485. doi:10.1017/S0266267108002101

de Ruiter, M. B., Veltman, D. J., Goudriaan, A. E., Oosterlaan, J., Sjoerds, Z., & van den Brink, W. (2009). Response perseveration and ventral prefrontal sensitivity to reward and punishment in male problem gamblers and smokers. Neuropsychopharmacology, 34(4), 1027-38. doi:10.1038/npp.2008.175

Rutledge, R. B., Dean, M., Caplin, A., & Glimcher, P. W. (2010). Testing the reward prediction error hypothesis with an axiomatic model. J Neurosci, 30(40), 13525-36. doi:10.1523/JNEUROSCI.1747-10.2010

Sabatinelli, D., Bradley, M. M., Fitzsimmons, J. R., & Lang, P. J. (2005). Parallel amygdala and inferotemporal activation reflect emotional intensity and fear relevance. NeuroImage, 24(4), 1265-70. doi:10.1016/j.neuroimage.2004.12.015

Sacks, O. (1998). The man who mistook his wife for a hat: And other clinical tales. Simon and Schuster.

Santos, J. P., Seixas, D., Brandão, S., & Moutinho, L. (2011). Investigating the role of the ventromedial prefrontal cortex in the assessment of brands. Front Neurosci, 5, 77. doi:10.3389/fnins.2011.00077

Sarkar, A. (2011). Romancing with a brand: A conceptual analysis of romantic consumer-brand relationship. Management & Marketing, 6(1).

Sato, K., Nariai, T., Tanaka, Y., Maehara, T., Miyakawa, N., Sasaki, S., . . . Ohno, K. (2005). Functional representation of the finger and face in the human somatosensory cortex: Intraoperative intrinsic optical imaging. NeuroImage, 25(4), 1292-1301.

Schaefer, M., & Rotte, M. (2007). Thinking on luxury or pragmatic brand products: Brain responses to different categories of culturally based brands. Brain Res, 1165, 98-104. doi:10.1016/j.brainres.2007.06.038

Schaffer, C. E., Davidson, R. J., & Saron, C. (1983). Frontal and parietal electroencephalogram asymmetry in depressed and nondepressed subjects. Biological Psychiatry, 18(7), 753-762.

Schmajuk, N. A., & DiCarlo, J. J. (1992). Stimulus configuration, classical conditioning, and hippocampal function. Psychological Review, 99(2), 268.

Seed, A., Clayton, N., Carruthers, P., Dickinson, A., Glimcher, P. W., Güntürkün, O., . . . Stevens, J. R. (2011). Planning, memory, and decision making. In Strüngmann Forum Report: Animal thinking: Contemporary issues in comparative cognition, edited by R. Menzel & J. Fischer (Vol. 8, pp. 121-147). Cambridge, MA: MIT Press.

Sehlmeyer, C., Schöning, S., Zwitserlood, P., Pfleiderer, B., Kircher, T., Arolt, V., & Konrad, C. (2009). Human fear conditioning and extinction in neuroimaging: A systematic review. PLoS One, 4(6), e5865. doi:10.1371/journal.pone.0005865

Serences, J. T. (2008). Value-based modulations in human visual cortex. Neuron, 60(6), 1169-81. doi:10.1016/j.neuron.2008.10.051

Sescousse, G., Barbalat, G., Domenech, P., & Dreher, J. -C. (2013). Imbalance in the sensitivity to different types of rewards in pathological gambling. Brain, awt126.

Sescousse, G., Redouté, J., & Dreher, J. C. (2010). The architecture of reward value coding in the human orbitofrontal cortex. J Neurosci, 30(39), 13095-104. doi:10.1523/JNEUROSCI.3501-10.2010

Seth, A. K. (2009). Functions of consciousness. Encyclopedia of Consciousness, 1, 279-293.

Shanahan, M. P., & Baars, B. (2005). Applying global workspace theory to the frame problem. Cognition, 98(2), 157-176.

Shapiro, K. L., Raymond, J. E., & Arnell, K. M. (1994). Attention to visual pattern information produces the attentional blink in rapid serial visual presentation. Journal of

Experimental Psychology: Human Perception and Performance, 20(2), 357.

Shapiro, K. L., Raymond, J. E., & Arnell, K. M. (1997). The attentional blink. Trends in Cognitive Sciences, 1(8), 291-296.

Shapiro, M. S., Siller, S., & Kacelnik, A. (2008). Simultaneous and sequential choice as a function of reward delay and magnitude: Normative, descriptive and process-based models tested in the european starling (sturnus vulgaris). Journal of Experimental Psychology: Animal Behavior Processes, 34(1), 75.

Shulman, K. I. (1997). Disinhibition syndromes, secondary mania and bipolar disorder in old age. Journal of Affective Disorders, 46(3), 175-182.

Silberstein, R. B., Harris, P. G., Nield, G. A., Pipingas, A. Frontal steady-state potential changes predict long term recognition memory performance. International Journal of Psychophysiology. 2000; 39:79-85.

Silberstein, R.B. Nield, G.E. Brain activity correlates of consumer brand choice shift associated with television advertising. Int. J. Advertising. 2008; 27: 359 – 380.

Silberstein, R.B. Nield, G.E. Measuring Emotion in Advertising Research. . Pulse. 2012; 3: 24-27.

Silberstein, R. B., Schier, M. A., Pipingas, A., Ciorciari, J., Wood, S. R. and Simpson D. G. Steady state visually evoked potential topography associated with a visual vigilance task. Brain Topography 1990; 3: 337-347.

Simon, H. A. (1977). Scientific discovery and the psychology of problem solving. In Models of discovery (pp. 286-303). Springer.

Smith, B. W., Mitchell, D. G., Hardin, M. G., Jazbec, S., Fridberg, D., Blair, R. J., & Ernst, M. (2009). Neural substrates of reward magnitude, probability, and risk during a wheel of fortune decision-making task. NeuroImage, 44(2), 600-9. doi:10.1016/j.neuroimage.2008.08.016

Smith, S. M., & Blankenship, S. E. (1991). Incubation and the persistence of fixation in problem solving. The American Journal of Psychology, 61-87.

Starkstein, S. E., & Robinson, R. G. (1997). Mechanism of disinhibition after brain lesions. The Journal of Nervous and Mental Disease, 185(2), 108-114.

ENDNOTES

[1] Subliminal perception is the term for the processing of sensory stimuli below the limit (sub-limen) of consciousness.

[2] Ramsøy, T.Z. & Skov, M. The insentience of brand equity - Two studies of consciousness and brands. Conference on Neuroeconomics/NeuroPsychoEconomics 2010.

[3] However this also increases the opportunity for local minima and potentially makes the numeric conditioning of the system worse, thus increasing the effects of model errors. Many experiments use simple models, reducing possible sources of error and decreasing the computation time to find a solution. Localisation algorithms make use of the given source and head models to find a likely location for an underlying focal field generator. An alternative methodology involves performing a so-called Independent Component Analysis first in order to sort out the individual sources, and then localising the separated sources individually. This method has been shown to improve the signal-to-noise ratio of the data by correctly separating non-neuronal noise sources from neuronal sources, and has shown promise in segregating focal neuronal sources.

Generally, localisation algorithms operate by successive refinement. The system is initialised with a first guess. Then a loop is entered, in which a forward model is used to generate the magnetic field that would result from the current guess, and the guess then adjusted to reduce the difference between this estimated field and the measured field. This process it repeated until a convergence between estimated and measured field is reached.

Another approach is to ignore the inverse problem, and use an estimation algorithm to localise sources. One such approach is the second-order technique known as Synthetic Aperture Magnetometry, which uses a linear weighting of the sensor channels to focus the array on a given target location. This approach, also known as "beamforming", has an advantage over more traditional source localisation techniques because most sources in the brain are distributed and cannot be well described with a point source such as a current dipole.

A solution can then be combined with Magnetic Resonance Imaging (MRI) images to create Magnetic Source Images (MSI). The two sets of data are combined by measuring the location of a common set of fiducial points marked during MRI with lipid markers and marked during MEG with electrified coils of wire that give off magnetic fields. The locations of the fiducial points in each data set are then used to define a common coordinate system so that superimposing ("co-registering") the functional MEG data onto the structural MRI data is possible.

A criticism of the use of this technique in clinical practice is that it produces coloured areas with definite boundaries superimposed upon an MRI scan: the untrained viewer may not realise that the colours do not represent a physiological certainty, because of the relatively low spatial resolution of MEG, but rather a probability cloud derived from statistical processes. However, when the magnetic source image corroborates other data, it can be of clinical utility.

[4] Bagdziunaite D, Nassri K, Clement J, Ramsøy TZ (2014) An added value of neuroscientific tools to understand consumers' in-store behaviour. EMAC 2014

[5] As found in the American Marketing Association Dictionary: https://www.ama.org/resources/Pages/Dictionary.aspx

[6] The article is currently in peer review, and going under the reference: Ramsøy TZ, Lins J, Have AS, Jacobsen C, Zeller C (in review) Undetected Actual Control Exacerbates the Illusion of Control

[7] Ramsøy TZ, Clement J, Loving PN, Ringberg T, Skov M (in review) Effects of Ovarian Phase on Women's Immediate Emotional and Cognitive Responses to Sexual Contents in Adverts

[8] Please note that this term does not refer to the concept of "mirror neurons", although the finding is quite similar. Mirror neurons are mainly referring to studies where one finds that regions of the motor cortex that are responsible for a particular action – such as reaching out and grasping an object – are also active when we see others doing the same action.

[9] Ramsøy TZ, Jacobsen C, Friis-Olivarius M, Bagdziunaite D, Skov M (in press) Predictive value of body posture and pupil dilation in assessing consumer preference and choice,

[10] Ramsøy, T.Z. & Skov, M. The insentience of brand equity - Two studies of consciousness and brands. Conference on Neuroeconomics/NeuroPsychoEconomics 2010.

[11] Presented at the 2014 NeuroPsychoEconomics conference: Niaz N, Jacobsen C, Zeller C, Lins J, Ramsøy TZ (2014) Happy and in charge. How moods affect the illusion of control.

[12] Presented at the 2014 and 26th Annual Convention for the Association for Psychological Science: Niaz N, Rasmussen TL, Christensen T, Ramsøy TZ (2014) Talk to me, nicely – How positive words improve work motivation.

[13] "Hippocampi" is the plural for "hippocampus"

[14] We treat classical and operant conditioning in a separate section of this chapter.

[15] You can read more about LTP on this excellent Wikipedia page, and for those really keen to understand some extra details, I'd recommend also reading about second-messenger systems.

[16] Twitmyer and Pavlov made their observations independently of each other, an incident of simultaneous discovery, but Pavlov's reports have by far been the most known and popularised version.

While Pavlov studied dogs, Twitmyer (based in Pennsylvania) made the accidental discovery when studying the knee-jerk reflex (aka the patellar tendon reflex). Here, he found that an auditory stimulus preceding the hammer that caused the knee jerk, would predictably produce the knee jerk response, even despite that the hammer never touched the knee.

[17] Published as a Master thesis by Helle Shin Andersen: http://studenttheses.cbs.dk/handle/10417/2977

[18] The publication is currently in peer review: Hulme OJ, Skov M, Siebner HR, Ramsøy TZ (forthcoming) Distributed Hippocampal Encoding of Associative Density

[19] Presented at the 204 NeuroPsychoEconomics conference: Bagdziunaite D, Jensen AS, Auning J, Clement J, Ramsøy TZ (2014) What counts most? How price, country of origin and nationality dynamically affect consumer preference.

[20] Forthcoming publication: Ramsøy TZ, Christensen MK, Skov M & Stahlhut C (in review) A Neural Predictor of Willingness to Pay.

[21] This is presented in two papers: One presented at the 2014 EMAC conference: Bagdziunaite D, Nassri K, Clement J, Ramsøy TZ (2014) An added value of neuroscientific tools to understand consumers' in-store behaviour, and one forthcoming publication: Ramsøy TZ, Nassri K, Storm MZ, Clement J, Nel K, Bagdziunaite D (forthcoming) Rapid attentional and neural "motivation" responses drive in-store purchases.

[22] The following case may, to some, be reminiscent of a much more famous case of Phineas Gage, who has been described in both the academic and popular literature for decades. I decide to use my own case simply because the case of Gage has been described so many times and so well before in books like Antonio Damasio's "Descarte's Error." I also find that using my own case makes my examples more relevant, rather than using a clinical case that is a century old.

[23] Forthcoming publication: Gelskov SA, Henningsson S, Madsen KH, Siebner HR, Ramsøy TZ (in review) Amygdala activity reflects the subjective decision space of gamble ratios

[24] Reported in: Gelskov SA, Madsen KH, Skimminge A, Henningsson S, Siebner HR, Ramsøy TZ (2010) Anterior insula evaluates the magnitude of potential losses. Society for Neuroscience 2010

[25] Reported in: Gelskov SA, Henningsson S, Madsen KH, Siebner HR, Ramsøy TZ (in review) The amygdala encodes the subjective decision space of gain-loss ratios

[26] Reported in the 2014 conference Micro-Foundations for Strategic Management Research: Embracing Individuals: Niaz N, Jensen NR, Wilke R, Ramsøy TZ (2014) Reinforcement Makes You Think Fast and Wrong. The Micro-foundations of Hubristic Behavior.

[27] Article in review: Ramsøy TZ, Zuraigat FQ, Zeller C, Jacobsen C, Christensen MK, Bechara A. The Emotional and Cognitive Bases of Compulsive Buying Disorder, and presented at the 2013 NeuroPsychoEconomics conference: Ramsøy TZ, Zuraigat FQ, Bagdziunaite D, Christensen MK, Bechara A. Arousal, Executive Control and Decision Making in Compulsive Buying Disorder.

[28] Article currently in review: Gelskov SA, Henningsson S, Madsen KH, Siebner HR, Ramsøy TZ. Hyper-sensitivity in dorsal striatum and dorsal prefrontal cortex in pathological gamblers during risky decision-making

[29] The following is an adaption of my text to the Wharton special issue on the future of marketing: Ramsøy TZ (2013) The brain buzz: harvesting the true impacts of neuromarketing. Wharton Future of Advertising Program – Advertising 2020

www.ingramcontent.com/pod-product-compliance
Lightning Source LLC
Chambersburg PA
CBHW041441210326
41599CB00004B/96